Understanding Cryptography

Christof Paar · Jan Pelzl

Understanding Cryptography

A Textbook for Students and Practitioners

Foreword by Bart Preneel

 Springer

Prof. Dr.-Ing. Christof Paar
Lehrstuhl für Kommunikationssicherheit
Fakultät für Elektrotechnik und
Informationstechnik
Ruhr-Universität Bochum
44780 Bochum
Germany
christof.paar@rub.de

Dr. Jan Pelzl
escrypt GmbH - Embedded Security
Zentrum für IT-Sicherheit
Lise-Meitner-Allee 4
44801 Bochum
Germany
jpelzl@escrypt.com

ISBN 978-3-642-44649-8   ISBN 978-3-642-04101-3 (eBook)
DOI 10.1007/978-3-642-04101-3
Springer Heidelberg Dordrecht London New York

Library of Congress Control Number: 2009940447

ACM Computing Classification (1998): E.3, K.4.4, K.6.5.

Cover design: KuenkelLopka GmbH

Printed on acid-free paper

Springer is part of Springer Science+Business Media (www.springer.com)

To

Flora, Maja, Noah and Sarah

as well as to

Karl, Greta and Nele

While writing this book we noticed that for some reason the names of our spouses and children are limited to five letters. As far as we know, this has no cryptographic relevance.

Foreword

Academic research in cryptology started in the mid-1970s; today it is a mature research discipline with an established professional organization (IACR, International Association for Cryptologic Research), thousands of researchers, and dozens of international conferences. Every year more than a thousand scientific papers are published on cryptology and its applications.

Until the 1970s, cryptography was almost exclusively found in diplomatic, military and government applications. During the 1980s, the financial and telecommunications industries deployed hardware cryptographic devices. The first mass-market cryptographic application was the digital mobile phone system of the late 1980s. Today, everyone uses cryptography on a daily basis: Examples include unlocking a car or garage door with a remote-control device, connecting to a wireless LAN, buying goods with a credit or debit card in a brick and mortar store or on the Internet, installing a software update, making a phone call via voice-over-IP, or paying for a ride on a public transport system. There is no doubt that emerging application areas such as e-health, car telematics and smart buildings will make cryptography even more ubiquitous.

Cryptology is a fascinating discipline at the intersection of computer science, mathematics and electrical engineering. As cryptology is moving fast, it is hard to keep up with all the developments. During the last 25 years, the theoretical foundations of the area have been strengthened; we now have a solid understanding of security definitions and of ways to prove constructions secure. Also in the area of applied cryptography we witness very fast developments: old algorithms are broken and withdrawn and new algorithms and protocols emerge.

While several excellent textbooks on cryptology have been published in the last decade, they tend to focus on readers with a strong mathematical background. Moreover, the exciting new developments and advanced protocols form a temptation to add ever more fancy material. It is the great merit of this textbook that it restricts itself to those topics that are relevant to practitioners today. Moreover, the mathematical background and formalism is limited to what is strictly necessary and it is introduced exactly in the place where it is needed. This "less is more" approach is very suitable to address the needs of newcomers in the field, as they get introduced step by step to the basic concepts and judiciously chosen algorithms and protocols. Each chapter contains very helpful pointers to further reading, for those who want to expand and deepen their knowledge.

Overall, I am very pleased that the authors have succeeded in creating a highly valuable introduction to the subject of applied cryptography. I hope that it can serve as a guide for practitioners to build more secure systems based on cryptography, and as a stepping stone for future researchers to explore the exciting world of cryptography and its applications.

Leuven, August 2009 *Bart Preneel*

Preface

Cryptography has crept into everything, from Web browsers and e-mail programs to cell phones, bank cards, cars and even into medical implants. In the near future we will see many new exciting applications for cryptography such as radio frequency identification (RFID) tags for anti-counterfeiting or car-to-car communications (we've worked on securing both of these applications). This is quite a change from the past, where cryptography had been traditionally confined to very specific applications, especially government communications and banking systems. As a consequence of the pervasiveness of crypto algorithms, an increasing number of people must understand how they work and how they can be applied in practice. This book addresses this issue by providing a comprehensive introduction to modern applied cryptography that is equally suited for students and practitioners in industry.

Our book provides the reader with a deep understanding of how modern cryptographic schemes work. We introduce the necessary mathematical concepts in a way that is accessible for every reader with a minimum background in college-level calculus. It is thus equally well suited as a textbook for undergraduate or beginning graduate classes, or as a reference book for practicing engineers and computer scientists who are interested in a solid understanding of modern cryptography.

The book has many features that make it a unique source for practitioners and students. We focused on practical relevance by introducing most crypto algorithms that are used in modern real-world applications. For every crypto scheme, up-to-date security estimations and key length recommendations are given. We also discuss the important issue of software and hardware implementation for every algorithm. In addition to crypto algorithms, we introduce topics such as important cryptographic protocols, modes of operation, security services and key establishment techniques. Many very timely topics, e.g., lightweight ciphers which are optimized for constrained applications (such as RFID tags or smart cards) or new modes of operations, are also contained in the book.

A discussion section at the end of each chapter with annotated references provides plenty of material for further reading. For classroom use, these sections are an excellent source for course projects. In particular, when used as a textbook, the companion website for the book is highly recommended:

> www.crypto-textbook.com

Readers will find many ideas for course projects, links to open-source software, test vectors, and much more information on contemporary cryptography. In addition, links to video lectures are provided.

How to Use the Book

The material in this book has evolved over many years and is "classroom proven". We've taught it both as a course for beginning graduate students and advanced undergraduate students and as a pure undergraduate course for students majoring in our IT security programs. We found that one can teach most of the book content in a two-semester course, with 90 minutes of lecture time plus 45 minutes of help session with exercises per week (total of 10 ECTS credits). In a typical US-style three-credit course, or in a one-semester European course, some of the material should be omitted. Here are some reasonable choices for a one-semester course:

Curriculum 1 Focus on the *application of cryptography*, e.g., in a computer science or electrical engineering program. This crypto course is a good addition to courses in computer networks or more advanced security courses: Chap. 1; Sects. 2.1–2.2; Chap. 4; Sect. 5.1; Chap. 6; Sects. 7.1–7.3; Sects. 8.1–8.4; Sects. 10.1–10.2; Chap. 11; Chap. 12; and Chap. 13.

Curriculum 2 Focus on *cryptographic algorithms and their mathematical background*, e.g., as an applied cryptography course in computer science, electrical engineering or in an (undergraduate) math program. This crypto course works also nicely as preparation for a more theoretical graduate courses in cryptography: Chap. 1; Chap. 2; Chap. 3; Chap. 4; Chap. 6; Chap. 7; Sects. 8.1–8.4; Chap. 9; Chap. 10; and Sects. 11.1–11.2.

Trained as engineers, we have worked in applied cryptography and security for more than 15 years and hope that the readers will have as much fun with this fascinating field as we've had!

Bochum,   *Christof Paar*
September 2009   *Jan Pelzl*

Acknowledgements

Writing this book would have been impossible without the help of many people. We hope we did not forget anyone in our list.

We are grateful for the excellent work of Daehyun Strobel and Pascal Wißmann, who provided most of the artwork in the book and never complained about our many changes. Axel Poschmann provided the section about the PRESENT block cipher, a very timely topic, and we are thankful for his excellent work. Help with technical questions was provided by Frederick Armknecht (stream ciphers), Roberto Avanzi (finite fields and elliptic curves), Alex May (number theory), Alfred Menezes and Neal Koblitz (history of elliptic curve cryptography), Matt Robshaw (AES), and Damian Weber (discrete logarithms).

Many thanks go the members of the Embedded Security group at the University of Bochum — Andrey Bogdanov, Benedikt Driessen, Thomas Eisenbarth, Tim Güneysu, Stefan Heyse, Markus Kasper, Timo Kasper, Amir Moradi and Daehyun Strobel — who did much of the technical proofreading and provided numerous suggestions for improving the presentation of the material. Special thanks to Daehyun for helping with examples and some advanced LaTeX work, and to Markus for his help with problems. Olga Paustjan's help with artwork and typesetting is also very much appreciated.

An earlier generation of doctoral students from our group — Sandeep Kumar, Kerstin Lemke-Rust, Andy Rupp, Kai Schramm, and Marko Wolf — helped to create an online course that covered similar material. Their work was very useful and was a great inspiration when writing the book.

Bart Preneel's willingness to provide the Foreword is a great honor for us and we would like to thank him at this point again. Last but not least, we thank the people from Springer for their support and encouragement. In particular, thanks to our editor Ronan Nugent and to Alfred Hofmann.

Table of Contents

Chapter 1
Introduction to Cryptography and Data Security

This section will introduce the most important terms of modern cryptology and will teach an important lesson about proprietary vs. openly known algorithms. We will also introduce modular arithmetic which is also of major importance in public-key cryptography.

In this chapter you will learn:

- The general rules of cryptography
- Key lengths for short-, medium- and long-term security
- The difference between different types of attacks against ciphers
- A few historical ciphers, and on the way we will learn about modular arithmetic, which is of major importance for modern cryptography as well
- Why one should only use well-established encryption algorithms

1.1 Overview of Cryptology (and This Book)

If we hear the word *cryptography* our first associations might be e-mail encryption, secure website access, smart cards for banking applications or code breaking during World War II, such as the famous attack against the German Enigma encryption machine (Fig. 1.1).

Fig. 1.1 The German Enigma encryption machine (reproduced with permission from the Deutsches Museum, Munich)

Cryptography seems closely linked to modern electronic communication. However, cryptography is a rather old business, with early examples dating back to about 2000 B.C., when non-standard "secret" hieroglyphics were used in ancient Egypt. Since Egyptian days cryptography has been used in one form or the other in many, if not most, cultures that developed written language. For instance, there are documented cases of secret writing in ancient Greece, namely the *scytale* of Sparta (Fig. 1.2), or the famous Caesar cipher in ancient Rome, about which we will learn later in this chapter. This book, however, strongly focuses on modern cryptographic

Fig. 1.2 Scytale of Sparta

methods and also teaches many data security issues and their relationship with cryptography.

Let's now have a look at the field of *cryptography* (Fig. 1.3). The first thing

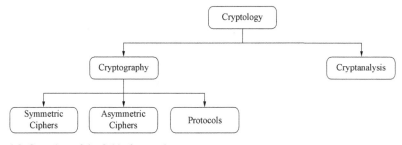

Fig. 1.3 Overview of the field of cryptology

that we notice is that the most general term is *cryptology* and not *cryptography*. Cryptology splits into two main branches:

Cryptography is the science of secret writing with the goal of hiding the meaning of a message.

Cryptanalysis is the science and sometimes art of *breaking* cryptosystems. You might think that code breaking is for the intelligence community or perhaps organized crime, and should not be included in a serious classification of a scientific discipline. However, most cryptanalysis is done by respectable researchers in academia nowadays. Cryptanalysis is of central importance for modern cryptosystems: without people who try to break our crypto methods, we will never know whether they are really secure or not. See Sect. 1.3 for more discussion about this issue.

Because cryptanalysis is the only way to assure that a cryptosystem is secure, it is an integral part of cryptology. Nevertheless, the focus of this book is on crypto**graphy**: We introduce most important practical crypto algorithms in detail. These are all crypto algorithms that have withstood cryptanalysis for a long time, in most cases for several decades. In the case of crypt**analysis** we will mainly restrict ourselves to providing state-of-the-art results with respect to breaking the crypto algorithms that are introduced, e.g., the factoring record for breaking the RSA scheme.

Let's now go back to Fig. 1.3. Cryptography itself splits into three main branches:

Symmetric Algorithms are what many people assume cryptography is about: two parties have an encryption and decryption method for which they share a secret key. All cryptography from ancient times until 1976 was exclusively based on symmetric methods. Symmetric ciphers are still in widespread use, especially for data encryption and integrity check of messages.

Asymmetric (or Public-Key) Algorithms In 1976 an entirely different type of cipher was introduced by Whitfield Diffie, Martin Hellman and Ralph Merkle. In public-key cryptography, a user possesses a secret key as in symmetric cryptography but also a public key. Asymmetric algorithms can be used for applications such as digital signatures and key establishment, and also for classical data encryption.

Cryptographic Protocols Roughly speaking, crypto protocols deal with the application of cryptographic algorithms. Symmetric and asymmetric algorithms can be viewed as building blocks with which applications such as secure Internet communication can be realized. The Transport Layer Security (TLS) scheme, which is used in every Web browser, is an example of a cryptographic protocol.

Strictly speaking, hash functions, which will be introduced in Chap. 11, form a third class of algorithms but at the same time they share some properties with symmetric ciphers.

In the majority of cryptographic applications in practical systems, symmetric and asymmetric algorithms (and often also hash functions) are all used together. This is sometimes referred to as *hybrid schemes*. The reason for using both families of algorithms is that each has specific strengths and weaknesses.

The main focus of this book is on symmetric and asymmetric algorithms, as well as hash functions. However, we will also introduce basic security protocols. In particular, we will introduce several key establishment protocols and what can be achieved with crypto protocols: confidentiality of data, integrity of data, authentication of data, user identification, etc.

1.2 Symmetric Cryptography

This section deals with the concepts of symmetric ciphers and it introduces the historic substitution cipher. Using the substitution cipher as an example, we will learn the difference between brute-force and analytical attacks.

1.2.1 Basics

Symmetric cryptographic schemes are also referred to as *symmetric-key*, *secret-key*, and *single-key* schemes or algorithms. Symmetric cryptography is best introduced with an easy to understand problem: There are two users, Alice and Bob, who want to communicate over an insecure *channel* (Fig. 1.4). The term channel might sound a bit abstract but it is just a general term for the communication link: This can be the Internet, a stretch of air in the case of mobile phones or wireless LAN communication, or any other communication media you can think of. The actual problem starts with the bad guy, Oscar[1], who has access to the channel, for instance, by hacking into an Internet router or by listening to the radio signals of a Wi-Fi communication. This type of unauthorized listening is called *eavesdropping*. Obviously, there are many situations in which Alice and Bob would prefer to communicate without Oscar listening. For instance, if Alice and Bob represent two offices of a car manufacturer, and they are transmitting documents containing the business strategy for the introduction of new car models in the next few years, these documents should

---

[1] The name Oscar was chosen to remind us of the word opponent.

not get into the hands of their competitors, or of foreign intelligence agencies for that matter.

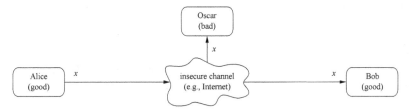

Fig. 1.4 Communication over an insecure channel

In this situation, symmetric cryptography offers a powerful solution: Alice encrypts her message x using a symmetric algorithm, yielding the ciphertext y. Bob receives the ciphertext and decrypts the message. Decryption is, thus, the inverse process of encryption (Fig. 1.5). What is the advantage? If we have a strong encryption algorithm, the ciphertext will look like random bits to Oscar and will contain no information whatsoever that is useful to him.

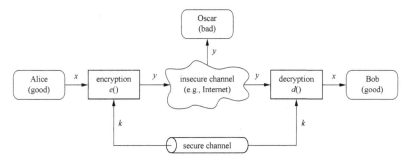

Fig. 1.5 Symmetric-key cryptosystem

The variables x, y and k in Fig. 1.5 are important in cryptography and have special names:

- x is called *plaintext* or *cleartext*,
- y is called *ciphertext*,
- k is called the *key*,
- the set of all possible keys is called the *key space*.

The system needs a secure channel for distribution of the key between Alice and Bob. The secure channel shown in Fig. 1.5 can, for instance, be a human who is transporting the key in a wallet between Alice and Bob. This is, of course, a somewhat cumbersome method. An example where this method works nicely is the pre-shared keys used in Wi-Fi Protected Access (WPA) encryption in wireless

LANs. Later in this book we will learn methods for establishing keys over insecure channels. In any case, the key has only to be transmitted once between Alice and Bob and can then be used for securing many subsequent communications.

One important and also counterintuitive fact in this situation is that both the encryption and the decryption algorithms are publicly known. It seems that keeping the *encryption algorithm* secret should make the whole system harder to break. However, secret algorithms also mean untested algorithms: The only way to find out whether an encryption method is strong, i.e., cannot be broken by a determined attacker, is to make it public and have it analyzed by other cryptographers. Please see Sect. 1.3 for more discussion on this topic. The only thing that should be kept secret in a sound cryptosystem is the key.

Remarks:

1. Of course, if Oscar gets hold of the key, he can easily decrypt the message since the algorithm is publicly known. Hence it is crucial to note that the problem of transmitting a message securely is reduced to the problems of transmitting a key secretly and of storing the key in a secure fashion.
2. In this scenario we only consider the problem of confidentiality, that is, of hiding the contents of the message from an eavesdropper. We will see later in this book that there are many other things we can do with cryptography, such as preventing Oscar from making unnoticed changes to the message (message integrity) or assuring that a message really comes from Alice (sender authentication).

1.2.2 Simple Symmetric Encryption: The Substitution Cipher

We will now learn one of the simplest methods for encrypting text, the *substitution (= replacement) cipher*. Historically this type of cipher has been used many times, and it is a good illustration of basic cryptography. We will use the substitution cipher for learning some important facts about key lengths and about different ways of attacking ciphers.

The goal of the substitution cipher is the encryption of text (as opposed to bits in modern digital systems). The idea is very simple: We substitute each letter of the alphabet with another one.

Example 1.1.

$$A \rightarrow k$$
$$B \rightarrow d$$
$$C \rightarrow w$$
$$\cdots$$

For instance, the pop group ABBA would be encrypted as kddk.

◇

We assume that we choose the substitution table completely randomly, so that an attacker is not able to guess it. Note that the substitution table is the key of this cryptosystem. As always in symmetric cryptography, the key has to be distributed between Alice and Bob in a secure fashion.

Example 1.2. Let's look at another ciphertext:

```
iq ifcc vqqr fb rdq vfllcq na rdq cfjwhwz hr bnnb
            hcc hwwhbsqvqbre hwq vhlq
```

◇

This does not seem to make too much sense and looks like decent cryptography. *However, the substitution cipher is not secure at all!* Let's look at ways of breaking the cipher.

First Attack: Brute-Force or Exhaustive Key Search

Brute-force attacks are based on a simple concept: Oscar, the attacker, has the ciphertext from eavesdropping on the channel and happens to have a short piece of plaintext, e.g., the header of a file that was encrypted. Oscar now simply decrypts the first piece of ciphertext with *all possible* keys. Again, the key for this cipher is the substitution table. If the resulting plaintext matches the short piece of plaintext, he knows that he has found the correct key.

Definition 1.2.1 Basic Exhaustive Key Search or Brute-force Attack

Let (x,y) denote the pair of plaintext and ciphertext, and let $K = \{k_1,...,k_\kappa\}$ be the key space of all possible keys k_i. A brute-force attack now checks for every $k_i \in K$ if

$$d_{k_i}(y) \stackrel{?}{=} x.$$

If the equality holds, a possible correct key is found; if not, proceed with the next key.

In practice, a brute-force attack can be more complicated because incorrect keys can give false positive results. We will address this issue in Sect. 5.2.

It is important to note that a brute-force attack against symmetric ciphers is always possible *in principle*. Whether it is feasible in practice depends on the key space, i.e., on the number of possible keys that exist for a given cipher. If testing all the keys on many modern computers takes too much time, i.e., several decades, the cipher is *computationally secure* against a brute-force attack.

Let's determine the key space of the substitution cipher: When choosing the replacement for the first letter A, we randomly choose one letter from the 26 letters of the alphabet (in the example above we chose k). The replacement for the next alphabet letter B was randomly chosen from the remaining 25 letters, etc. Thus there exist the following number of different substitution tables:

$$\text{key space of the substitution cipher} = 26 \cdot 25 \cdots 3 \cdot 2 \cdot 1 = 26! \approx 2^{88}$$

Even with hundreds of thousands of high-end PCs such a search would take several decades! Thus, we are tempted to conclude that the substitution cipher is secure. But this is incorrect because there is another, more powerful attack.

Second Attack: Letter Frequency Analysis

First we note that the brute-force attack from above treats the cipher as a black box, i.e., we do not analyze the internal structure of the cipher. The substitution cipher can easily be broken by such an analytical attack.

The major weakness of the cipher is that each plaintext symbol always maps to the same ciphertext symbol. That means that the statistical properties of the plaintext are preserved in the ciphertext. If we go back to the second example we observe that the letter q occurs most frequently in the text. From this we know that q must be the substitution for one of the frequent letters in the English language.

For practical attacks, the following properties of language can be exploited:

1. Determine the frequency of every ciphertext letter. The frequency distribution, often even of relatively short pieces of encrypted text, will be close to that of the given language in general. In particular, the most frequent letters can often easily be spotted in ciphertexts. For instance, in English E is the most frequent letter (about 13%), T is the second most frequent letter (about 9%), A is the third most frequent letter (about 8%), and so on. Table 1.1 lists the letter frequency distribution of English.
2. The method above can be generalized by looking at pairs or triples, or quadruples, and so on of ciphertext symbols. For instance, in English (and some other European languages), the letter Q is almost always followed by a U. This behavior can be exploited to detect the substitution of the letter Q and the letter U.
3. If we assume that word separators (blanks) have been found (which is only sometimes the case), one can often detect frequent short words such as THE, AND, etc. Once we have identified one of these words, we immediately know three letters (or whatever the length of the word is) for the entire text.

In practice, the three techniques listed above are often combined to break substitution ciphers.

Example 1.3. If we analyze the encrypted text from Example 1.2, we obtain:

```
WE WILL MEET IN THE MIDDLE OF THE LIBRARY AT NOON
          ALL ARRANGEMENTS ARE MADE
```

Table 1.1 Relative letter frequencies of the English language

Letter	Frequency	Letter	Frequency
A	0.0817	N	0.0675
B	0.0150	O	0.0751
C	0.0278	P	0.0193
D	0.0425	Q	0.0010
E	0.1270	R	0.0599
F	0.0223	S	0.0633
G	0.0202	T	0.0906
H	0.0609	U	0.0276
I	0.0697	V	0.0098
J	0.0015	W	0.0236
K	0.0077	X	0.0015
L	0.0403	Y	0.0197
M	0.0241	Z	0.0007

◇

Lesson learned Good ciphers should hide the statistical properties of the encrypted plaintext. The ciphertext symbols should appear to be random. Also, a large key space alone is not sufficient for a strong encryption function.

1.3 Cryptanalysis

This section deals with recommended key lengths of symmetric ciphers and different ways of attacking crypto algorithms. It is stressed that a cipher should be secure even if the attacker knows the details of the algorithm.

1.3.1 General Thoughts on Breaking Cryptosystems

If we ask someone with some technical background what breaking ciphers is about, he/she will most likely say that code breaking has to do with heavy mathematics, smart people and large computers. We have images in mind of the British code breakers during World War II, attacking the German Enigma cipher with extremely smart mathematicians (the famous computer scientist Alan Turing headed the efforts) and room-sized electro-mechanical computers. However, in practice there are also other methods of code breaking. Let's look at different ways of breaking cryptosystems *in the real world* (Fig. 1.6).

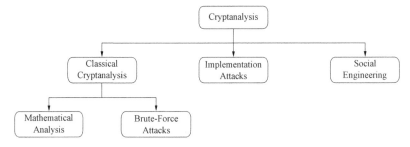

Fig. 1.6 Overview of cryptanalysis

Classical Cryptanalysis

Classical cryptanalysis is understood as the science of recovering the plaintext x from the ciphertext y, or, alternatively, recovering the key k from the ciphertext y. We recall from the earlier discussion that cryptanalysis can be divided into analytical attacks, which exploit the internal structure of the encryption method, and brute-force attacks, which treat the encryption algorithm as a black box and test all possible keys.

Implementation Attacks

Side-channel analysis can be used to obtain a secret key, for instance, by measuring the electrical power consumption of a processor which operates on the secret key. The power trace can then be used to recover the key by applying signal processing techniques. In addition to power consumption, electromagnetic radiation or the run-time behavior of algorithms can give information about the secret key and are, thus, useful side channels.[2] Note also that implementation attacks are mostly relevant against cryptosystems to which an attacker has physical access, such as smart cards. In most Internet-based attacks against remote systems, implementation attacks are usually not a concern.

Social Engineering Attacks

Bribing, blackmailing, tricking or classical espionage can be used to obtain a secret key by involving humans. For instance, forcing someone to reveal his/her secret key, e.g., by holding a gun to his/her head can be quite successful. Another, less violent, attack is to call people whom we want to attack on the phone, and say: "This is

---

[2] Before you switch on the digital oscilloscope in your lab in order to reload your Geldkarte (the Geldkarte is the electronic wallet function integrated in most German bank cards) to the maximum amount of €200: Modern smart cards have built-in countermeasures against side channel attacks and are very hard to break.

the IT department of your company. For important software updates we need your password". It is always surprising how many people are naïve enough to actually give out their passwords in such situations.

This list of attacks against cryptographic system is certainly not exhaustive. For instance, buffer overflow attacks or malware can also reveal secret keys in software systems. You might think that many of these attacks, especially social engineering and implementation attacks, are "unfair," but there is little fairness in real-world cryptography. If people want to break your IT system, they are already breaking the rules and are, thus, unfair. The major point to learn here is:

An attacker always looks for the weakest link in your cryptosystem. That means we have to choose strong algorithms *and* we have to make sure that social engineering and implementation attacks are not practical.

Even though both implementation attacks and social engineering attacks can be quite powerful in practice, this book mainly assumes attacks based on mathematical cryptanalysis.

Solid cryptosystems should adhere to *Kerckhoffs' Principle*, postulated by Auguste Kerckhoffs in 1883:

Definition 1.3.1 Kerckhoffs' Principle
A cryptosystem should be secure even if the attacker (Oscar) knows all details about the system, with the exception of the secret key. In particular, the system should be secure when the attacker knows the encryption and decryption algorithms.

Important Remark: Kerckhoffs' Principle is counterintuitive! It is extremely tempting to design a system which appears to be more secure because we keep the details hidden. This is called *security by obscurity*. However, experience and military history has shown time and again that such systems are almost always weak, and they are very often broken easily as soon as the secret design has been reverse-engineered or leaked out through other means. An example is the Content Scrambling System (CSS) for DVD content protection, which was broken easily once it was reverse-engineered. This is why a cryptographic scheme must remain secure even if its description becomes available to an attacker.

1.3.2 How Many Key Bits Are Enough?

During the 1990s there was much public discussion about the key length of ciphers. Before we provide some guidelines, there are two crucial aspects to remember:

1. The discussion of key lengths for symmetric crypto algorithms is only relevant if a brute-force attack is the best known attack. As we saw in Sect. 1.2.2 during the security analysis of the substitution cipher, if there is an analytical attack that

works, a large key space does not help at all. Of course, if there is the possibility of social engineering or implementation attacks, a long key also does not help.
2. The key lengths for symmetric and asymmetric algorithms are dramatically different. For instance, an 80-bit symmetric key provides roughly the same security as a 1024-bit RSA (RSA is a popular asymmetric algorithm) key.

Both facts are often misunderstood, especially in the semitechnical literature.

Table 1.2 gives a rough indication of the security of symmetric ciphers *with respect to brute-force attacks*. As described in Sect. 1.2.2, a large key space is a necessary but not sufficient condition for a secure symmetric cipher. The cipher must also be strong against analytical attacks.

Table 1.2 Estimated time for successful brute-force attacks on symmetric algorithms with different key lengths

Key length	Security estimation
56–64 bits	short term: a few hours or days
112–128 bits	long term: several decades in the absence of quantum computers
256 bits	long term: several decades, even with quantum computers that run the currently known quantum computing algorithms

Foretelling the Future Of course, predicting the future tends to be tricky: We can't really foresee new technical or theoretical developments with certainty. As you can imagine, it is very hard to know what kinds of computers will be available in the year 2030. For medium-term predictions, *Moore's Law* is often assumed. Roughly speaking, Moore's Law states that computing power doubles every 18 months while the costs stay constant. This has the following implications in cryptography: If today we need one month and computers worth $1,000,000 to break a cipher X, then:

■ The cost for breaking the cipher will be $500,000 in 18 months (since we only have to buy half as many computers),
■ $250,000 in 3 years,
■ $125,000 in 4.5 years, and so on.

It is important to stress that Moore's Law is an exponential function. In 15 years, i.e., after 10 iterations of computer power doubling, we can do $2^{10} = 1024$ as many computations for the same money we would need to spend today. Stated differently, we only need to spend about 1/1000th of today's money to do the same computation. In the example above that means that we can break cipher X in 15 years within one month at a cost of about $1,000,000/1024 \approx \$1000$. Alternatively, with $1,000,000, an attack can be accomplished within 45 minutes in 15 years from now. Moore's Law behaves similarly to a bank account with a 50% interest rate: The compound interest grows very, very quickly. Unfortunately, there are few trustworthy banks which offer such an interest rate.

1.4 Modular Arithmetic and More Historical Ciphers

In this section we use two historical ciphers to introduce modular arithmetic with integers. Even though the historical ciphers are no longer relevant, modular arithmetic is extremely important in modern cryptography, especially for asymmetric algorithms. Ancient ciphers date back to Egypt, where substitution ciphers were used. A very popular special case of the substitution cipher is the *Caesar cipher*, which is said to have been used by Julius Caesar to communicate with his army. The Caesar cipher simply shifts the letters in the alphabet by a constant number of steps. When the end of the alphabet is reached, the letters repeat in a cyclic way, similar to numbers in modular arithmetic.

To make computations with letters more practicable, we can assign each letter of the alphabet a number. By doing so, an encryption with the Caesar cipher simply becomes a (modular) addition with a fixed value. Instead of just adding constants, a multiplication with a constant can be applied as well. This leads us to the *affine cipher*.

Both the Caesar cipher and the affine cipher will now be discussed in more detail.

1.4.1 Modular Arithmetic

Almost all crypto algorithms, both symmetric ciphers and asymmetric ciphers, are based on arithmetic within a finite number of elements. Most number sets we are used to, such as the set of natural numbers or the set of real numbers, are infinite. In the following we introduce modular arithmetic, which is a simple way of performing arithmetic in a finite set of integers.

Let's look at an example of a finite set of integers from everyday life:

Example 1.4. Consider the hours on a clock. If you keep adding one hour, you obtain:

$$1h, 2h, 3h, \ldots, 11h, 12h, 1h, 2h, 3h, \ldots, 11h, 12h, 1h, 2h, 3h, \ldots$$

Even though we keep adding one hour, we never leave the set.

◇

Let's look at a general way of dealing with arithmetic in such finite sets.

Example 1.5. We consider the set of the nine numbers:

$$\{0, 1, 2, 3, 4, 5, 6, 7, 8\}$$

We can do regular arithmetic as long as the results are smaller than 9. For instance:

$$2 \times 3 = 6$$
$$4 + 4 = 8$$

But what about $8+4$? Now we try the following rule: Perform regular integer arithmetic and divide the result by 9. We then consider **only the remainder** rather than the original result. Since $8+4 = 12$, and $12/9$ has a remainder of 3, we write:

$$8+4 \equiv 3 \text{ mod } 9$$

◇

We now introduce an exact definition of the modulo operation:

Definition 1.4.1 Modulo Operation
Let $a, r, m \in \mathbb{Z}$ (where $\mathbb{Z}$ is a set of all integers) and $m > 0$. We write

$$a \equiv r \text{ mod } m$$

if m divides $a - r$.
m is called the modulus *and r is called the* remainder.

There are a few implications from this definition which go beyond the casual rule "divide by the modulus and consider the remainder." We discuss these implications below.

Computation of the Remainder

It is always possible to write $a \in \mathbb{Z}$, such that

$$a = q \cdot m + r \quad \text{for} \quad 0 \leq r < m \tag{1.1}$$

Since $a - r = q \cdot m$ (m divides $a - r$) we can now write: $a \equiv r \text{ mod } m$. Note that $r \in \{0, 1, 2, \ldots, m-1\}$.

Example 1.6. Let $a = 42$ and $m = 9$. Then

$$42 = 4 \cdot 9 + 6$$

and therefore $42 \equiv 6 \text{ mod } 9$.
◇

The Remainder Is Not Unique

It is somewhat surprising that for every given modulus m and number a, there are (infinitely) many valid remainders. Let's look at another example:

Example 1.7. We want to reduce 12 modulo 9. Here are several results which are correct according to the definition:

- $12 \equiv 3 \bmod 9$,   3 is a valid remainder since $9|(12-3)$
- $12 \equiv 21 \bmod 9$,   21 is a valid remainder since $9|(21-3)$
- $12 \equiv -6 \bmod 9$, -6 is a valid remainder since $9|(-6-3)$

where the "$x|y$" means "x divides y". There is a system behind this behavior. The set of numbers

$$\{\ldots, -24, -15, -6, 3, 12, 21, 30, \ldots\}$$

form what is called an *equivalence class*. There are eight other equivalence classes for the modulus 9:

$$\{\ldots, -27, -18, -9, 0, 9, 18, 27, \ldots\}$$
$$\{\ldots, -26, -17, -8, 1, 10, 19, 28, \ldots\}$$
$$\vdots$$
$$\{\ldots, -19, -10, -1, 8, 17, 26, 35, \ldots\}$$

◇

All Members of a Given Equivalence Class Behave Equivalently

For a given modulus m, it does not matter which element from a class we choose for a given computation. This property of equivalent classes has major practical implications. If we have involved computations with a fixed modulus — which is usually the case in cryptography — we are free to choose the class element that results in the easiest computation. Let's look first at an example:

Example 1.8. The core operation in many practical public-key schemes is an exponentiation of the form $x^e \bmod m$, where x, e, m are very large integers, say, 2048 bits each. Using a toy-size example, we can demonstrate two ways of doing modular exponentiation. We want to compute $3^8 \bmod 7$. The first method is the straightforward approach, and for the second one we switch between equivalent classes.

- $3^8 = 6561 \equiv 2 \bmod 7$, since $6561 = 937 \cdot 7 + 2$
  Note that we obtain the fairly large intermediate result 6561 even though we know that our final result cannot be larger than 6.
- Here is a much smarter method: First we perform two partial exponentiations:

$$3^8 = 3^4 \cdot 3^4 = 81 \cdot 81$$

We can now replace the intermediate results 81 by another member of the same equivalence class. The smallest positive member modulo 7 in the class is 4 (since $81 = 11 \cdot 7 + 4$). Hence:

$$3^8 = 81 \cdot 81 \equiv 4 \cdot 4 = 16 \bmod 7$$

From here we obtain the final result easily as $16 \equiv 2 \bmod 7$.

Note that we could perform the second method without a pocket calculator since the numbers never become larger than 81. For the first method, on the other hand, dividing 6561 by 7 is mentally already a bit challenging. As a general rule we should remember that it is almost always of computational advantage to apply the modulo reduction as soon as we can in order to keep the numbers small.

$\diamond$

Of course, the final result of any modulo computation is always the same, no matter how often we switch back and forth between equivalent classes.

Which Remainder Do We Choose?

By agreement, we usually choose r in Eq. (1.1) such that:

$$0 \leq r \leq m - 1.$$

However, mathematically it does not matter which member of an equivalent class we use.

1.4.2 Integer Rings

After studying the properties of modulo reduction we are now ready to define in more general terms a structure that is based on modulo arithmetic. Let's look at the mathematical construction that we obtain if we consider the set of integers from zero to $m - 1$ together with the operations addition and multiplication:

Definition 1.4.2 Ring
The integer ring $\mathbb{Z}_m$ *consists of:*

1. The set $\mathbb{Z}_m = \{0, 1, 2, \ldots, m - 1\}$
2. Two operations "+" and "×" for all $a, b \in \mathbb{Z}_m$ *such that:*
   1. $a + b \equiv c \bmod m$, $(c \in \mathbb{Z}_m)$
   2. $a \times b \equiv d \bmod m$, $(d \in \mathbb{Z}_m)$

Let's first look at an example for a small integer ring.

Example 1.9. Let $m = 9$, i.e., we are dealing with the ring $\mathbb{Z}_9 = \{0, 1, 2, 3, 4, 5, 6, 7, 8\}$. Let's look at a few simple arithmetic operations:

$$6 + 8 = 14 \equiv 5 \bmod 9$$
$$6 \times 8 = 48 \equiv 3 \bmod 9$$

◇

More about rings and finite fields which are related to rings is discussed in Sect. 4.2. At this point, the following properties of rings are important:

■ We can add and multiply any two numbers and the result is always in the ring. A ring is said to be *closed*.
■ Addition and multiplication are *associative*, e.g., $a + (b + c) = (a + b) + c$, and $a \cdot (b \cdot c) = (a \cdot b) \cdot c$ for all $a, b, c \in \mathbb{Z}_m$.
■ There is the *neutral element 0 with respect to addition*, i.e., for every element $a \in \mathbb{Z}_m$ it holds that $a + 0 \equiv a \bmod m$.
■ For any element a in the ring, there is always the negative element $-a$ such that $a + (-a) \equiv 0 \bmod m$, i.e., the *additive inverse* always exists.
■ There is the *neutral element 1 with respect to multiplication*, i.e., for every element $a \in \mathbb{Z}_m$ it holds that $a \times 1 \equiv a \bmod m$.
■ The *multiplicative inverse* exists only for some, but not for all, elements. Let $a \in \mathbb{Z}$, the inverse a^{-1} is defined such that

$$a \cdot a^{-1} \equiv 1 \bmod m$$

If an inverse exists for a, we can divide by this element since $b/a \equiv b \cdot a^{-1} \bmod m$.
■ It takes some effort to *find* the inverse (usually employing the Euclidean algorithm, which is taught in Sect. 6.3). However, there is an easy way of telling whether an inverse for a given element a *exists* or not:
An element $a \in \mathbb{Z}$ has a multiplicative inverse a^{-1} if and only if $\gcd(a, m) = 1$, where gcd is the *greatest common divisor*, i.e., the largest integer that divides both numbers a and m. The fact that two numbers have a gcd of 1 is of great importance in number theory, and there is a special name for it: if $\gcd(a, m) = 1$, then a and m are said to be *relatively prime* or coprime.

Example 1.10. Let's see whether the multiplicative inverse of 15 exists in $\mathbb{Z}_{26}$. Because

$$\gcd(15, 26) = 1$$

the inverse must exist. On the other hand, since

$$\gcd(14, 26) = 2 \neq 1$$

the multiplicative inverse of 14 does not exist in $\mathbb{Z}_{26}$.
◇

Another ring property is that $a \times (b + c) = (a \times b) + (a \times c)$ for all $a, b, c \in \mathbb{Z}_m$, i.e., the *distributive law* holds. In summary, roughly speaking, we can say that the ring $\mathbb{Z}_m$ is the set of integers $\{0, 1, 2, \ldots, m - 1\}$ in which we can add, subtract, multiply, and sometimes divide.

As mentioned earlier, the ring $\mathbb{Z}_m$, and thus integer arithmetic with the modulo operation, is of central importance to modern public-key cryptography. In practice, the integers involved have a length of 150–4096 bits so that efficient modular computations are a crucial aspect.

1.4.3 Shift Cipher (or Caesar Cipher)

We now introduce another historical cipher, the *shift cipher*. It is actually a special case of the substitution cipher and has a very elegant mathematical description.

The shift cipher itself is extremely simple: We simply shift every plaintext letter by a fixed number of positions in the alphabet. For instance, if we shift by 3 positions, A would be substituted by d, B by e, etc. The only problem arises towards the end of the alphabet: what should we do with X, Y, Z? As you might have guessed, they should "wrap around". That means X should become a, Y should become b, and Z is replaced by c. Allegedly, Julius Caesar used this cipher with a three-position shift.

The shift cipher also has an elegant description using modular arithmetic. For the mathematical statement of the cipher, the letters of the alphabet are encoded as numbers, as depicted in Table 1.3.

Table 1.3 Encoding of letters for the shift cipher

A	B	C	D	E	F	G	H	I	J	K	L	M
0	1	2	3	4	5	6	7	8	9	10	11	12
N	O	P	Q	R	S	T	U	V	W	X	Y	Z
13	14	15	16	17	18	19	20	21	22	23	24	25

Both the plaintext letters and the ciphertext letters are now elements of the ring $\mathbb{Z}_{26}$. Also, the key, i.e., the number of shift positions, is also in $\mathbb{Z}_{26}$ since more than 26 shifts would not make sense (27 shifts would be the same as 1 shift, etc.). The encryption and decryption of the shift cipher follows now as:

Definition 1.4.3 Shift Cipher
Let $x, y, k \in \mathbb{Z}_{26}$.
Encryption: $e_k(x) \equiv x + k \bmod 26$
Decryption: $d_k(y) \equiv y - k \bmod 26$

Example 1.11. Let the key be $k = 17$, and the plaintext is:

$$\text{ATTACK} = x_1, x_2, \ldots, x_6 = 0, 19, 19, 0, 2, 10.$$

The ciphertext is then computed as

$$y_1, y_2, \ldots, y_6 = 17, 10, 10, 17, 19, 1 = \text{rkkrtb}$$

◇

As you can guess from the discussion of the substitution cipher earlier in this book, the shift cipher is not secure at all. There are two ways of attacking it:

1. Since there are only 26 different keys (shift positions), one can easily launch a brute-force attack by trying to decrypt a given ciphertext with all possible 26 keys. If the resulting plaintext is readable text, you have found the key.
2. As for the substitution cipher, one can also use letter frequency analysis.

1.4.4 Affine Cipher

Now, we try to improve the shift cipher by generalizing the encryption function. Recall that the actual encryption of the shift cipher was the addition of the key $y_i = x_i + k \bmod 26$. The *affine cipher* encrypts by multiplying the plaintext by one part of the key followed by addition of another part of the key.

Definition 1.4.4 Affine Cipher
Let $x, y, a, b \in \mathbb{Z}_{26}$.
Encryption: $e_k(x) = y \equiv a \cdot x + b \bmod 26$
Decryption: $d_k(y) = x \equiv a^{-1} \cdot (y - b) \bmod 26$
with the key: $k = (a, b)$, which has the restriction: $\gcd(a, 26) = 1$.

The decryption is easily derived from the encryption function:

$$a \cdot x + b \equiv y \bmod 26$$
$$a \cdot x \equiv (y - b) \bmod 26$$
$$x \equiv a^{-1} \cdot (y - b) \bmod 26$$

The restriction $\gcd(a, 26) = 1$ stems from the fact that the key parameter a needs to be inverted for decryption. We recall from Sect. 1.4.2 that an element a and the modulus must be relatively prime for the inverse of a to exist. Thus, a must be in the set:

$$a \in \{1, 3, 5, 7, 9, 11, 15, 17, 19, 21, 23, 25\} \tag{1.2}$$

But how do we find a^{-1}? For now, we can simply compute it by trial and error: For a given a we simply try all possible values a^{-1} until we obtain:

$$a \cdot a^{-1} \equiv 1 \bmod 26$$

For instance, if $a = 3$, then $a^{-1} = 9$ since $3 \cdot 9 = 27 \equiv 1 \bmod 26$. Note that a^{-1} also always fulfills the condition $\gcd(a^{-1}, 26) = 1$ since the inverse of a^{-1} always exists. In fact, the inverse of a^{-1} is a itself. Hence, for the trial-and-error determination of a^{-1} one only has to check the values given in Eq. (1.2).

Example 1.12. Let the key be $k = (a, b) = (9, 13)$, and the plaintext be

$$\texttt{ATTACK} = x_1, x_2, \ldots, x_6 = 0, 19, 19, 0, 2, 10.$$

The inverse a^{-1} of a exists and is given by $a^{-1} = 3$. The ciphertext is computed as

$$y_1, y_2, \ldots, y_6 = 13, 2, 2, 13, 5, 25 = \texttt{nccnfz}$$

⋄

Is the affine cipher secure? No! The key space is only a bit larger than in the case of the shift cipher:

$$\text{key space} = (\#\text{values for } a) \times (\#\text{values for } b)$$
$$= 12 \times 26 = 312$$

A key space with 312 elements can, of course, still be searched exhaustively, i.e., brute-force attacked, in a fraction of a second with current desktop PCs. In addition, the affine cipher has the same weakness as the shift and substitution cipher: The mapping between plaintext letters and ciphertext letters is fixed. Hence, it can easily be broken with letter frequency analysis.

The remainder of this book deals with strong cryptographic algorithms which are of practical relevance.

1.5 Discussion and Further Reading

This book addresses practical aspects of cryptography and data security and is intended to be used as an introduction; it is suited for classroom use, distance learning and self-study. At the end of each chapter, we provide a discussion section in which we briefly describe topics for readers interested in further study of the material.

About This Chapter: Historical Ciphers and Modular Arithmetic This chapter introduced a few historical ciphers. However, there are many, many more, ranging from ciphers in ancient times to WW II encryption methods. To readers who wish to learn more about historical ciphers and the role they played over the centuries, the books by Bauer [13], Kahn [97] and Singh [157] are highly recommended. Besides making fascinating bedtime reading, these books help one to understand the role that military and diplomatic intelligence played in shaping world history. They also help to show modern cryptography in a larger context.

The mathematics introduced in this chapter, modular arithmetic, belongs to the field of number theory. This is a fascinating subject area which is, unfortunately, historically viewed as a "branch of mathematics without applications". Thus, it is rarely taught outside mathematics curricula. There is a wealth of books on number theory. Among the classic introductory books are references [129, 148]. A particularly accessible book written for non-mathematications is [156].

Research Community and General References Even though cryptography has matured considerably over the last 30 years, it is still a relatively young field compared to other disciplines, and every year brings many new developments and discoveries. Many research results are published at events organized by the International Association for Cryptologic Research (IACR). The proceedings of the three IACR conferences CRYPTO, EUROCRYPT, and ASIACRYPT as well as the IACR workshops Cryptographic Hardware and Embedded Systems (CHES), Fast Software Encryption (FSE), Public Key Cryptography (PKC) and Theory of Cryptograpy (TCC), are excellent sources for tracking the recent developments in the field of cryptology at large. Two important conferences which deal with the larger issue of security (of which cryptography is one aspect) are the IEEE Symposium on Security and Privacy and the USENIX Security forum. All of the events listed take place annually.

There are several good books on cryptography. As reference sources, the *Handbook of Applied Cryptography* [120] and the more recent *Encyclopedia of Cryptography and Security* [168] are highly recommended; both make excellent additions to this textbook.

Provable Security Due to our focus on practical cryptography, this book omits most aspects related to the theoretical foundations of crypto algorithms and protocols. Especially in modern cryptographic research, there is a strong desire to provide statements about cryptographic schemes which are provable in a strict mathematical sense. For this, the goals of both a security system and the adversary are described in a formal model. Often, proofs are achieved by reducing the security of a system to certain assumptions, e.g., that factorization of integers is hard or that a hash function is collision free.

The field of provable security is quite large. We list now some important subareas. A recent survey on the specific area of provable public-key encryption is given in [55]. Provable security is closely related to *cryptographic foundations*, which studies the general assumptions and approaches needed. For instance, the interrelationship between certain presumably hard problems (e.g., integer factorization and discrete logarithm) are studied. The standard references are [81, 83]. *Zero-knowledge proofs* are concerned with proving a certain knowledge towards another party without revealing the secret. They were originally motivated by proving an entity's identity without revealing a password or key. However, they are typically not used that way any more. An early reference is [139], and a more recent tutorial is given in [82]. *Multiparty computation* can be used to compute answers such as the outcome of an election or determining the highest bid in an auction based on encrypted data. The interesting part is that when the protocol is completed the participants know only their own input and the answer but nothing about the encrypted data of the other participants. Good reference sources are [112] and [83, Chap. 7].

A few times this book also touches upon provable security, for instance the relationship between Diffie–Hellman key exchange and the Diffie–Hellman problem (cf. Sect. 8.4), the block cipher based hash functions in Sect. 11.3.2 or the security of the HMAC message authentication scheme in Sect. 12.2.

As a word of caution, it should be mentioned that even though very practical results have been derived from research in the provable security of crypto schemes, many findings are only of limited practical value. Also, the whole field is not without controversy [84, 102].

Secure System Design Cryptography is often an important tool for building a secure system, but on the other hand secure system design encompasses many other aspects. Security systems are intended to protect something valuable, e.g., information, monetary values, personal property, etc. The main objective of secure system design is to make breaking the system more costly than the value of the protected assets, where the "cost" should be measured in monetary value but also in more abstract terms such as effort or reputation. Generally speaking, adding security to a system often narrows its usability.

In order to approach the problem systematically, several general frameworks exist. They typically require that assets and corresponding security needs have to be defined, and that the attack potential and possible attack paths must be evaluated. Finally, adequate countermeasures have to be specified in order to realize an appropriate level of security for a particular application or environment.

There are standards which can be used for evaluation and help to define a secure system. Among the more prominent ones are ISO/IEC [94] (15408, 15443-1, 15446, 19790, 19791, 19792, 21827), the Common Criteria for Information Technology Security Evaluation [46], the German IT-Grundschutzhandbuch [37], FIPS PUBS [77] and many more.

1.6 Lessons Learned

- Never ever develop your own crypto algorithm unless you have a team of experienced cryptanalysts checking your design.
- Do not use unproven crypto algorithms (i.e., symmetric ciphers, asymmetric ciphers, hash functions) or unproven protocols.
- Attackers always look for the weakest point of a cryptosystem. For instance, a large key space by itself is no guarantee for a cipher being secure; the cipher might still be vulnerable against analytical attacks.
- Key lengths for symmetric algorithms in order to thwart exhaustive key-search attacks are:
  - □ 64 bits: insecure except for data with extremely short-term value.
  - □ 112–128 bits: long-term security of several decades, including attacks by intelligence agencies unless they possess quantum computers. Based on our current knowledge, attacks are only feasible with quantum computers (which do not exist and perhaps never will).
  - □ 256 bit: as above, but possibly against attacks by quantum computers.

■ Modular arithmetic is a tool for expressing historical encryption schemes, such as the affine cipher, in a mathematically elegant way.

Problems

1.1. The ciphertext below was encrypted using a substitution cipher. Decrypt the ciphertext without knowledge of the key.

```
lrvmnir bpr sumvbwvr jx bpr lmiwv yjeryrkbi jx qmbm wi
bpr xjvni mkd ymibrut jx irhx wi bpr riirkvr jx
ymbinlmtmipw utn qmumbr dj w ipmhh but bj rhnvwdmbr bpr
yjeryrkbi jx bpr qmbm mvvjudwko bj yt wkbrusurbmbwjk
lmird jk xjubt trmui jx ibndt

wb wi kjb mk rmit bmiq bj rashmwk rmvp yjeryrkb mkd wbi
iwokwxwvmkvr mkd ijyr ynib urymwk nkrashmwkrd bj ower m
vjyshrbr rashmkmbwjk jkr cjnhd pmer bj lr fnmhwxwrd mkd
wkiswurd bj invp mk rabrkb bpmb pr vjnhd urmvp bpr ibmbr
jx rkhwopbrkrd ywkd vmsmlhr jx urvjokwgwko ijnkdhrii
ijnkd mkd ipmsrhrii ipmsr w dj kjb drry ytirhx bpr xwkmh
mnbpjuwbt lnb yt rasruwrkvr cwbp qmbm pmi hrxb kj djnlb
bpmb bpr xjhhjcwko wi bpr sujsru msshwvmbwjk mkd
wkbrusurbmbwjk w jxxru yt bprjuwri wk bpr pjsr bpmb
riirkvr jx jqwkmcmk qmumbr cwhh urymwk wkbmvb
```

1. Compute the relative frequency of all letters A . . . Z in the ciphertext. You may want to use a tool such as the open-source program CrypTool [50] for this task. However, a paper and pencil approach is also still doable.
2. Decrypt the ciphertext with the help of the relative letter frequency of the English language (see Table 1.1 in Sect. 1.2.2). Note that the text is relatively short and that the letter frequencies in it might not perfectly align with that of general English language from the table.
3. Who wrote the text?

1.2. We received the following ciphertext which was encoded with a shift cipher:
```
xultpaajcxitltlxaarpjhtiwtgxktghidhipxciwtvgtpilpit
ghlxiwiwtxgqadds.
```

1. Perform an attack against the cipher based on a letter frequency count: How many letters do you have to identify through a frequency count to recover the key? What is the cleartext?
2. Who wrote this message?

1.3. We consider the long-term security of the Advanced Encryption Standard (AES) with a key length of 128-bit with respect to exhaustive key-search attacks. AES is perhaps the most widely used symmetric cipher at this time.

1. Assume that an attacker has a special purpose application specific integrated circuit (ASIC) which checks $5 \cdot 10^8$ keys per second, and she has a budget of $1 million. One ASIC costs $50, and we assume 100% overhead for integrating

the ASIC (manufacturing the printed circuit boards, power supply, cooling, etc.). How many ASICs can we run in parallel with the given budget? How long does an average key search take? Relate this time to the age of the Universe, which is about 10^{10} years.
2. We try now to take advances in computer technology into account. Predicting the future tends to be tricky but the estimate usually applied is Moore's Law, which states that the computer power doubles every 18 months while the costs of integrated circuits stay constant. How many years do we have to wait until a key-search machine can be built for breaking AES with 128 bit with an average search time of 24 hours? Again, assume a budget of $1 million (do not take inflation into account).

1.4. We now consider the relation between passwords and key size. For this purpose we consider a cryptosystem where the user enters a key in the form of a password.

1. Assume a password consisting of 8 letters, where each letter is encoded by the ASCII scheme (7 bits per character, i.e., 128 possible characters). What is the size of the key space which can be constructed by such passwords?
2. What is the corresponding key length in bits?
3. Assume that most users use only the 26 lowercase letters from the alphabet instead of the full 7 bits of the ASCII-encoding. What is the corresponding key length in bits in this case?
4. At least how many characters are required for a password in order to generate a key length of 128 bits in case of letters consisting of

   a. 7-bit characters?
   b. 26 lowercase letters from the alphabet?

1.5. As we learned in this chapter, modular arithmetic is the basis of many cryptosystems. As a consequence, we will address this topic with several problems in this and upcoming chapters.

Let's start with an easy one: Compute the result without a calculator.

1. $15 \cdot 29 \bmod 13$
2. $2 \cdot 29 \bmod 13$
3. $2 \cdot 3 \bmod 13$
4. $-11 \cdot 3 \bmod 13$

The results should be given in the range from $0, 1, \ldots,$ modulus-1. Briefly describe the relation between the different parts of the problem.

1.6. Compute without a calculator:

1. $1/5 \bmod 13$
2. $1/5 \bmod 7$
3. $3 \cdot 2/5 \bmod 7$

1.7. We consider the ring $\mathbb{Z}_4$. Construct a table which describes the addition of all elements in the ring with each other:

$$
\begin{array}{c|cccc}
+ & 0 & 1 & 2 & 3 \\
\hline
0 & 0 & 1 & 2 & 3 \\
1 & 1 & 2 & \cdots & \\
2 & \cdots & & & \\
3 & & & &
\end{array}
$$

1. Construct the multiplication table for $\mathbb{Z}_4$.
2. Construct the addition and multiplication tables for $\mathbb{Z}_5$.
3. Construct the addition and multiplication tables for $\mathbb{Z}_6$.
4. There are elements in $\mathbb{Z}_4$ and $\mathbb{Z}_6$ without a multiplicative inverse. Which elements are these? Why does a multiplicative inverse exist for all nonzero elements in $\mathbb{Z}_5$?

1.8. What is the multiplicative inverse of 5 in $\mathbb{Z}_{11}$, $\mathbb{Z}_{12}$, and $\mathbb{Z}_{13}$? You can do a trial-and-error search using a calculator or a PC.

With this simple problem we want now to stress the fact that the inverse of an integer in a given ring depends completely on the ring considered. That is, if the modulus changes, the inverse changes. Hence, it doesn't make sense to talk about an inverse of an element unless it is clear what the modulus is. This fact is crucial for the RSA cryptosystem, which is introduced in Chap. 7. The extended Euclidean algorithm, which can be used for computing inverses efficiently, is introduced in Sect. 6.3.

1.9. Compute x as far as possible without a calculator. Where appropriate, make use of a smart decomposition of the exponent as shown in the example in Sect. 1.4.1:

1. $x = 3^2 \bmod 13$
2. $x = 7^2 \bmod 13$
3. $x = 3^{10} \bmod 13$
4. $x = 7^{100} \bmod 13$
5. $7^x = 11 \bmod 13$

The last problem is called a *discrete logarithm* and points to a hard problem which we discuss in Chap. 8. The security of many public-key schemes is based on the hardness of solving the discrete logarithm for large numbers, e.g., with more than 1000 bits.

1.10. Find all integers n between $0 \le n < m$ that are relatively prime to m for $m = 4, 5, 9, 26$. We denote the *number* of integers n which fulfill the condition by $\phi(m)$, e.g. $\phi(3) = 2$. This function is called "Euler's phi function". What is $\phi(m)$ for $m = 4, 5, 9, 26$?

1.11. This problem deals with the affine cipher with the key parameters $a = 7$, $b = 22$.

1. Decrypt the text below:
   `falszztysyjzyjkywjrztyjztyynaryjkyswarztyegyyj`
2. Who wrote the line?

1.12. Now, we want to extend the affine cipher from Sect. 1.4.4 such that we can encrypt and decrypt messages written with the full German alphabet. The German alphabet consists of the English one together with the three umlauts, Ä, Ö, Ü, and the (even stranger) "double s" character ß. We use the following mapping from letters to integers:

A ↔ 0	B ↔ 1	C ↔ 2	D ↔ 3	E ↔ 4	F ↔ 5
G ↔ 6	H ↔ 7	I ↔ 8	J ↔ 9	K ↔ 10	L ↔ 11
M ↔ 12	N ↔ 13	O ↔ 14	P ↔ 15	Q ↔ 16	R ↔ 17
S ↔ 18	T ↔ 19	U ↔ 20	V ↔ 21	W ↔ 22	X ↔ 23
Y ↔ 24	Z ↔ 25	Ä ↔ 26	Ö ↔ 27	Ü ↔ 28	ß ↔ 29

1. What are the encryption and decryption equations for the cipher?
2. How large is the key space of the affine cipher for this alphabet?
3. The following ciphertext was encrypted using the key $(a = 17, b = 1)$. What is the corresponding plaintext?

  ä u ß w ß

4. From which village does the plaintext come?

1.13. In an attack scenario, we assume that the attacker Oscar manages somehow to provide Alice with a few pieces of plaintext that she encrypts. Show how Oscar can break the affine cipher by using two pairs of plaintext–ciphertext, (x_1, y_1) and (x_2, y_2). What is the condition for choosing x_1 and x_2?

Remark: In practice, such an assumption turns out to be valid for certain settings, e.g., encryption by Web servers, etc. This attack scenario is, thus, very important and is denoted as a *chosen plaintext attack*.

1.14. An obvious approach to increase the security of a symmetric algorithm is to apply the same cipher twice, i.e.:

$$y = e_{k2}(e_{k1}(x))$$

As is often the case in cryptography, things are very tricky and results are often different from the expected and/ or desired ones. In this problem we show that a double encryption with the affine cipher is only as secure as single encryption! Assume two affine ciphers $e_{k1} = a_1 x + b_1$ and $e_{k2} = a_2 x + b_2$.

1. Show that there is a single affine cipher $e_{k3} = a_3 x + b_3$ which performs exactly the same encryption (and decryption) as the combination $e_{k2}(e_{k1}(x))$.
2. Find the values for a_3, b_3 when $a_1 = 3, b_1 = 5$ and $a_2 = 11, b_2 = 7$.
3. For verification: (1) encrypt the letter K first with e_{k1} and the result with e_{k2}, and (2) encrypt the letter K with e_{k3}.
4. Briefly describe what happens if an exhaustive key-search attack is applied to a double-encrypted affine ciphertext. Is the effective key space increased?

Remark: The issue of multiple encryption is of great practical importance in the case of the Data Encryption Standard (DES), for which multiple encryption (in particular, triple encryption) does increase security considerably.

Chapter 2
Stream Ciphers

If we look at the types of cryptographic algorithms that exist in a little bit more detail, we see that the symmetric ciphers can be divided into stream ciphers and block ciphers, as shown in Fig. 2.1.

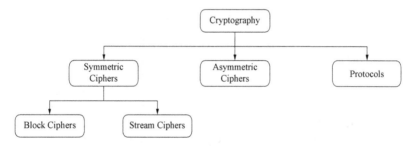

Fig. 2.1 Main areas within cryptography

This chapter gives an introduction to stream ciphers:

■ The pros and cons of stream ciphers
■ Random and pseudorandom number generators
■ A truly unbreakable cipher: the One-Time Pad (OTP)
■ Linear feedback shift registers and Trivium, a modern stream cipher

2.1 Introduction

2.1.1 Stream Ciphers vs. Block Ciphers

Symmetric cryptography is split into block ciphers and stream ciphers, which are easy to distinguish. Figure 2.2 depicts the operational differences between stream (Fig. 2.2a) and block (Fig. 2.2b) ciphers when we want to encrypt b bits at a time, where b is the width of the block cipher.

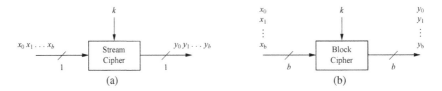

Fig. 2.2 Principles of encrypting b bits with a stream (a) and a block (b) cipher

A description of the principles of the two types of symmetric ciphers follows.

Stream ciphers encrypt bits individually. This is achieved by adding a bit from a *key stream* to a plaintext bit. There are synchronous stream ciphers where the key stream depends only on the key, and asynchronous ones where the key stream also depends on the ciphertext. If the dotted line in Fig. 2.3 is present, the stream cipher is an asynchronous one. Most practical stream ciphers are synchronous ones and Sect. 2.3 of this chapter will deal with them. An example of an asynchronous stream cipher is the cipher feedback (CFB) mode introduced in Sect. 5.1.4.

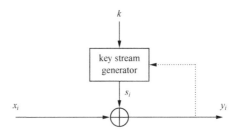

Fig. 2.3 Synchronous and asynchronous stream ciphers

Block ciphers encrypt an entire block of plaintext bits at a time with the same key. This means that the encryption of any plaintext bit in a given block depends on every other plaintext bit in the same block. In practice, the vast majority of block ciphers either have a block length of 128 bits (16 bytes) such as the advanced encryption standard (AES), or a block length of 64 bits (8 bytes) such as the data encryption standard (DES) or triple DES (3DES) algorithm. All of these ciphers are introduced in later chapters.

This chapter gives an introduction to stream ciphers. Before we go into more detail, it will be helpful to learn some useful facts about stream ciphers vs. block ciphers:

1. In practice, in particular for encrypting computer communication on the Internet, block ciphers are used more often than stream ciphers.
2. Because stream ciphers tend to be small and fast, they are particularly relevant for applications with little computational resources, e.g., for cell phones or other small embedded devices. A prominent example for a stream cipher is the A5/1 cipher, which is part of the GSM mobile phone standard and is used for voice encryption. However, stream ciphers are sometimes also used for encrypting Internet traffic, especially the stream cipher RC4.
3. Traditionally, it was assumed that stream ciphers tended to encrypt more efficiently than block ciphers. *Efficient* for software-optimized stream ciphers means that they need fewer processor instructions (or processor cycles) to encrypt one bit of plaintext. For hardware-optimized stream ciphers, *efficient* means they need fewer gates (or smaller chip area) than a block cipher for encrypting at the same data rate. However, modern block ciphers such as AES are also very efficient in software. Moreover, for hardware, there are also highly efficient block ciphers, such as PRESENT, which are as efficient as very compact stream ciphers.

2.1.2 Encryption and Decryption with Stream Ciphers

As mentioned above, stream ciphers encrypt plaintext bits individually. The question now is: How does encryption of an individual bit work? The answer is surprisingly simple: Each bit x_i is encrypted by adding a secret key stream bit s_i modulo 2.

Definition 2.1.1 Stream Cipher Encryption and Decryption
The plaintext, the ciphertext and the key stream consist of individual bits,
i.e., $x_i, y_i, s_i \in \{0, 1\}$.
Encryption: $y_i = e_{s_i}(x_i) \equiv x_i + s_i \bmod 2$
Decryption: $x_i = d_{s_i}(y_i) \equiv y_i + s_i \bmod 2$

Since encryption and decryption functions are both simple additions modulo 2, we can depict the basic operation of a stream cipher as shown in Fig. 2.4. Note that we use a circle with an addition sign as the symbol for modulo 2 addition.

Just looking at the formulae, there are three points about the stream cipher encryption and decryption function which we should clarify:

1. Encryption and decryption are the same functions!

Fig. 2.4 Encryption and decryption with stream ciphers

2. Why can we use a simple modulo 2 addition as encryption?
3. What is the nature of the key stream bits s_i?

The following discussion of these three items will give us already an understanding of some important stream cipher properties.

Why Are Encryption and Decryption the Same Function?

The reason for the similarity of the encryption and decryption function can easily be shown. We must prove that the decryption function actually produces the plaintext bit x_i again. We know that ciphertext bit y_i was computed using the encryption function $y_i \equiv x_i + s_i \bmod 2$. We insert this encryption expression in the decryption function:

$$
\begin{aligned}
d_{s_i}(y_i) &\equiv y_i + s_i \bmod 2 \\
&\equiv (x_i + s_i) + s_i \bmod 2 \\
&\equiv x_i + s_i + s_i \bmod 2 \\
&\equiv x_i + 2s_i \bmod 2 \\
&\equiv x_i + 0 \bmod 2 \\
&\equiv x_i \bmod 2 \quad Q.E.D.
\end{aligned}
$$

The trick here is that the expression ($2s_i \bmod 2$) has always the value zero since $2 \equiv 0 \bmod 2$. Another way of understanding this is as follows: If s_i has either the value 0, in which case $2s_i = 2 \cdot 0 \equiv 0 \bmod 2$. If $s_i = 1$, we have $2s_i = 2 \cdot 1 = 2 \equiv 0 \bmod 2$.

Why Is Modulo 2 Addition a Good Encryption Function?

A mathematical explanation for this is given in the context of the One-Time Pad in Sect. 2.2.2. However, it is worth having a closer look at addition modulo 2. If we do arithmetic modulo 2, the only possible values are 0 and 1 (because if you divide by 2, the only possible remainders are 0 and 1). Thus, we can treat arithmetic modulo 2 as Boolean functions such as AND gates, OR gates, NAND gates, etc. Let's look at the truth table for modulo 2 addition:

This should look familiar to most readers: It is the truth table of the *exclusive-OR*, also called *XOR*, gate. This is in important fact: **Modulo 2 addition is equivalent to**

$$\begin{array}{cc|c} x_i & s_i & y_i \equiv x_i + s_i \bmod 2 \\ \hline 0 & 0 & 0 \\ 0 & 1 & 1 \\ 1 & 0 & 1 \\ 1 & 1 & 0 \end{array}$$

the XOR operation. The XOR operation plays a major role in modern cryptography and will be used many times in the remainder of this book.

The question now is, why is the XOR operation so useful, as opposed to, for instance, the AND operation? Let's assume we want to encrypt the plaintext bit $x_i = 0$. If we look at the truth table we find that we are on either the 1st or 2nd line of the truth table:

Table 2.1 Truth table of the XOR operation

$$\begin{array}{cc|c} x_i & s_i & y_i \\ \hline \mathbf{0} & \mathbf{0} & \mathbf{0} \\ \mathbf{0} & \mathbf{1} & \mathbf{1} \\ 1 & 0 & 1 \\ 1 & 1 & 0 \end{array}$$

Depending on the key bit, the ciphertext y_i is either a zero ($s_i = 0$) or one ($s_i = 1$). If the key bit s_i behaves perfectly randomly, i.e., it is unpredictable and has exactly a 50% chance to have the value 0 or 1, then both possible ciphertexts also occur with a 50% likelihood. Likewise, if we encrypt the plaintext bit $x_i = 1$, we are on line 3 or 4 of the truth table. Again, depending on the value of the key stream bit s_i, there is a 50% chance that the ciphertext is either a 1 or a 0.

We just observed that the XOR function is perfectly balanced, i.e., by observing an output value, there is exactly a 50% chance for any value of the input bits. This distinguishes the XOR gate from other Boolean functions such as the OR, AND or NAND gate. Moreover, AND and NAND gates are not invertible. Let's look at a very simple example for stream cipher encryption.

Example 2.1. Alice wants to encrypt the letter A, where the letter is given in ASCII code. The ASCII value for A is $65_{10} = 1000001_2$. Let's furthermore assume that the first key stream bits are $(s_0, \ldots, s_6) = 0101100$.

Alice	**Oscar**	**Bob**
$x_0, \ldots, x_6 = 1000001 = \text{A}$		
$\oplus$		
$s_0, \ldots, s_6 = 0101100$		
$y_0, \ldots, y_6 = 1101101 = \text{m}$		
	$\xrightarrow{\text{m}=1101101}$	
		$y_0, \ldots, y_6 = 1101101$
		$\oplus$
		$s_0, \ldots, s_6 = 0101100$
		$x_0, \ldots, x_6 = 1000001 = \text{A}$

Note that the encryption by Alice turns the uppercase A into the lower case letter
m. Oscar, the attacker who eavesdrops on the channel, only sees the ciphertext letter
m. Decryption by Bob *with the same key stream* reproduces the plaintext A again.
◇

So far, stream ciphers look unbelievably easy: One simply takes the plaintext,
performs an XOR operation with the key and obtains the ciphertext. On the receiving
side, Bob does the same. The "only" thing left to discuss is the last question from
above.

What Exactly Is the Nature of the Key Stream?

It turns out that the generation of the values s_i, which are called the *key stream*, is
the central issue for the security of stream ciphers. In fact, the security of a stream
cipher *completely depends on the key stream*. The key stream bits s_i are *not* the key
bits themselves. So, how do we get the key stream? Generating the key stream is
pretty much what stream ciphers are about. This is a major topic and is discussed
later in this chapter. However, we can already guess that a central requirement for
the key stream bits should be that they appear like a random sequence to an attacker.
Otherwise, an attacker Oscar could guess the bits and do the decryption by himself.
Hence, we first need to learn more about random numbers.

Historical Remark Stream ciphers were invented in 1917 by Gilbert Vernam, even
though they were not called stream ciphers back at that time. He built an elec-
tromechanical machine which automatically encrypted teletypewriter communication. The plaintext was fed into the machine as one paper tape, and the key stream
as a second tape. This was the first time that encryption and transmission was au-
tomated in one machine. Vernam studied electrical engineering at Worcester Poly-
technic Institute (WPI) in Massachusetts where, by coincidence, one of the authors
of this book was a professor in the 1990s. Stream ciphers are sometimes referred to
as Vernam ciphers. Occasionally, one-time pads are also called Vernam ciphers. For
further reading on Vernam's machine, the book by Kahn [97] is recommended.

2.2 Random Numbers and an Unbreakable Stream Cipher

2.2.1 Random Number Generators

As we saw in the previous section, the actual encryption and decryption of stream
ciphers is extremely simple. The security of stream ciphers hinges entirely on a
"suitable" key stream $s_0, s_1, s_2, \ldots$. Since randomness plays a major role, we will first
learn about the three types of random number generators (RNG) that are important
for us.

True Random Number Generators (TRNG)

True random number generators (TRNGs) are characterized by the fact that their output cannot be reproduced. For instance, if we flip a coin 100 times and record the resulting sequence of 100 bits, it will be virtually impossible for anyone on Earth to generate the same 100 bit sequence. The chance of success is $1/2^{100}$, which is an extremely small probability. TRNGs are based on physical processes. Examples include coin flipping, rolling of dice, semiconductor noise, clock jitter in digital circuits and radioactive decay. In cryptography, TRNGs are often needed for generating session keys, which are then distributed between Alice and Bob, and for other purposes.

(General) Pseudorandom Number Generators (PRNG)

Pseudorandom number generators (PRNGs) generate sequences which are *computed* from an initial seed value. Often they are computed recursively in the following way:

$$s_0 = \text{seed}$$
$$s_{i+1} = f(s_i), \quad i = 0, 1, \ldots$$

A generalization of this are generators of the form $s_{i+1} = f(s_i, s_{i-1}, \ldots, s_{i-t})$, where t is a fixed integer. A popular example is the *linear congruential generator*:

$$s_0 = \text{seed}$$
$$s_{i+1} \equiv a\,s_i + b \bmod m, \quad i = 0, 1, \ldots$$

where a, b, m are integer constants. Note that PRNGs are not random in a true sense because they can be computed and are thus completely deterministic. A widely used example is the *rand()* function used in ANSI C. It has the parameters:

$$s_0 = 12345$$
$$s_{i+1} \equiv 1103515245\,s_i + 12345 \bmod 2^{31}, \quad i = 0, 1, \ldots$$

A common requirement of PRNGs is that they possess good statistical properties, meaning their output approximates a sequence of true random numbers. There are many mathematical tests, e.g., the chi-square test, which can verify the statistical behavior of PRNG sequences. Note that there are many, many applications for pseudorandom numbers outside cryptography. For instance, many types of simulations or testing, e.g., of software or of VLSI chips, need random data as input. That is the reason why a PRNG is included in the ANSI C specification.

Cryptographically Secure Pseudorandom Number Generators (CSPRNG)

Cryptographically secure pseudorandom number generators (CSPRNGs) are a special type of PRNG which possess the following additional property: A CSPRNG is PRNG which is *unpredictable*. Informally, this means that given n output bits of the key stream $s_i, s_{i+1}, \ldots, s_{i+n-1}$, where n is some integer, it is computationally infeasible to compute the subsequent bits $s_{i+n}, s_{i+n+1}, \ldots$. A more exact definition is that given n consecutive bits of the key stream, there is no polynomial time algorithm that can predict the next bit s_{n+1} with better than 50% chance of success. Another property of CSPRNG is that given the above sequence, it should be computationally infeasible to compute any preceding bits $s_{i-1}, s_{i-2}, \ldots$.

Note that the need for unpredictability of CSPRNGs is unique to cryptography. In virtually all other situations where pseudorandom numbers are needed in computer science or engineering, unpredictability is not needed. As a consequence, the distinction between PRNG and CSPRN and their relevance for stream ciphers is often not clear to non-cryptographers. Almost all PRNG that were designed without the clear purpose of being stream ciphers are not CSPRNGs.

2.2.2 The One-Time Pad

In the following we discuss what happens if we use the three types of random numbers as generators for the key stream sequence $s_0, s_1, s_2, \ldots$ of a stream cipher. Let's first define what a perfect cipher should be:

Definition 2.2.1 Unconditional Security
A cryptosystem is unconditionally or information-theoretically secure if it cannot be broken even with infinite computational resources.

Unconditional security is based on information theory and assumes no limit on the attacker's computational power. This looks like a pretty straightforward definition. It is in fact straightforward, but the requirements for a cipher to be unconditionally secure are tremendous. Let's look at it using a gedankenexperiment: Assume we have a symmetric encryption algorithm (it doesn't matter whether it's a block cipher or stream cipher) with a key length of 10,000 bits, and the only attack that works is an exhaustive key search, i.e, a brute-force attack. From the discussion in Sect. 1.3.2, we recall that 128 bits are more than enough for long-term security. So, is a cipher with 10,000 bits unconditionally secure? The answer is simple: No! Since an attacker can have *infinite* computational resources, we can simply assume that the attacker has 2^{10000} computers available and every computer checks exactly one key. This will give us a correct key in one time step. Of course, there is no way that 2^{10000} computer can ever be built, the number is too large. (It is estimated that there are "only" about 2^{266} atoms in the known universe.) The cipher would merely be *computationally secure* but not unconditionally.

All this said, we now show a way to build an unconditionally secure cipher that is quite simple. This cipher is called the One-Time Pad.

Definition 2.2.2 One-Time Pad (OTP)
A stream cipher for which

1. *the key stream $s_0, s_1, s_2, \ldots$ is generated by a true random number generator, and*
2. *the key stream is only known to the legitimate communicating parties, and*
3. *every key stream bit s_i is only used once*

is called a one-time pad. The one-time pad is unconditionally secure.

It is easy to show why the OTP is unconditionally secure. Here is a sketch of a proof. For every ciphertext bit we get an equation of this form:

$$y_0 \equiv x_0 + s_0 \bmod 2$$
$$y_1 \equiv x_1 + s_1 \bmod 2$$
$$\vdots$$

Each individual relation is a linear equation modulo 2 with two unknowns. They are impossible to solve. If the attacker knows the value for y_0 (0 or 1), he cannot determine the value of x_0. In fact, the solutions $x_0 = 0$ and $x_0 = 1$ are exactly equally likely if s_0 stems from a truly random source and there is 50% chance that it has the value 0 and 1. The situation is identical for the second equation and all subsequent ones. Note that the situation is different if the values s_i are not truly random. In this case, there is some functional relationship between them, and the equations shown above are not independent. Even though it might still be hard to solve the system of equations, it is not provably secure!

So, now we have a simple cipher which is perfectly secure. There are rumors that the red telephone between the White House and the Kremlin was encrypted using an OTP during the Cold War. Obviously there must be a catch since OTPs are not used for Web browsers, e-mail encryption, smart cards, mobile phones, or other important applications. Let's look at the implications of the three requirements in Defintion 2.2.2. The first requirement means that we need a TRNG. That means we need a device, e.g., based on white noise of a semiconductor, that generates truly random bits. Since standard PCs do not have TRNG, this requirement might not be that convenient but can certainly be met. The second requirement means that Alice has to get the random bits securely to Bob. In practice that could mean that Alice burns the true random bits on a CD ROM and sends them securely, e.g., with a trusted courier, to Bob. Still doable, but not great. The third requirement is probably the most impractical one: Key stream bits cannot be re-used. *This implies that we need one key bit for every bit of plaintext.* Hence, our key is as long as the plaintext! This is probably the major drawback of the OTP. Even if Alice and Bob share a CD with 1 MByte of true random numbers, we run quickly into limits. If they send a single email with an attachment of 1 MByte, they could encrypt and decrypt it, but after that they would need to exchange a true random key stream again.

For these reasons OTPs are rarely used in practice. However, they give us a great design idea for secure ciphers: If we XOR truly random bits and plaintext, we get ciphertext that can certainly not be broken by an attacker. We will see in the next section how we can use this fact to build practical stream ciphers.

2.2.3 Towards Practical Stream Ciphers

In the previous section we saw that OTPs are unconditionally secure, but that they have drawbacks which make them impractical. What we try to do with practical stream ciphers is to replace the truly random key stream bits by a pseudorandom number generator where the key k serves as a seed. The principle of practical stream ciphers is shown in Fig. 2.5.

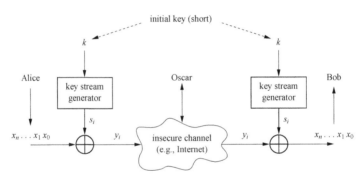

Fig. 2.5 Practical stream ciphers

Before we turn to stream ciphers used in the real world, it should be stressed that practical stream ciphers are not unconditionally secure. In fact, *all* known practical crypto algorithms (stream ciphers, block ciphers, public-key algorithms) are not unconditionally secure. The best we can hope for is *computational security*, which we define as follows:

Definition 2.2.3 Computational Security
A cryptosystem is computationally secure *if the best known algorithm for breaking it requires at least* t *operations.*

This seems like a reasonable definition, but there are still several problems with it. First, often we do not know what the best algorithm for a given attack is. A prime example is the RSA public-key scheme, which can be broken by factoring large integers. Even though many factoring algorithms are known, we do not know whether there exist any better ones. Second, even if a lower bound on the complexity of one attack is known, we do not know whether any other, more powerful attacks are possible. We saw this in Sect. 1.2.2 during the discussion about the substitution cipher: Even though we know the exact computational complexity for an exhaustive key search, there exist other more powerful attacks. The best we can do in practice is to design crypto schemes for which it is *assumed* that they are computationally secure. For symmetric ciphers this usually means one hopes that there is no attack method with a complexity better than an exhaustive key search.

Let's go back to Fig. 2.5. This design emulates ("behaves like") a one-time pad. It has the major advantage over the OTP that Alice and Bob only need to exchange a secret key that is at most a few 100 bits long, and that does not have to be as long as the message we want to encrypt. We now have to think carefully about the properties of the key stream $s_0, s_1, s_2, \ldots$ that is generated by Alice and Bob. Obviously, we need some type of random number generator to derive the key stream. First, we note that we cannot use a TRNG since, by definition, Alice and Bob will not be able to generate the same key stream. Instead we need deterministic, i.e., pseudorandom, number generators. We now look at the other two generators that were introduced in the previous section.

Building Key Streams from PRNGs

Here is an idea that seems promising (but in fact is pretty bad): Many PRNGs possess good statistical properties, which are necessary for a strong stream cipher. If we apply statistical tests to the key stream sequence, the output should pretty much behave like the bit sequence generated by tossing a coin. So it is tempting to assume that a PRNG can be used to generate the key stream. But all of this is not sufficient for a stream cipher since our opponent, Oscar, is smart. Consider the following attack:

Example 2.2. Let's assume a PRNG based on the linear congruential generator:

$$S_0 = \text{seed}$$
$$S_{i+1} \equiv A S_i + B \bmod m, \quad i = 0, 1, \ldots$$

where we choose m to be 100 bits long and $S_i, A, B \in \{0, 1, \ldots, m-1\}$. Note that this PRNG can have excellent statistical properties if we choose the parameters carefully. The modulus m is part of the encryption scheme and is publicly known. The secret key comprises the values (A, B) and possibly the seed S_0, each with a length of 100. That gives us a key length of 200 bit, which is more than sufficient to protect against a brute-force attack. Since this is a stream cipher, Alice can encrypt:

$$y_i \equiv x_i + s_i \bmod 2$$

where s_i are the bits of the binary representation of the PRNG output symbols S_j.

But Oscar can easily launch an attack. Assume he knows the first 300 bits of plaintext (this is only 300/8=37.5 byte), e.g., file header information, or he guesses part of the plaintext. Since he certainly knows the ciphertext, he can now compute the first 300 bits of key stream as:

$$s_i \equiv y_i + x_i \bmod m \ , \ i = 1, 2, \ldots, 300$$

These 300 bits immediately give the first three output symbols of the PRNG: $S_1 = (s_1, \ldots, s_{100})$, $S_2 = (s_{101}, \ldots, s_{200})$ and $S_3 = (s_{201}, \ldots, s_{300})$. Oscar can now generate two equations:

$$S_2 \equiv A S_1 + B \bmod m$$
$$S_3 \equiv A S_2 + B \bmod m$$

This is a system of linear equations over $\mathbb{Z}_m$ with two unknowns A and B. But those two values are the key, and we can immediately solve the system, yielding:

$$A \equiv (S_2 - S_3)/(S_1 - S_2) \bmod m$$
$$B \equiv S_2 - S_1(S_2 - S_3)/(S_1 - S_2) \bmod m$$

In case $\gcd((S_1 - S_2), m)) \neq 1$ we get multiple solutions since this is an equation system over $\mathbb{Z}_m$. However, with a fourth piece of known plaintext the key can uniquely be detected in almost all cases. Alternatively, Oscar simply tries to encrypt the message with each of the multiple solutions found. Hence, in summary: if we know a few pieces of plaintext, we can compute the key and decrypt the entire ciphertext!

◇

This type of attack is why the notation of CSPRNG was invented.

Building Key Streams from CSPRNGs

What we need to do to prevent the attack above is to use a CSPRNG, which assures that the key stream is unpredictable. We recall that this means that given the first n output bits of the key stream $s_1, s_2, \ldots, s_n$, it is computationally infeasible to compute the bits $s_{n+1}, s_{n+2}, \ldots$. Unfortunately, pretty much all pseudorandom number generators that are used for applications outside cryptography are *not* cryptographically secure. Hence, in practice, we need to use specially designed pseudorandom number generators for stream ciphers.

The question now is how practical stream ciphers actually look. There are many proposals for stream ciphers out in the literature. They can roughly be classified as ciphers either optimized for software implementation or optimized for hardware implementation. In the former case, the ciphers typically require few CPU instructions to compute one key stream bit. In the latter case, they tend to be based on operations which can easily be realized in hardware. A popular example is shift registers with feedback, which are discussed in the next section. A third class of stream ciphers is realized by using block ciphers as building blocks. The cipher feedback mode, output feedback mode and counter mode to be introduced in Chap. 5 are examples of stream ciphers derived from block ciphers.

It could be argued that the state-of-the-art in block cipher design is more advanced than stream ciphers. Currently it seems to be easier for scientists to design "secure" block ciphers than stream ciphers. Subsequent chapters deal in great detail with the two most popular and standardized block ciphers, DES and AES.

2.3 Shift Register-Based Stream Ciphers

As we have learned so far, practical stream ciphers use a stream of key bits $s_1, s_2, \ldots$ that are generated by the key stream generator, which should have certain properties. An elegant way of realizing long pseudorandom sequences is to use linear feedback shift registers (LFSRs). LFSRs are easily implemented in hardware and many, but certainly not all, stream ciphers make use of LFSRs. A prominent example is the A5/1 cipher, which is standardized for voice encryption in GSM. As we will see, even though a plain LFSR produces a sequence with good statistical properties, it is cryptographically weak. However, combinations of LFSRs, such as A5/1 or the cipher Trivium, can make secure stream ciphers. It should be stressed that there are many ways for constructing stream ciphers. This section only introduces one of several popular approaches.

2.3.1 Linear Feedback Shift Registers (LFSR)

An LFSR consists of clocked storage elements (*flip-flops*) and a *feedback path*. The number of storage elements gives us the *degree* of the LFSR. In other words, an LFSR with m flip-flops is said to be of degree m. The feedback network computes the input for the last flip-flop as XOR-sum of certain flip-flops in the shift register.

Example 2.3. **Simple LFSR** We consider an LFSR of degree $m = 3$ with flip-flops FF_2, FF_1, FF_0, and a feedback path as shown in Fig. 2.6. The internal state bits are denoted by s_i and are shifted by one to the right with each clock tick. The rightmost state bit is also the current output bit. The leftmost state bit is computed in the feedback path, which is the XOR sum of some of the flip-flop values in the previous clock period. Since the XOR is a linear operation, such circuits are called linear feedback shift registers. If we assume an initial state of $(s_2 = 1, s_1 = 0, s_0 = 0)$, Table 2.2 gives the complete sequence of states of the LFSR. Note that the rightmost column is the output of the LFSR. One can see from this example that the LFSR

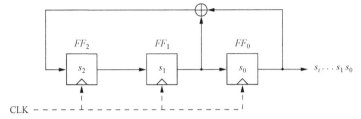

Fig. 2.6 Linear feedback shift register of degree 3 with initial values s_2, s_1, s_0

Table 2.2 Sequence of states of the LFSR

clk	FF_2	FF_1	$FF_0 = s_i$
0	1	0	0
1	0	1	0
2	1	0	1
3	1	1	0
4	1	1	1
5	0	1	1
6	0	0	1
7	1	0	0
8	0	1	0

starts to repeat after clock cycle 6. This means the LFSR output has period of length 7 and has the form:

$$0010111 \; 0010111 \; 0010111 \; \ldots$$

There is a simple formula which determines the functioning of this LFSR. Let's look at how the output bits s_i are computed, assuming the initial state bits s_0, s_1, s_2:

$$s_3 \equiv s_1 + s_0 \bmod 2$$
$$s_4 \equiv s_2 + s_1 \bmod 2$$
$$s_5 \equiv s_3 + s_2 \bmod 2$$
$$\vdots$$

In general, the output bit is computed as:

$$s_{i+3} \equiv s_{i+1} + s_i \bmod 2$$

where $i = 0, 1, 2, \ldots$

◇

This was, of course, a simple example. However, we could already observe many important properties. We will now look at general LFSRs.

A Mathematical Description of LFSRs

The general form of an LFSR of degree m is shown in Fig. 2.7. It shows m flip-flops and m possible feedback locations, all combined by the XOR operation. Whether a feedback path is active or not, is defined by the *feedback coefficient* $p_0, p_1, \ldots, p_{m-1}$:

- If $p_i = 1$ (closed switch), the feedback is active.
- If $p_i = 0$ (open switch), the corresponding flip-flop output is not used for the feedback.

With this notation, we obtain an elegant mathematical description for the feedback path. If we *multiply* the output of flip-flop i by its coefficient p_i, the result is either the output value if $p_i = 1$, which corresponds to a closed switch, or the value zero if $p_i = 0$, which corresponds to an open switch. The values of the feedback coefficients are crucial for the output sequence produced by the LFSR.

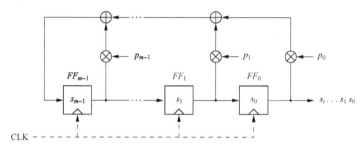

Fig. 2.7 General LFSR with feedback coefficients p_i and initial values $s_{m-1}, \ldots, s_0$

Let's assume the LFSR is initially loaded with the values $s_0, \ldots, s_{m-1}$. The next output bit of the LFSR s_m, which is also the input to the leftmost flip-flop, can be computed by the XOR-sum of the products of flip-flop outputs and corresponding feedback coefficient:

$$s_m \equiv s_{m-1}p_{m-1} + \cdots + s_1 p_1 + s_0 p_0 \bmod 2$$

The next LFSR output can be computed as:

$$s_{m+1} \equiv s_m p_{m-1} + \cdots + s_2 p_1 + s_1 p_0 \bmod 2$$

In general, the output sequence can be described as:

$$s_{i+m} \equiv \sum_{j=0}^{m-1} p_j \cdot s_{i+j} \bmod 2; \quad s_i, p_j \in \{0,1\}; \ i = 0, 1, 2, \ldots \tag{2.1}$$

Clearly, the output values are given through a combination of some previous output values. LFSRs are sometimes referred to as *linear recurrences*.

Due to the finite number of recurring states, the output sequence of an LFSR repeats periodically. This was also illustrated in Example 2.3. Moreover, an LFSR can produce output sequences of different lengths, depending on the feedback coefficients. The following theorem gives us the *maximum length* of an LFSR as function of its degree.

Theorem 2.3.1 *The maximum sequence length generated by an LFSR of degree m is* $2^m - 1$.

It is easy to show that this theorem holds. The *state* of an LFSR is uniquely determined by the m internal register bits. Given a certain state, the LFSR deterministically assumes its next state. Because of this, as soon as an LFSR assumes a previous state, it starts to repeat. Since an m-bit state vector can only assume $2^m - 1$ nonzero states, the maximum sequence length before repetition is $2^m - 1$. Note that the all-zero state must be excluded. If an LFSR assumes this state, it will get "stuck" in it, i.e., it will never be able to leave it again. Note that only certain configurations $(p_0, \ldots, p_{m-1})$ yield maximum length LFSRs. We give a small example for this below.

Example 2.4. LFSR with maximum-length output sequence
Given an LFSR of degree $m = 4$ and the feedback path $(p_3 = 0, p_2 = 0, p_1 = 1, p_0 = 1)$, the output sequence of the LFSR has a period of $2^m - 1 = 15$, i.e., it is a maximum-length LFSR.
$\diamond$

Example 2.5. LFSR with non-maximum output sequence
Given an LFSR of degree $m = 4$ and $(p_3 = 1, p_2 = 1, p_1 = 1, p_0 = 1)$, then the output sequence has period of 5; therefore, it is not a maximum-length LFSR. $\diamond$

The mathematical background of the properties of LFSR sequences is beyond the scope of this book. However, we conclude this introduction to LFSRs with some additional facts. LFSRs are often specified by polynomials using the following notation: An LFSR with a feedback coefficient vector $(p_{m-1}, \ldots, p_1, p_0)$ is represented by the polynomial

$$P(x) = x^m + p_{m-1}x^{m-1} + \ldots + p_1x + p_0$$

For instance, the LFSR from the example above with coefficients $(p_3 = 0, p_2 = 0, p_1 = 1, p_0 = 1)$ can alternatively be specified by the polynomial $x^4 + x + 1$. This seemingly odd notation as a polynomial has several advantages. For instance, maximum-length LFSRs have what is called *primitive polynomials*. Primitive polynomials are a special type of irreducible polynomial. Irreducible polynomials are roughly comparable with prime numbers, i.e., their only factors are 1 and the polynomial itself. Primitive polynomials can relatively easily be computed. Hence, maximum-length LFSRs can easily be found. Table 2.3 shows one primitive polynomial for every value of m in the range from $m = 2, 3, \ldots, 128$. As an example, the notation $(0,2,5)$ refers to the polynomial $1+x^2+x^5$. Note that there are many primitive polynomials for every given degree m. For instance, there exist 69,273,666 different primitive polynomials of degree $m=31$.

Table 2.3 Primitive polynomials for maximum-length LFSRs

(0,1,2)	(0,1,3,4,24)	(0,1,46)	(0,1,5,7,68)	(0,2,3,5,90)	(0,3,4,5,112)
(0,1,3)	(0,3,25)	(0,5,47)	(0,2,5,6,69)	(0,1,5,8,91)	(0,2,3,5,113)
(0,1,4)	(0,1,3,4,26)	(0,2,3,5,48)	(0,1,3,5,70)	(0,2,5,6,92)	(0,2,3,5,114)
(0,2,5)	(0,1,2,5,27)	(0,4,5,6,49)	(0,1,3,5,71)	(0,2,93)	(0,5,7,8,115)
(0,1,6)	(0,1,28)	(0,2,3,4,50)	(0,3,9,10,72)	(0,1,5,6,94)	(0,1,2,4,116)
(0,1,7)	(0,2,29)	(0,1,3,6,51)	(0,2,3,4,73)	(0,11,95)	(0,1,2,5,117)
(0,1,3,4,8)	(0,1,30)	(0,3,52)	(0,1,2,6,74)	(0,6,9,10,96)	(0,2,5,6,118)
(0,1,9)	(0,3,31)	(0,1,2,6,53)	(0,1,3,6,75)	(0,6,97)	(0,8,119)
(0,3,10)	(0,2,3,7,32)	(0,3,6,8,54)	(0,2,4,5,76)	(0,3,4,7,98)	(0,1,3,4,120)
(0,2,11)	(0,1,3,6,33)	(0,1,2,6,55)	(0,2,5,6,77)	(0,1,3,6,99)	(0,1,5,8,121)
(0,3,12)	(0,1,3,4,34)	(0,2,4,7,56)	(0,1,2,7,78)	(0,2,5,6,100)	(0,1,2,6,122)
(0,1,3,4,13)	(0,2,35)	(0,4,57)	(0,2,3,4,79)	(0,1,6,7,101)	(0,2,123)
(0,5,14)	(0,2,4,5,36)	(0,1,5,6,58)	(0,2,4,9,80)	(0,3,5,6,102)	(0,37,124)
(0,1,15)	(0,1,4,6,37)	(0,2,4,7,59)	(0,4,81)	(0,9,103)	(0,5,6,7,125)
(0,1,3,5,16)	(0,1,5,6,38)	(0,1,60)	(0,4,6,9,82)	(0,1,3,4,104)	(0,2,4,7,126)
(0,3,17)	(0,4,39)	(0,1,2,5,61)	(0,2,4,7,83)	(0,4,105)	(0,1,127)
(0,3,18)	(0,3,4,5,40)	(0,3,5,6,62)	(0,5,84)	(0,1,5,6,106)	(0,1,2,7,128)
(0,1,2,5,19)	(0,3,41)	(0,1,63)	(0,1,2,8,85)	(0,4,7,9,107)	
(0,3,20)	(0,1,2,5,42)	(0,1,3,4,64)	(0,2,5,6,86)	(0,1,4,6,108)	
(0,2,21)	(0,3,4,6,43)	(0,1,3,4,65)	(0,1,5,7,87)	(0,2,4,5,109)	
(0,1,22)	(0,5,44)	(0,3,66)	(0,8,9,11,88)	(0,1,4,6,110)	
(0,5,23)	(0,1,3,4,45)	(0,1,2,5,67)	(0,3,5,6,89)	(0,2,4,7,111)	

2.3.2 Known-Plaintext Attack Against Single LFSRs

As indicated by its name, LFSRs are linear. Linear systems are governed by linear relationships between their inputs and outputs. Since linear dependencies can relatively easily be analyzed, this can be a major advantage, e.g., in communication systems. However, a cryptosystem where the key bits only occur in linear relationships makes a highly insecure cipher. We will now investigate how the linear behavior of a LFSR leads to a powerful attack.

If we use an LFSR as a stream cipher, the secret key k is the feedback coefficient vector $(p_{m-1}, \ldots, p_1, p_0)$. An attack is possible if the attacker Oscar knows some plaintext and the corresponding ciphertext. We further assume that Oscar knows the degree m of the LFSR. The attack is so efficient that he can easily try a large number of possible m values, so that this assumption is not a major restriction. Let the known plaintext be given by $x_0, x_1, \ldots, x_{2m-1}$ and the corresponding ciphertext by $y_0, y_1, \ldots, y_{2m-1}$. With these $2m$ pairs of plaintext and ciphertext bits, Oscar reconstructs the first $2m$ key stream bits:

$$s_i \equiv x_i + y_i \bmod 2; \quad i = 0, 1, \ldots, 2m-1.$$

The goal is now to find the key which is given by the feedback coefficients p_i.

Eq. (2.1) is a description of the relationship of the unknown key bits p_i and the key stream output. We repeat the equation here for convenience:

$$s_{i+m} \equiv \sum_{j=0}^{m-1} p_j \cdot s_{i+j} \bmod 2; \quad s_i, p_j \in \{0,1\}; \quad i = 0, 1, 2, \ldots$$

Note that we get a different equation for every value of i. Moreover, the equations are linearly independent. With this knowledge, Oscar can generate m equations for the first m values of i:

$$
\begin{aligned}
i &= 0, & s_m &\equiv p_{m-1}s_{m-1} + \ldots + p_1 s_1 + p_0 s_0 & \bmod 2 \\
i &= 1, & s_{m+1} &\equiv p_{m-1}s_m + \ldots + p_1 s_2 + p_0 s_1 & \bmod 2 \\
&\vdots & \vdots &\quad\vdots\;\; \vdots & \vdots \\
i &= m-1, & s_{2m-1} &\equiv p_{m-1}s_{2m-2} + \ldots + p_1 s_m + p_0 s_{m-1} & \bmod 2
\end{aligned}
\tag{2.2}
$$

He has now m linear equations in m unknowns $p_0, p_1, \ldots, p_{m-1}$. This system can easily be solved by Oscar using Gaussian elimination, matrix inversion or any other algorithm for solving systems of linear equations. Even for large values of m, this can be done easily with a standard PC.

This situation has major consequences: **as soon as Oscar knows $2m$ output bits of an LFSR of degree m, the p_i coefficients can exactly be constructed by merely solving a system of linear equations**. Once he has computed these feedback coefficients, he can "build" the LFSR and load it with any m consecutive output bits that he already knows. Oscar can now clock the LFSR and produce the entire output sequence. Because of this powerful attack, LFSRs by themselves are extremely insecure! They are a good example of a PRNG with good statistical properties but with terrible cryptographical ones. Nevertheless, all is not lost. There are many stream ciphers which use *combinations* of several LFSRs to build strong cryptosystems. The cipher Trivium in the next section is an example.

2.3.3 Trivium

Trivium is a relatively new stream cipher which uses an 80-bit key. It is based on a combination of three shift registers. Even though these are feedback shift registers, there are nonlinear components used to derive the output of each register, unlike the LFSRs that we studied in the previous section.

Description of Trivium

As shown in Fig. 2.8, at the heart of Trivium are three shift registers, A, B and C. The lengths of the registers are 93, 84 and 111, respectively. The XOR-sum of all three register outputs forms the key stream s_i. A specific feature of the cipher is that

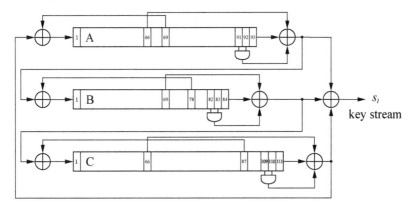

Fig. 2.8 Internal structure of the stream cipher Trivium

the output of each register is connected to the input of another register. Thus, the registers are arranged in circle-like fashion. The cipher can be viewed as consisting of one circular register with a total length of $93 + 84 + 111 = 288$. Each of the three registers has similar structure as described below.

The input of each register is computed as the XOR-sum of two bits:

- The output bit of another register according to Fig. 2.8. For instance, the output of register A is part of the input of register B.
- One register bit at a specific location is fed back to the input. The positions are given in Table 2.4. For instance, bit 69 of register A is fed back to its input.

The output of each register is computed as the XOR-sum of three bits:

- The rightmost register bit.
- One register bit at a specific location is fed forward to the output. The positions are given in Table 2.4. For instance, bit 66 of register A is fed to its output.
- The output of a logical AND function whose input is two specific register bits. Again, the positions of the AND gate inputs are given in Table 2.4.

Table 2.4 Specification of Trivium

	register length	feedback bit	feedforward bit	AND inputs
A	93	69	66	91, 92
B	84	78	69	82, 83
C	111	87	66	109, 110

Note that the AND operation is equal to multiplication in modulo 2 arithmetic. If we multiply two unknowns, and the register contents are the unknowns that an attacker wants to recover, the resulting equations are no longer linear as they contain products of two unknowns. Thus, the feedforward paths involving the AND operation are crucial for the security of Trivium as they prevent attacks that exploit the linearity of the cipher, as the one applicable to plain LFSRs shown in the previous section.

Encryption with Trivium

Almost all modern stream ciphers have two input parameters: a key k and an initialization vector IV. The former is the regular key that is used in every symmetric crypto system. The IV serves as a randomizer and should take a new value for every encryption session. It is important to note that the IV does not have to be kept secret, it merely must change for every session. Such values are often referred to as *nonces*, which stands for "number used once". Its main purpose is that two key streams produced by the cipher should be different, even though the key has not changed. If this were not the case, the following attack becomes possible. If an attacker has known plaintext from a first encryption, he can compute the corresponding key stream. The second encryption using the same key stream can now immediately be deciphered. Without a changing IV, stream cipher encryption is highly deterministic. Methods for generating IVs are discussed in Sect. 5.1.2. Let's look at the details of running Trivium:

Initialization Initially, an 80-bit IV is loaded into the 80 leftmost locations of register A, and an 80-bit key is loaded in the 80 leftmost locations of register B. All other register bits are set to zero with the exception of the three rightmost bits of register C, i.e., bits c_{109}, c_{110} and c_{111}, which are set to 1.

Warm-up Phase In the first phase, the cipher is clocked $4 \times 288 = 1152$ times. No cipher output is generated.

Encryption Phase The bits produced hereafter, i.e., starting with the output bit of cycle 1153, form the key stream.

The warm-up phase is needed for randomizing the cipher sufficiently. It makes sure that the key stream depends on both the key k and the IV.

An attractive feature of Trivium is its compactness, especially if implemented in hardware. It mainly consists of a 288-bit shift register and a few Boolean operations. It is estimated that a hardware implementation of the cipher occupies and area of between about 3500 and 5500 gate equivalences, depending on the degree of parallelization. (A gate equivalence is the chip area occupied by a 2-input NAND gate.) For instance, an implementation with 4000 gates computes the key stream at a rate of 16 bits/clock cycle. This is considerably smaller than most block ciphers such as AES and is very fast. If we assume that this hardware design is clocked at a moderate 125 MHz, the encryption rate would be $16\text{bit} \times 125\text{MHz} = 2$ Gbit/sec. In software, it is estimated that computing 8 output bits takes 12 cycles on a 1.5 GHz Intel CPU, resulting in a theoretical encryption rate of 1 Gbit/sec.

Even though there are no known attacks at the time of writing, one should keep in mind that Trivium is a relatively new cipher and attacks in the future are certainly a possibility. In the past, many other stream ciphers were found to be not secure. More information on Trivium can be found in [164].

2.4 Discussion and Further Reading

Established Stream Ciphers Even though many stream ciphers have been proposed over the years, there are considerably fewer well-investigated ones. The security of many proposed stream ciphers is unknown, and many stream ciphers have been broken. In the case of software-oriented stream ciphers, arguably the best-investigated ones are RC4 [144] and SEAL [120, Sect. 6.4.1]. Note that there are some known weaknesses in RC4, even though it is still secure in practice if it is used correctly [142]. The SEAL cipher, on the other hand, is patented.

In the case of hardware-oriented ciphers, there is a wealth of LFSR-based algorithms. Many proposed ciphers have been broken; see references [8, 85] for an introduction. Among the best-studied ones are the A5/1 and A5/2 algorithms which are used in GSM mobile networks for voice encryption between cell phones and base stations. A5/1, which is the cipher used in most industrialized nations, had originally been kept secret but was reverse-engineered and published on the Internet in 1998. The cipher is borderline secure today [22], whereas the weaker A5/2 has much more serious flaws [11]. Neither of the two ciphers is recommended based on today's understanding of cryptanalysis. For 3GPP mobile communication, a different cipher A5/3 (also named *KASUMI*) is used, but it is a block cipher.

This somewhat pessimistic outlook on the state-of-the-art in stream ciphers changed with the eSTREAM project, described below.

eSTREAM Project The *eSTREAM project* had the explicit goal to advance the state-of-the-art knowledge about stream cipher design. As part of this objective, new stream ciphers that might become suitable for widespread adoption were investigated. eSTREAM was organized by the European Network of Excellence in Cryptography (ECRYPT). The call for stream ciphers was first issued in November 2004 and ended in 2008. The ciphers were divided into two "profiles", depending on the intended application:

- Profile 1: Stream ciphers for software applications with high throughput requirements.
- Profile 2: Stream ciphers for hardware applications with restricted resources such as limited storage, gate count, or power consumption.

Some cryptographers had emphasized the importance of including an authentication method, and hence two further profiles were also included to deal with ciphers that also provide authentication.

A total of 34 candidates were submitted to eSTREAM. At the end of the project four software-oriented ("Profile 1") ciphers were found to have desirable properties: *HC-128*, *Rabbit*, *Salsa20/12* and *SOSEMANUK*. With respect to hardware-oriented ciphers ("Profile 2"), the following three ciphers were selected: *Grain v1*, *MICKEY v2* and *Trivium*. Note that all of these are relatively new ciphers and only time will tell whether they are really cryptographically strong. The algorithm description, source code and the results of the four-year evaluation process are available online [69], and the official book provides more detailed information [146].

It is important to keep in mind that ECRYPT is not a standardization body, so the status of the eSTREAM finalist ciphers cannot be compared to that of AES at the end of its selection process (cf. Sect. 4.1).

True Random Number Generation We introduced in this chapter different classes of RNGs, and found that cryptographically secure pseudorandom number generators are of central importance for stream ciphers. For other cryptographic applications, true random number generators are important. For instance, true random numbers are needed for the generation of cryptographic keys which are then to be distributed. Many ciphers and modes of operation rely on initial values that are often generated from TRNGs. Also, many protocols require nonces (numbers used only once), which may stem from a TRNG. All TRNGs need to exploit some entropy source, i.e., some process which behaves truly randomly. Many TRNG designs have been proposed over the years. They can coarsely be classified as approaches that use specially designed hardware as an entropy source or as TRNGs that exploit external sources of randomness. Examples of the former are circuits with random behavior, e.g., that are based on semiconductor noise or on several uncorrelated oscillators. Reference [104, Chap. 5] contains a good survey. Examples of the latter ones are computer systems which measure the times between key strokes or the arrival times of packets at a network interface. In all these cases, one has to be extremely careful to make sure that the noise source in fact has enough entropy. There are many examples of TRNG designs which turned out to have poor random behavior and which constitute a serious security weakness, depending on how they are used. There are tools available that test the statistical properties of TRNG output sequences [56, 125]. There are also standards with which TRNGs can be formally evaluated [80].

2.5 Lessons Learned

- Stream ciphers are less popular than block ciphers in most domains such as Internet security. There are exceptions, for instance, the popular stream cipher RC4.
- Stream ciphers sometimes require fewer resources, e.g., code size or chip area, for implementation than block ciphers, and they are attractive for use in constrained environments such as cell phones.
- The requirements for a *cryptographically secure* pseudorandom number generator are far more demanding than the requirements for pseudorandom number generators used in other applications such as testing or simulation.

- The One-Time Pad is a provable secure symmetric cipher. However, it is highly impractical for most applications because the key length has to equal the message length.
- Single LFSRs make poor stream ciphers despite their good statistical properties. However, careful combinations of several LFSR can yield strong ciphers.

Problems

2.1. The stream cipher described in Definition 2.1.1 can easily be generalized to work in alphabets other than the binary one. For manual encryption, an especially useful one is a stream cipher that operates on letters.

1. Develop a scheme which operates with the letters A, B,..., Z, represented by the numbers 0,1,...,25. What does the key (stream) look like? What are the encryption and decryption functions?
2. Decrypt the following cipher text:
   bsaspp kkuosp
   which was encrypted using the key:
   rsidpy dkawoa
3. How was the young man murdered?

2.2. Assume we store a one-time key on a CD-ROM with a capacity of 1 Gbyte. Discuss the *real-life* implications of a One-Time-Pad (OTP) system. Address issues such as life cycle of the key, storage of the key during the life cycle/after the life cycle, key distribution, generation of the key, etc.

2.3. Assume an OTP-like encryption with a short key of 128 bit. This key is then being used periodically to encrypt large volumes of data. Describe how an attack works that breaks this scheme.

2.4. At first glance it seems as though an exhaustive key search is possible against an OTP system. Given is a short message, let's say 5 ASCII characters represented by 40 bit, which was encrypted using a 40-bit OTP. Explain *exactly* why an exhaustive key search will not succeed even though sufficient computational resources are available. This is a paradox since we know that the OTP is unconditionally secure. That is, explain why a brute-force attack does not work.

Note: You have to resolve the paradox! That means answers such as "The OTP is unconditionally secure and therefore a brute-force attack does not work" are not valid.

2.5. We will now analyze a pseudorandom number sequence generated by a LFSR characterized by $(c_2 = 1, c_1 = 0, c_0 = 1)$.

1. What is the sequence generated from the initialization vector $(s_2 = 1, s_1 = 0, s_0 = 0)$?
2. What is the sequence generated from the initialization vector $(s_2 = 0, s_1 = 1, s_0 = 1)$?
3. How are the two sequences related?

2.6. Assume we have a stream cipher whose period is quite short. We happen to know that the period is 150–200 bit in length. We assume that we do *not* know anything else about the internals of the stream cipher. In particular, we should not assume that it is a simple LFSR. For simplicity, assume that English text in ASCII format is being encrypted.

Describe in detail how such a cipher can be attacked. Specify exactly what Oscar has to know in terms of plaintext/ciphertext, and how he can decrypt all ciphertext.

2.7. Compute the first two output bytes of the LFSR of degree 8 and the feedback polynomial from Table 2.3 where the initialization vector has the value FF in hexadecimal notation.

2.8. In this problem we will study LFSRs in somewhat more detail. LFSRs come in three flavors:

■ LFSRs which generate a maximum-length sequence. These LFSRs are based on *primitive polynomials*.
■ LFSRs which do not generate a maximum-length sequence but whose sequence length is independent of the initial value of the register. These LFSRs are based on *irreducible polynomials* that are not primitive. Note that all primitive polynomials are also irreducible.
■ LFSRs which do not generate a maximum-length sequence and whose sequence length depends on the initial values of the register. These LFSRs are based on *reducible polynomials*.

We will study examples in the following. Determine *all* sequences generated by

1. $x^4 + x + 1$
2. $x^4 + x^2 + 1$
3. $x^4 + x^3 + x^2 + x + 1$

Draw the corresponding LFSR for each of the three polynomials. Which of the polynomials is primitive, which is only irreducible, and which one is reducible? Note that the lengths of all sequences generated by each of the LFSRs should add up to $2^m - 1$.

2.9. Given is a stream cipher which uses a single LFSR as key stream generator. The LFSR has a degree of 256.

1. How many plaintext/ciphertext bit pairs are needed to launch a successful attack?
2. Describe all steps of the attack in detail and develop the formulae that need to be solved.
3. What is the key in this system? Why doesn't it make sense to use the initial contents of the LFSR as the key or as part of the key?

2.10. We conduct a known-plaintext attack on an LFSR-based stream cipher. We know that the plaintext sent was:
1001 0010 0110 1101 1001 0010 0110
By tapping the channel we observe the following stream:
1011 1100 0011 0001 0010 1011 0001

1. What is the degree m of the key stream generator?
2. What is the initialization vector?
3. Determine the feedback coefficients of the LFSR.

4. Draw a circuit diagram and verify the output sequence of the LFSR.

2.11. We want to perform an attack on another LFSR-based stream cipher. In order to process letters, each of the 26 uppercase letters and the numbers 0, 1, 2, 3, 4, 5 are represented by a 5-bit vector according to the following mapping:

$$A \leftrightarrow 0 = 00000_2$$

$$\vdots$$

$$Z \leftrightarrow 25 = 11001_2$$
$$0 \leftrightarrow 26 = 11010_2$$

$$\vdots$$

$$5 \leftrightarrow 31 = 11111_2$$

We happen to know the following facts about the system:

- The degree of the LFSR is $m = 6$.
- Every message starts with the header `WPI`.

We observe now on the channel the following message (the fourth letter is a zero):
`j5a0edj2b`

1. What is the initialization vector?
2. What are the feedback coefficients of the LFSR?
3. Write a program in your favorite programming language which generates the whole sequence, and find the whole plaintext.
4. Where does the thing after `WPI` live?
5. What type of attack did we perform?

2.12. Assume the *IV* and the key of Trivium each consist of 80 all-zero bits. Compute the first 70 bits $s_1, \ldots, s_{70}$ during the warm-up phase of Trivium. Note that these are only internal bits which are not used for encryption since the warm-up phase lasts for 1152 clock cycles.

Chapter 3
The Data Encryption Standard (DES) and Alternatives

The *Data Encryption Standard (DES)* has been by far the most popular block cipher for most of the last 30 years. Even though it is nowadays not considered secure against a determined attacker because the DES key space is too small, it is still used in legacy applications. Furthermore, encrypting data three times in a row with DES — a process referred to as *3DES* or *triple DES* — yields a very secure cipher which is still widely used today (Section 3.5 deals with 3DES.) Perhaps what is more important, since DES is by far the best-studied symmetric algorithm, its design principles have inspired many current ciphers. Hence, studying DES helps us to understand many other symmetric algorithms.

In this chapter you will learn:

- The design process of DES, which is very helpful for understanding the technical and political evolution of modern cryptography
- Basic design ideas of block ciphers, including confusion and diffusion, which are important properties of all modern block ciphers
- The internal structure of DES, including Feistel networks, S-boxes and the key schedule
- Security analysis of DES
- Alternatives to DES, including 3DES

3.1 Introduction to DES

In 1972 a mildly revolutionary act was performed by the US National Bureau of Standards (NBS), which is now called *National Institute of Standards and Technology (NIST)*: the NBS initiated a request for proposals for a standardized cipher in the USA. The idea was to find a single secure cryptographic algorithm which could be used for a variety of applications. Up to this point in time governments had always considered cryptography, and in particular cryptanalysis, so crucial for national security that it had to be kept secret. However, by the early 1970s the demand for encryption for commercial applications such as banking had become so pressing that it could not be ignored without economic consequences.

The NBS received the most promising candidate in 1974 from a team of cryptographers working at IBM. The algorithm IBM submitted was based on the cipher *Lucifer*. Lucifer was a family of ciphers developed by Horst Feistel in the late 1960s, and was one of the first instances of block ciphers operating on digital data. Lucifer is a Feistel cipher which encrypts blocks of 64 bits using a key size of 128 bits. In order to investigate the security of the submitted ciphers, the NBS requested the help of the *National Security Agency (NSA)*, which did not even admit its existence at that point in time. It seems certain that the NSA influenced changes to the cipher, which was rechristened DES. One of the changes that occurred was that DES is specifically designed to withstand differential cryptanalysis, an attack not known to the public until 1990. It is not clear whether the IBM team developed the knowledge about differential cryptanalysis by themselves or whether they were guided by the NSA. Allegedly, the NSA also convinced IBM to reduce the Lucifer key length of 128 bit to 56 bit, which made the cipher much more vulnerable to brute-force attacks.

The NSA involvement worried some people because it was feared that a secret trapdoor, i.e., a mathematical property with which DES could be broken but which is only known to NSA, might have been the real reason for the modifications. Another major complaint was the reduction of the key size. Some people conjectured that the NSA would be able to search through a key space of 2^{56}, thus breaking it by brute-force. In later decades, most of these concerns turned out to be unfounded. Section 3.5 provides more information about real and perceived security weaknesses of DES.

Despite of all the criticism and concerns, in 1977 the NBS finally released all specifications of the modified IBM cipher as the *Data Encryption Standard (FIPS PUB 46)* to the public. Even though the cipher is described down to the bit level in the standard, the motivation for parts of the DES design (the so-called design criteria), especially the choice of the substitution boxes, were never officially released.

With the rapid increase in personal computers in the early 1980s and all specifications of DES being publicly available, it become easier to analyze the inner structure of the cipher. During this period, the civilian cryptography research community also grew and DES underwent major scrutiny. However, no serious weaknesses were found until 1990. Originally, DES was only standardized for 10 years, until 1987. Due to the wide use of DES and the lack of security weaknesses, the NIST reaffirmed the federal use of the cipher until 1999, when it was finally replaced by the *Advanced Encryption Standard (AES)*.

3.1.1 Confusion and Diffusion

Before we start with the details of DES, it is instructive to look at primitive operations which can be applied in order to achieve strong encryption. According to the famous information theorist Claude Shannon, there are two primitive operations with which strong encryption algorithms can be built:

1. **Confusion** is an encryption operation where the relationship between key and ciphertext is obscured. Today, a common element for achieving confusion is substitution, which is found in both DES and AES.
2. **Diffusion** is an encryption operation where the influence of one plaintext symbol is spread over many ciphertext symbols with the goal of hiding statistical properties of the plaintext. A simple diffusion element is the bit permutation, which is used frequently within DES. AES uses the more advanced Mixcolumn operation.

Ciphers which only perform confusion, such as the Shift Cipher (cf. Sect. 1.4.3) or the World War II encryption machine Enigma, are not secure. Neither are ciphers which only perform diffusion. However, through the concatenation of such operations, a strong cipher can be built. The idea of concatenating several encryption operation was also proposed by Shannon. Such ciphers are known as *product ciphers*. All of today's block ciphers are product ciphers as they consist of rounds which are applied repeatedly to the data (Fig. 3.1).

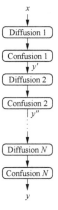

Fig. 3.1 Principle of an N round product cipher, where each round performs a confusion and diffusion operation

Modern block ciphers possess excellent diffusion properties. On a cipher level this means that changing of one bit of plaintext results *on average* in the change of half the output bits, i.e., the second ciphertext looks statistically independent of the first one. This is an important property to keep in mind when dealing with block ciphers. We demonstrate this behavior with the following simple example.

Example 3.1. Let's assume a small block cipher with a block length of 8 bits. Encryption of two plaintexts x_1 and x_2, which differ only by one bit, should roughly result in something as shown in Fig. 3.2.

$$x_1 = 0010\ 1011$$
$$\boxed{\text{Block Cipher}}$$
$$x_2 = 0000\ 1011$$
$$y_1 = 1011\ 1001$$
$$y_2 = 0110\ 1100$$

Fig. 3.2 Principle of diffusion of a block cipher

Note that modern block ciphers have block lengths of 64 or 128 bit but they show exactly the same behavior if one input bit is flipped.

◇

3.2 Overview of the DES Algorithm

DES is a cipher which encrypts blocks of length of 64 bits with a key of size of 56 bits (Fig. 3.3).

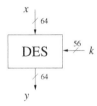

Fig. 3.3 DES block cipher

DES is a symmetric cipher, i.e., the same same key is used for encryption and decryption. DES is, like virtually all modern block ciphers, an iterative algorithm. For each block of plaintext, encryption is handled in 16 rounds which all perform the identical operation. Figure 3.4 shows the round structure of DES. In every round a different subkey is used and all subkeys k_i are derived from the main key k.

Let's now have a more detailed view on the internals of DES, as shown in Fig. 3.5.The structure in the figure is called a *Feistel network*. It can lead to very strong ciphers if carefully designed. Feistel networks are used in many, but certainly not in all, modern block ciphers. (In fact, AES is not a Feistel cipher.) In addition to its potential cryptographic strength, one advantage of Feistel networks is that encryption and decryption are almost the same operation. Decryption requires

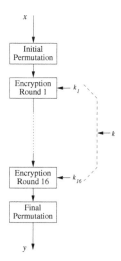

Fig. 3.4 Iterative structure of DES

only a reversed key schedule, which is an advantage in software and hardware implementations. We discuss the Feistel network in the following.

After the initial bitwise permutation IP of a 64-bit plaintext x, the plaintext is split into two halves L_0 and R_0. These two 32-bit halves are the input to the Feistel network, which consists of 16 rounds. The right half R_i is fed into the function f. The output of the f function is XORed (as usually denoted by the symbol $\oplus$) with the left 32-bit half L_i. Finally, the right and left half are swapped. This process repeats in the next round and can be expressed as:

$$L_i = R_{i-1},$$
$$R_i = L_{i-1} \oplus f(R_{i-1}, k_i)$$

where i= 1,...,16. After round 16, the 32-bit halves L_{16} and R_{16} are swapped again, and the final permutation IP^{-1} is the last operation of DES. As the notation suggests, the final permutation IP^{-1} is the inverse of the initial permutation IP. In each round, a round key k_i is derived from the main 56-bit key using what is called the key schedule.

It is crucial to note that the Feistel structure really only encrypts (decrypts) half of the input bits per each round, namely the left half of the input. The right half is copied to the next round unchanged. In particular, the right half is *not encrypted* with the f function. In order to get a better understanding of the working of Feistel cipher, the following interpretation is helpful: Think of the f function as a pseudorandom generator with the two input parameters R_{i-1} and k_i. The output of the pseudorandom generator is then used to encrypt the left half L_{i-1} with an XOR operation. As we saw in Chap. 2, if the output of the f function is not predictable for an attacker, this results in a strong encryption method.

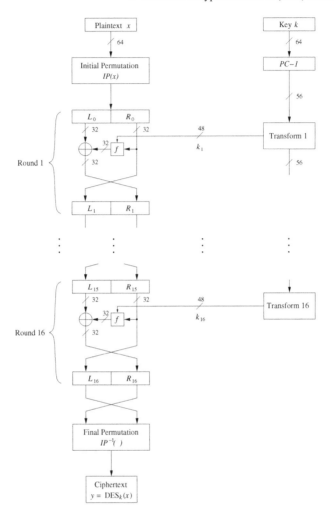

Fig. 3.5 The Feistel structure of DES

The two aforementioned basic properties of ciphers, i.e., confusion and diffusion, are realized within the f-function. In order to thwart advanced analytical attacks, the f-function must be designed extremely carefully. Once the f-function has been designed securely, the security of a Feistel cipher increases with the number of key bits used and the number of rounds.

Before we discuss all components of DES in detail, here is an algebraic description of the Feistel network for the mathematically inclined reader. The Feistel structure of each round bijectively maps a block of 64 input bits to 64 output bits (i.e., every possible input is mapped uniquely to exactly one output, and vice versa). This mapping remains bijective for some arbitrary function f, i.e., even if the embedded function f is not bijective itself. In the case of DES, the function f is in fact a surjective (many-to-one) mapping. It uses nonlinear building blocks and maps 32 input bits to 32 output bits using a 48-bit round key k_i, with $1 \leq i \leq 16$.

3.3 Internal Structure of DES

The structure of DES as depicted in Fig. 3.5 shows the internal functions which we will discuss in this section. The building blocks are the initial and final permutation, the actual DES rounds with its core, the f-function, and the key schedule.

3.3.1 Initial and Final Permutation

As shown in Figs. 3.6 and 3.7, the *initial permutation IP* and the *final permutation IP*$^{-1}$ are bitwise permutations. A bitwise permutation can be viewed as simple crosswiring. Interestingly, permutations can be very easily implemented in hardware but are not particularly fast in software. Note that both permutations do not increase the security of DES at all. The exact rationale for the existence of these two permutations is not known, but it seems likely that their original purpose was to arrange the plaintext, ciphertext and bits in a bytewise manner to make data fetches easier for 8-bit data busses, which were the state-of-the-art register size in the early 1970s.

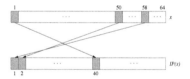

Fig. 3.6 Examples for the bit swaps of the initial permutation

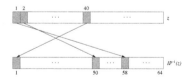

Fig. 3.7 Examples for the bit swaps of the final permutation

The details of the transformation *IP* are given in Table 3.1. This table, like all other tables in this chapter, should be read from left to right, top to bottom. The table indicates that input bit 58 is mapped to output position 1, input bit 50 is mapped to the second output position, and so forth. The final permutation IP^{-1} performs the inverse operation of IP as shown in Table 3.2.

Table 3.1 Initial permutation IP

IP
58 50 42 34 26 18 10 2
60 52 44 36 28 20 12 4
62 54 46 38 30 22 14 6
64 56 48 40 32 24 16 8
57 49 41 33 25 17 9 1
59 51 43 35 27 19 11 3
61 53 45 37 29 21 13 5
63 55 47 39 31 23 15 7

Table 3.2 Final permutation IP^{-1}

IP^{-1}
40 8 48 16 56 24 64 32
39 7 47 15 55 23 63 31
38 6 46 14 54 22 62 30
37 5 45 13 53 21 61 29
36 4 44 12 52 20 60 28
35 3 43 11 51 19 59 27
34 2 42 10 50 18 58 26
33 1 41 9 49 17 57 25

3.3.2 The f-Function

As mentioned earlier, the f-function plays a crucial role for the security of DES. In round i it takes the right half R_{i-1} of the output of the previous round and the current round key k_i as input. The output of the f-function is used as an XOR-mask for encrypting the left half input bits L_{i-1}.

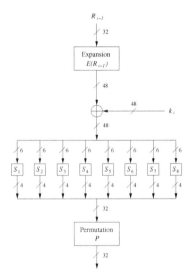

Fig. 3.8 Block diagram of the f-function

The structure of the f-function is shown in Fig. 3.8. First, the 32-bit input is expanded to 48 bits by partitioning the input into eight 4-bit blocks and by expanding each block to 6 bits. This happens in the E-box, which is a special type of permutation. The first block consists of the bits $(1,2,3,4)$, the second one of $(5,6,7,8)$, etc. The expansion to six bits can be seen in Fig. 3.9.

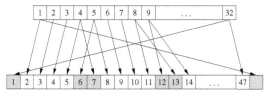

Fig. 3.9 Examples for the bit swaps of the expansion function E

As can be seen from the Table 3.3, exactly 16 of the 32 input bits appear twice in the output. However, an input bit never appears twice in the same 6-bit output block. The expansion box increases the diffusion behavior of DES since certain input bits influence two different output locations.

Table 3.3 Expansion permutation E

E					
32	1	2	3	4	5
4	5	6	7	8	9
8	9	10	11	12	13
12	13	14	15	16	17
16	17	18	19	20	21
20	21	22	23	24	25
24	25	26	27	28	29
28	29	30	31	32	1

Next, the 48-bit result of the expansion is XORed with the round key k_i, and the eight 6-bit blocks are fed into eight different *substition boxes*, which are often referred to as *S-boxes*. Each S-box is a lookup table that maps a 6-bit input to a 4-bit output. Larger tables would have been cryptographically better, but they also become much larger; eight 4-by-6 tables were probably close the maximum size which could be fit on a single integrated circuit in 1974. Each S-box contains $2^6 = 64$ entries, which are typically represented by a table with 16 columns and 4 rows. Each entry is a 4-bit value. All S-boxes are listed in Tables 3.4 to 3.11. Note that all S-boxes are different. The tables are to be read as indicated in Fig. 3.10: the most significant bit (MSB) and the least significant bit (LSB) of each 6-bit input select the row of the table, while the four inner bits select the column. The integers $0,1,\ldots,15$ of each entry in the table represent the decimal notation of a 4-bit value.

Example 3.2. The S-box input $b = (100101)_2$ indicates the row $11_2 = 3$ (i.e., fourth row, numbering starts with 00_2) and the column $0010_2 = 2$ (i.e., the third column). If the input b is fed into S-box 1, the output is $S_1(37 = 100101_2) = 8 = 1000_2$.

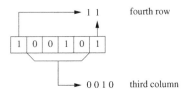

Fig. 3.10 Example of the decoding of the input 100101_2 by S-box 1

◇

Table 3.4 S-box S_1

S_1	0	1	2	3	4	5	6	7	8	9	10	11	12	13	14	15
0	14	04	13	01	02	15	11	08	03	10	06	12	05	09	00	07
1	00	15	07	04	14	02	13	01	10	06	12	11	09	05	03	08
2	04	01	14	08	13	06	02	11	15	12	09	07	03	10	05	00
3	15	12	08	02	04	09	01	07	05	11	03	14	10	00	06	13

Table 3.5 S-box S_2

S_2	0	1	2	3	4	5	6	7	8	9	10	11	12	13	14	15
0	15	01	08	14	06	11	03	04	09	07	02	13	12	00	05	10
1	03	13	04	07	15	02	08	14	12	00	01	10	06	09	11	05
2	00	14	07	11	10	04	13	01	05	08	12	06	09	03	02	15
3	13	08	10	01	03	15	04	02	11	06	07	12	00	05	14	09

Table 3.6 S-box S_3

S_3	0	1	2	3	4	5	6	7	8	9	10	11	12	13	14	15
0	10	00	09	14	06	03	15	05	01	13	12	07	11	04	02	08
1	13	07	00	09	03	04	06	10	02	08	05	14	12	11	15	01
2	13	06	04	09	08	15	03	00	11	01	02	12	05	10	14	07
3	01	10	13	00	06	09	08	07	04	15	14	03	11	05	02	12

The S-boxes are the core of DES in terms of cryptographic strength. They are the only nonlinear element in the algorithm and provide confusion. Even though the entire specification of DES was released by NBS/NIST in 1977, the motivation for the choice of the S-box tables was never completely revealed. This often gave rise

Table 3.7 S-box S_4

S_4	0	1	2	3	4	5	6	7	8	9	10	11	12	13	14	15
0	07	13	14	03	00	06	09	10	01	02	08	05	11	12	04	15
1	13	08	11	05	06	15	00	03	04	07	02	12	01	10	14	09
2	10	06	09	00	12	11	07	13	15	01	03	14	05	02	08	04
3	03	15	00	06	10	01	13	08	09	04	05	11	12	07	02	14

Table 3.8 S-box S_5

S_5	0	1	2	3	4	5	6	7	8	9	10	11	12	13	14	15
0	02	12	04	01	07	10	11	06	08	05	03	15	13	00	14	09
1	14	11	02	12	04	07	13	01	05	00	15	10	03	09	08	06
2	04	02	01	11	10	13	07	08	15	09	12	05	06	03	00	14
3	11	08	12	07	01	14	02	13	06	15	00	09	10	04	05	03

Table 3.9 S-box S_6

S_6	0	1	2	3	4	5	6	7	8	9	10	11	12	13	14	15
0	12	01	10	15	09	02	06	08	00	13	03	04	14	07	05	11
1	10	15	04	02	07	12	09	05	06	01	13	14	00	11	03	08
2	09	14	15	05	02	08	12	03	07	00	04	10	01	13	11	06
3	04	03	02	12	09	05	15	10	11	14	01	07	06	00	08	13

Table 3.10 S-box S_7

S_7	0	1	2	3	4	5	6	7	8	9	10	11	12	13	14	15
0	04	11	02	14	15	00	08	13	03	12	09	07	05	10	06	01
1	13	00	11	07	04	09	01	10	14	03	05	12	02	15	08	06
2	01	04	11	13	12	03	07	14	10	15	06	08	00	05	09	02
3	06	11	13	08	01	04	10	07	09	05	00	15	14	02	03	12

Table 3.11 S-box S_8

S_8	0	1	2	3	4	5	6	7	8	9	10	11	12	13	14	15
0	13	02	08	04	06	15	11	01	10	09	03	14	05	00	12	07
1	01	15	13	08	10	03	07	04	12	05	06	11	00	14	09	02
2	07	11	04	01	09	12	14	02	00	06	10	13	15	03	05	08
3	02	01	14	07	04	10	08	13	15	12	09	00	03	05	06	11

to speculation, in particular with respect to the possible existence of a secret back door or some other intentionally constructed weakness, which could be exploited by the NSA. However, now we know that the S-boxes were designed according to the criteria listed below.

1. Each S-box has six input bits and four output bits.
2. No single output bit should be too close to a linear combination of the input bits.
3. If the lowest and the highest bits of the input are fixed and the four middle bits are varied, each of the possible 4-bit output values must occur exactly once.
4. If two inputs to an S-box differ in exactly one bit, their outputs must differ in at least two bits.

5. If two inputs to an S-box differ in the two middle bits, their outputs must differ in at least two bits.
6. If two inputs to an S-box differ in their first two bits and are identical in their last two bits, the two outputs must be different.
7. For any nonzero 6-bit difference between inputs, no more than 8 of the 32 pairs of inputs exhibiting that difference may result in the same output difference.
8. A collision (zero output difference) at the 32-bit output of the eight S-boxes is only possible for three adjacent S-boxes.

Note that some of these design criteria were not revealed until the 1990s. More information about the issue of the secrecy of the design criteria is found in Sect. 3.5.

The S-boxes are the most crucial elements of DES because they introduce a *nonlinearity* to the cipher, i.e.,

$$S(a) \oplus S(b) \neq S(a \oplus b).$$

Without a nonlinear building block, an attacker could express the DES input and output with a system of linear equations where the key bits are the unknowns. Such systems can easily be solved, a fact that was used in the LFSR attack in Sect. 2.3.2. However, the S-boxes were carefully designed to also thwart advanced mathematical attacks, in particular *differential cryptanalysis*. Interestingly, differential cryptanalysis was first discovered in the research community in 1990. At this point, the IBM team declared that the attack was known to the designers at least 16 years earlier, and that DES was especially designed to withstand differential cryptanalysis.

Finally, the 32-bit output is permuted bitwise according to the P permutation, which is given in Table 3.12. Unlike the initial permutation IP and its inverse IP^{-1}, the permutation P introduces diffusion because the four output bits of each S-box are permuted in such a way that they affect several different S-boxes in the following round. The diffusion caused by the expansion, S-boxes and the permutation P guarantees that every bit at the end of the fifth round is a function of every plaintext bit and every key bit. This behavior is known as the *avalanche effect*.

Table 3.12 The permutation P within the f-function

P							
16	7	20	21	29	12	28	17
1	15	23	26	5	18	31	10
2	8	24	14	32	27	3	9
19	13	30	6	22	11	4	25

3.3.3 Key Schedule

The *key schedule* derives 16 round keys k_i, each consisting of 48 bits, from the original 56-bit key. Another term for round key is subkey. First, note that the DES input key is often stated as 64-bit, where every eighth bit is used as an odd parity bit over the preceding seven bits. It is not quite clear why DES was specified that way. In any case, the eight parity bits are **not** actual key bits and do not increase the security. DES is a 56-bit cipher, not a 64-bit one.

As shown in Fig. 3.11, the 64-bit key is first reduced to 56 bits by ignoring every eighth bit, i.e., the parity bits are stripped in the initial $PC-1$ permutation. Again, the parity bits certainly do not increase the key space! The name $PC-1$ stands for "permuted choice one". The exact bit connections that are realized by $PC-1$ are given in Table 3.13.

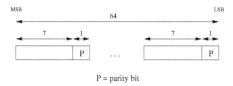

P = parity bit

Fig. 3.11 Location of the eight parity bits for a 64-bit input key

Table 3.13 Initial key permutation $PC-1$

$PC-1$							
57	49	41	33	25	17	9	1
58	50	42	34	26	18	10	2
59	51	43	35	27	19	11	3
60	52	44	36	63	55	47	39
31	23	15	7	62	54	46	38
30	22	14	6	61	53	45	37
29	21	13	5	28	20	12	4

The resulting 56-bit key is split into two halves C_0 and D_0, and the actual key schedule starts as shown in Fig. 3.12. The two 28-bit halves are cyclically shifted, i.e., rotated, left by one or two bit positions depending on the round i according to the following rules:

- In rounds $i = 1, 2, 9, 16$, the two halves are rotated left by one bit.
- In the other rounds where $i \neq 1, 2, 9, 16$, the two halves are rotated left by two bits.

Note that the rotations only take place within either the left or the right half. The total number of rotation positions is $4 \cdot 1 + 12 \cdot 2 = 28$. This leads to the interesting property that $C_0 = C_{16}$ and $D_0 = D_{16}$. This is very useful for the decryption key schedule where the subkeys have to be generated in reversed order, as we will see in Sect. 3.4.

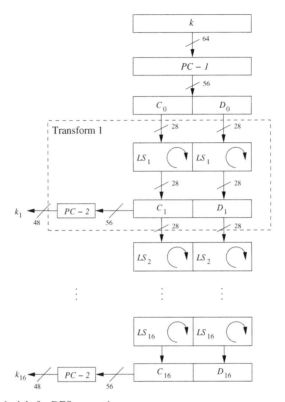

Fig. 3.12 Key schedule for DES encryption

To derive the 48-bit round keys k_i, the two halves are permuted bitwise again with $PC - 2$, which stands for "permuted choice 2". $PC - 2$ permutes the 56 input bits coming from C_i and D_i and ignores 8 of them. The exact bit-connections of $PC - 2$ are given in Table 3.14.

Table 3.14 Round key permutation $PC - 2$

$PC - 2$							
14	17	11	24	1	5	3	28
15	6	21	10	23	19	12	4
26	8	16	7	27	20	13	2
41	52	31	37	47	55	30	40
51	45	33	48	44	49	39	56
34	53	46	42	50	36	29	32

Note that every round key is a selection of 48 permuted bits of the input key k. The key schedule is merely a method of realizing the 16 permutations systematically. Especially in hardware, the key schedule is very easy to implement. The key schedule is also designed so that each of the 56 key bits is used in different round keys; each bit is used in approximately 14 of the 16 round keys.

3.4 Decryption

One advantage of DES is that decryption is essentially the same function as encryption. This is because DES is based on a Feistel network. Figure 3.13 shows a block diagram for DES decryption. Compared to encryption, only the key schedule is reversed, i.e., in decryption round 1, subkey 16 is needed; in round 2, subkey 15; etc. Thus, when in decryption mode, the key schedule algorithm has to generate the round keys as the sequence $k_{16}, k_{15}, \ldots, k_1$.

Reversed Key Schedule

The first question that we have to clarify is how, given the initial DES key k, can we easily generate k_{16}? Note that we saw above that $C_0 = C_{16}$ and $D_0 = D_{16}$. Hence k_{16} can be directly derived after $PC - 1$.

$$
\begin{aligned}
k_{16} &= PC - 2(C_{16}, D_{16}) \\
&= PC - 2(C_0, D_0) \\
&= PC - 2(PC - 1(k))
\end{aligned}
$$

To compute k_{15} we need the intermediate variables C_{15} and D_{15}, which can be derived from C_{16}, D_{16} through cyclic *right shifts (RS)*:

$$
\begin{aligned}
k_{15} &= PC - 2(C_{15}, D_{15}) \\
&= PC - 2(RS_2(C_{16}), RS_2(D_{16})) \\
&= PC - 2(RS_2(C_0), RS_2(D_0))
\end{aligned}
$$

The subsequent round keys $k_{14}, k_{13}, \ldots, k_1$ are derived via right shifts in a similar fashion. The number of bits shifted right for each round key in decryption mode

- In decryption round 1, the key is not rotated.
- In decryption rounds 2, 9, and 16 the two halves are rotated right by one bit.
- In the other rounds 3, 4, 5, 6, 7, 8, 10, 11, 12, 13, 14 and 15 the two halves are rotated right by two bits.

Figure 3.14 shows the reversed key schedule for decryption.

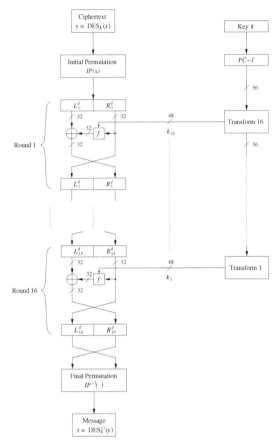

Fig. 3.13 DES decryption

Decryption in Feistel Networks

We have not addressed the core question: Why is the decryption function essentially the same as the encryption function? The basic idea is that the decryption function reverses the DES encryption in a round-by-round manner. That means that decryption round 1 reverses encryption round 16, decryption round 2 reverses encryption round 15, and so on. Let's first look at the initial stage of decryption by looking at Fig. 3.13. Note that the right and left halves are swapped in the last round of DES:

$$(L_0^d, R_0^d) = IP(Y) = IP(IP^{-1}(R_{16}, L_{16})) = (R_{16}, L_{16})$$

And thus:

$$L_0^d = R_{16}$$
$$R_0^d = L_{16} = R_{15}$$

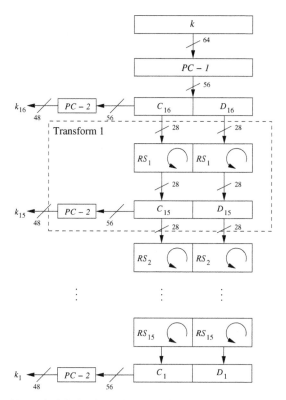

Fig. 3.14 Reversed key schedule for decryption of DES

Note that all variables in the decryption routine are marked with the superscript d, whereas the encryption variables do not have superscripts. The derived equation simply says that the input of the first round of decryption is the output of the last round of encryption because final and initial permutations cancel each other out. We will now show that the first decryption round reverses the last encryption round. For this, we have to express the output values (L_1^d, R_1^d) of the first decryption round 1 in terms of the input values of the last encryption round (L_{15}, R_{15}). The first one is easy:

$$L_1^d = R_0^d = L_{16} = R_{15}$$

We now look at how R_1^d is computed:

$$R_1^d = L_0^d \oplus f(R_0^d, k_{16}) = R_{16} \oplus f(L_{16}, k_{16})$$
$$R_1^d = [L_{15} \oplus f(R_{15}, k_{16})] \oplus f(R_{15}, k_{16})$$
$$R_1^d = L_{15} \oplus [f(R_{15}, k_{16}) \oplus f(R_{15}, k_{16})] = L_{15}$$

The crucial step is shown in the last equation above: An identical output of the f-function is XORed twice to L_{15}. These operations cancel each other out, so that

$R_1^d = L_{15}$. Hence, after the first decryption round, we in fact have computed the same values we had *before* the last encryption round. Thus, the first decryption round reverses the last encryption round. This is an iterative process which continues in the next 15 decryption rounds and that can be expressed as:

$$L_i^d = R_{16-i},$$
$$R_i^d = L_{16-i}$$

where $i = 0, 1, \ldots, 16$. In particular, after the last decryption round:

$$L_{16}^d = R_{16-16} = R_0$$
$$R_{16}^d = L_0$$

Finally, at the end of the decryption process, we have to reverse the initial permutation:

$$IP^{-1}(R_{16}^d, L_{16}^d) = IP^{-1}(L_0, R_0) = IP^{-1}(IP(x)) = x$$

where x is the plaintext that was the input to the DES encryption.

3.5 Security of DES

As we discussed in Sect. 1.2.2, ciphers can be attacked in several ways. With respect to cryptographic attacks, we distinguish between exhaustive key search or brute-force attacks, and analytical attacks. The latter was demonstrated with the LFSR attack in Sect. 2.3.2, where we could easily break a stream cipher by solving a system of linear equations. Shortly after DES was proposed, two major criticisms against the cryptographic strength of DES centered around two arguments:

1. The key space is too small, i.e., the algorithm is vulnerable against brute-force attacks.
2. The design criteria of the S-boxes was kept secret and there might have existed an analytical attack that exploits mathematical properties of the S-boxes, but which is only known to the DES designers.

We discuss both types of attacks below. However, we also state the main conclusion about DES security already here: Despite very intensive cryptanalysis over the lifetime of DES, current analytical attacks are not very efficient. However, DES can relatively easily be broken with an exhaustive key-search attack and, thus, plain DES is not suited for most applications any more.

3.5.1 Exhaustive Key Search

The first criticism is nowadays certainly justified. The original cipher proposed by IBM had a key length of 128 bits and it is suspicious that it was reduced to 56 bits. The official statement that a cipher with a shorter key length made it easier to implement the DES algorithm on a single chip in 1974 does not sound too convincing. For clarification, let's recall the principle of an exhaustive key search (or brute-force attack):

Definition 3.5.1 DES Exhaustive key search
Input: *at least one pair of plaintext–ciphertext* (x, y)
Output: *k, such that* $y = DES_k(x)$
Attack: *Test all* 2^{56} *possible keys until the following condition is fulfilled:*

$$DES_{k_i}^{-1}(y) \stackrel{?}{=} x \; , \; i = 0, 1, \ldots, 2^{56} - 1.$$

Note that there is a small chance of $1/2^{16}$ that an incorrect key is found, i.e., a key k which decrypts only the one ciphertext y correctly but not subsequent ciphertexts. If one wants to rule out this possibility, an attacker must check such a key candidate with a second plaintext–ciphertext pair. More about this is found in Sect. 5.2.

Regular computers are not particularly well suited to perform the 2^{56} key tests necessary, but special-purpose key-search machines are an option. It seems highly likely that large (government) institutions have long been able to build such *brute-force crackers*, which can break DES in a matter of days. In 1977, Whitfield Diffie and Martin Hellman [59] estimated that it was possible to build an exhaustive key-search machine for approximately $20,000,000. Even though they later stated that their cost estimate had been too optimistic, it was clear from the beginning that a cracker could be built with sufficient funding.

At the rump session of the CRYPTO 1993 conference, Michael Wiener proposed the design of a very efficient key-search machine which used pipelining techniques. An update of his proposal can be found in [174]. He estimated the cost of his design at approximately $1,000,000, and the time required to find the key at 1.5 days. This was a proposal only, and the machine was not built. In 1998, however, the EFF (Electronic Frontier Foundation) built the hardware machine *Deep Crack*, which performed a brute-force attack against DES in 56 hours. Figure 3.15 shows a photo of Deep Crack. The machine consisted of 1800 integrated circuits, where each had 24 key-test units. The average search time of Deep Crack was 15 days, and the machine was built for less than $250,000. The successful break with Deep Crack was considered the official demonstration that DES is no longer secure against determined attacks by many people. Please note that this break does not imply that a weak algorithm had been in use for more than 20 years. It was only possible to build Deep Crack at such a relatively low price because digital hardware had become cheap. In the 1980s it would have been impossible to build a DES cracker without spending many millions of dollars. It can be speculated that only government agencies were willing to spend such an amount of money for code breaking.

Fig. 3.15 Deep Crack — the hardware exhaustive key-search machine that broke DES in 1998 (reproduced with permission from Paul Kocher)

DES brute-force attacks also provide an excellent case study for the continuing decrease in hardware costs. In 2006, the *COPACOBANA (Cost-Optimized Parallel Code-Breaker)* machine was built based on commercial integrated circuits by a team of researchers from the Universities of Bochum and Kiel in Germany (the authors of this book were heavily involved in this effort). COPACOBANA allows one to break DES with an average search time of less than 7 days. The interesting part of this undertaking is that the machine could be built with hardware costs in the $10,000 range. Figure 3.16 shows a picture of COPACOBANA.

Fig. 3.16 COPACOBANA — A cost-optimized parallel code breaker

In summary, a key size of 56 bits is too short to encrypt confidential data nowadays. Hence, single DES should only be used for applications where only short-term security is needed — say, a few hours — or where the value of the encrypted data is very low. However, variants of DES, in particular 3DES, are still secure.

3.5.2 Analytical Attacks

As was shown in the first chapter, analytical attacks can be very powerful. Since the introduction of DES in the mid-1970s, many excellent researchers in academia (and without doubt many excellent researchers in intelligence agencies) tried to find weaknesses in the structure of DES which allowed them to break the cipher. It is a major triumph for the designers of DES that no weakness was found until 1990. In this year, Eli Biham and Adi Shamir discovered what is called *differential cryptanalysis (DC)*. This is a powerful attack which is *in principle* applicable to any block cipher. However, it turned out that the DES S-boxes are particularly resistant against this attack. In fact, one member of the original IBM design team declared after the discovery of DC that they had been aware of the attack at the time of design. Allegedly, the reason why the S-box design criteria were not made public was that the design team did not want to make such a powerful attack public. If this claim is true — and all circumstances support it — it means that the IBM and NSA team was 15 years ahead of the research community. It should be noted, however, that in the 1970s and 1980s relatively few people did active research in cryptography.

In 1993 a related but distinct analytical attack was published by Mitsuru Matsui, which was named *linear cryptanalysis (LC)*. Similar to differential cryptanalysis, the effectiveness of this attack also heavily depends on the structure of the S-boxes.

What is the practical relevance of these two analytical attacks against DES? It turns out that an attacker needs 2^{47} plaintext–ciphertext pairs for a successful DC attack. This assumes particularly chosen plaintext blocks; for random plaintext 2^{55} pairs are needed! In the case of LC, an attacker needs 2^{43} plaintext–ciphertext pairs. All these numbers seem highly impractical for several reasons. First, an attacker needs to know an extremely large number of plaintexts, i.e., pieces of data which are supposedly encrypted and thus hidden from the attacker. Second, collecting and storing such an amount of data takes a long time and requires considerable memory resources. Third, the attack only recovers one key. (This is actually one of many arguments for introducing key freshness in cryptographic applications.) As a result of all these arguments, it does not seem likely that DES can be broken with either DC or LC in real-world systems. However, both DC and LC are very powerful attacks which are applicable to many other block ciphers. Table 3.15 provides an overview of proposed and realized attacks against DES over the last three decades. Some entries refer to what is known as the DES Challenges. Starting in 1997, several DES-breaking challenges were organized by the company RSA Security.

3.6 Implementation in Software and Hardware

In the following, we provide a brief assessment of DES implementation properties in software and hardware. When we talk about software, we refer to DES implementations running on desktop CPUs or embedded microprocessors like smart cards

Table 3.15 History of full-round DES attacks

Date	Proposed or implemented attacks
1977	W. Diffie and M. Hellman propose cost estimate for key-search machine
1990	E. Biham and A. Shamir propose differential cryptanalysis, which requires 2^{47} chosen plaintexts
1993	M. Wiener proposes detailed hardware design for key-search machine with an average search time of 36 h and estimated cost of $1,000,000
1993	M. Matsui proposes linear cryptanalysis, which requires 2^{43} chosen ciphertexts
Jun. 1997	DES Challenge I broken through brute-force; distributed effort on the Internet took 4.5 months
Feb. 1998	DES Challenge II–1 broken through brute-force; distributed effort on the Internet took 39 days
Jul. 1998	DES Challenge II–2 broken through brute-force; Electronic Frontier Foundation built the Deep Crack key-search machine for about $250,000. The attack took 56 h (15 days average)
Jan. 1999	DES Challenge III broken through brute-force by distributed Internet effort combined with Deep Crack and a total search time of 22 hours
Apr. 2006	Universities of Bochum and Kiel built COPACOBANA key-search machine based on low-cost FPGAs for approximately $10,000. Average search time is 7 days

or cell phones. Hardware refers to DES implementations running on ICs such as application-specific integrated circuits (ASICs) or field programmable gate arrays (FPGAs).

Software

A straightforward software implementation which follows the data flow of most DES descriptions, such as the one presented in this chapter, results in a very poor performance. This is due to the fact that many of the atomic DES operations involve bit permutation, in particular the E and P permutation, which are slow in software. Similarly, small S-boxes such as used in DES are efficient in hardware but only moderately efficient on modern CPUs. There have been numerous methods proposed for accelerating DES software implementations. The general idea is to use tables with precomputed values of several DES operations, e.g., of several S-boxes and the permutation. Optimized implementations require about 240 cycles for encrypting one block on a 32-bit CPU. On a 2-GHz CPU this translates into a theoretical throughput of about 533 Mbits/s. 3DES, which is considerably more secure than single DES, runs at almost exactly 1/3 of the DES speed. Note that nonoptimized implementations are considerably slower, often below 100 Mbit/s.

A notable method for accelerating software implementations of DES is *bit-slicing*, developed by Eli Biham [20]. On a 300-MHz DEC Alpha workstation an encryption rate of 137 Mbit/sec has been reported, which was much faster than a standard DES implementation at that time. The limitation of bit-slicing, however, is that several blocks are encrypted in parallel, which can be a drawback for certain modes of operation such as Cipher Block Chaining (CBC) and Output Feedback (OFB) mode (cf. Chap. 5).

Hardware

One design criterion for DES was its efficiency in hardware. Permutations such as the E, P, IP and IP^{-1} permutations are very easy to implement in hardware, as they only require wiring but no logic. The small 6-by-4 S-boxes are also relatively easily realizable in hardware. Typically, they are implemented with Boolean logic, i.e., logic gates. On average, one S-box requires about 100 gates.

An area-efficient implementation of a single DES round can be done with less than 3000 gates. If a high throughput is desired, DES can be implemented extremely fast by fitting multiple rounds in one circuit, e.g., by using pipelining. On modern ASICs and FPGAs throughput rates of several 100 Gbit/sec are possible. On the other end of the performance spectrum, very small implementations with fewer than 3000 gates even fit onto lowcost radio frequency identification (RFID) chips.

3.7 DES Alternatives

There exist a wealth of other block ciphers. Even though there are many ciphers which have security weaknesses or which are not well investigated, there are also many block ciphers which appear very strong. In the following a brief list of ciphers is given which can be of interest depending on the application needs.

3.7.1 The Advanced Encryption Standard (AES) and the AES Finalist Ciphers

By now, the algorithm of choice for many, many applications has become the Advanced Encryption Standard (AES), which will be introduced in detail in the following chapter. AES is with its three key lengths of 128, 192 and 256 bit secure against brute-force attacks for several decades, and there are no analytical attacks with any reasonable chance of success known.

AES was the result of an open competition, and in the last stage of the selection process there were four other finalist algorithms. These are the block ciphers *Mars*, *RC6*, *Serpent* and *Twofish*. All of them are cryptographically strong and quite fast, especially in software. Based on today's knowledge, they can all be recommended. Mars, Serpent and Twofish can be used royalty-free.

3.7.2 Triple DES (3DES) and DESX

An alternative to AES or the AES finalist algorithms is *triple DES*, often denoted as
3DES. 3DES consists of three subsequent DES encryptions

$$y = DES_{k_3}(DES_{k_2}(DES_{k_1}(x)))$$

with different keys, as shown in Fig. 3.17.

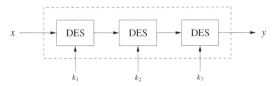

Fig. 3.17 Triple DES (3DES)

3DES seems resistant to both brute-force attacks and any analytical attack imag-
inable at the moment. See Chap. 5 for more information on double and triple encryption. Another version of 3DES is

$$y = DES_{k_3}(DES_{k_2}^{-1}(DES_{k_1}(x))).$$

The advantage here is that 3DES performs single DES encryption if $k_3 = k_2 = k_1$,
which is sometimes desired in implementations that should also support single DES
for legacy reasons. 3DES is very efficient in hardware but not particularly in software. It is popular in financial applications as well as for protecting biometric information in electronic passports.

A different approach for strengthening DES is to use key whitening. For this, two
additional 64-bit keys k_1 and k_2 are XORed to the plaintext and ciphertext, respectively, prior to and after the DES algorithm. This yields the following encryption
scheme:

$$y = DES_{k,k_1,k_2}(x) = DES_k(x \oplus k_1) \oplus k_2$$

This surprisingly simple modification makes DES much more resistant against exhaustive key searches. More about key whitening is said in Sect. 5.3.3.

3.7.3 Lightweight Cipher PRESENT

Over the last few years, several new block algorithms which are classified as
"lightweight ciphers" have been proposed. Lightweight commonly refers to algorithms with a very low implementation complexity, especially in hardware. Trivium
(Sect. 2.3.3) is an example of a lightweight stream cipher. A promising block cipher
candidate is *PRESENT*, which was designed specifically for applications such as

RFID tags or other pervasive computing applications that are extremely power or cost constrained. (One of the book authors participated in the design of PRESENT.)

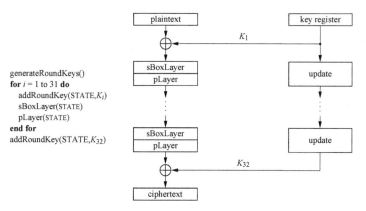

generateRoundKeys()
for $i = 1$ to 31 **do**
    addRoundKey(STATE,K_i)
    sBoxLayer(STATE)
    pLayer(STATE)
end for
addRoundKey(STATE,K_{32})

Fig. 3.18 Internal structure and pseudocode of the block cipher PRESENT

Unlike DES, PRESENT is not based on a Feistel network. Instead it is a substitution-permutation network (SP-network) and consists of 31 rounds. The block length is 64 bits, and two key lengths of 80 and 128 bits are supported. Each of the 31 rounds consists of an XOR operation to introduce a round key K_i for $1 \leq i \leq 32$, where K_{32} is used after round 31, a nonlinear substitution layer (sBoxLayer) and a linear bitwise permutation (pLayer). The nonlinear layer uses a single 4-bit S-box S, which is applied 16 times in parallel in each round. The key schedule generates 32 round keys from the user supplied key. The encryption routine of the cipher is described in pseudocode in Fig. 3.18, and each stage is now specified in turn.

addRoundKey At the beginning of each round, the round key K_i is XORed to the current STATE.

sBoxLayer PRESENT uses a single 4-bit to 4-bit S-box. This is a direct consequence of the pursuit of hardware efficiency, since such an S-Box allows a much more compact implementation than, e.g., an 8-bit S-box. The S-box entries in hexadecimal notation are given in Table 3.16.

Table 3.16 The PRESENT S-box in hexadecimal notation

x	0	1	2	3	4	5	6	7	8	9	A	B	C	D	E	F
$S[x]$	C	5	6	B	9	0	A	D	3	E	F	8	4	7	1	2

The 64 bit data path $b_{63} \ldots b_0$ is referred to as *state*. For the sBoxLayer the current state is considered as sixteen 4-bit words $w_{15} \ldots w_0$, where $w_i = b_{4*i+3}||b_{4*i+2}|| b_{4*i+1}||b_{4*i}$ for $0 \leq i \leq 15$, and the output are the 16 words $S[w_i]$.

pLayer Just like DES, the mixing layer was chosen as a bit permutation, which can be implemented extremely compactly in hardware. The bit permutation used in PRESENT is given by Table 3.17. Bit i of STATE is moved to bit position $P(i)$.

Table 3.17 The permutation layer of PRESENT

i	0	1	2	3	4	5	6	7	8	9	10	11	12	13	14	15
$P(i)$	0	16	32	48	1	17	33	49	2	18	34	50	3	19	35	51
i	16	17	18	19	20	21	22	23	24	25	26	27	28	29	30	31
$P(i)$	4	20	36	52	5	21	37	53	6	22	38	54	7	23	39	55
i	32	33	34	35	36	37	38	39	40	41	42	43	44	45	46	47
$P(i)$	8	24	40	56	9	25	41	57	10	26	42	58	11	27	43	59
i	48	49	50	51	52	53	54	55	56	57	58	59	60	61	62	63
$P(i)$	12	28	44	60	13	29	45	61	14	30	46	62	15	31	47	63

The bit permutation is quite regular and can in fact be expressed in the following way:

$$P(i) = \begin{cases} i \cdot 16 \mod 63, & i \in \{0, \ldots, 62\} \\ 63, & i = 63. \end{cases}$$

Key Schedule We describe in the following the key schedule for PRESENT with an 80-bit key. Since the main applications of PRESENT are low-cost systems, this key length is in most cases appropriate. (Details of the key schedule for PRESENT-128 can be found in [29].) The user-supplied key is stored in a key register K and is represented as $k_{79}k_{78} \ldots k_0$. At round i the 64-bit round key $K_i = \kappa_{63}\kappa_{62} \ldots \kappa_0$ consists of the 64 leftmost bits of the current contents of register K. Thus at round i we have:

$$K_i = \kappa_{63}\kappa_{62} \ldots \kappa_0 = k_{79}k_{78} \ldots k_{16}$$

The first subkey K_1 is a direct copy of 64 bit of the user supplied key. For the following subkeys $K_2, \ldots, K_{32}$ the key register $K = k_{79}k_{78} \ldots k_0$ is updated as follows:

1. $[k_{79}k_{78} \ldots k_1k_0] = [k_{18}k_{17} \ldots k_{20}k_{19}]$
2. $[k_{79}k_{78}k_{77}k_{76}] = S[k_{79}k_{78}k_{77}k_{76}]$
3. $[k_{19}k_{18}k_{17}k_{16}k_{15}] = [k_{19}k_{18}k_{17}k_{16}k_{15}] \oplus \text{round\_counter}$

Thus, the key schedule consists of three operations: (1) the key register is rotated by 61 bit positions to the left, (2) the leftmost four bits are passed through the PRESENT S-box, and (3) the `round_counter` value i is XORed with bits $k_{19}k_{18}k_{17}k_{16}k_{15}$ of K, where the least significant bit of `round_counter` is on the right. This counter is a simple integer which takes the values $(00001, 00010, \ldots , 11111)$. Note that for the derivation of K_2 the counter value 00001 is used; for K_3, the counter value 00010; and so on.

Implementation As a result of the aggressively hardware-optimized design of PRESENT, its software performance is not very competitive relative to modern ciphers like AES. An optimized software implementation on a Pentium III CPU in

C achieves a throughput of about 60 Mbit/s at a frequency of 1 GHz. However, it performs quite well on small microprocessors, which are common in inexpensive consumer products.

PRESENT-80 can be implemented in hardware with area requirements of approximately 1600 gate equivalences [147], where the encryption of one 64-bit plaintext block requires 32 clock cycles. As an example, at a clock rate of 1 MHz, which is quite typical on low-cost devices, a throughput of 2 Mbit/s is achieved, which is sufficient for most such applications. It is possible to realize the cipher with as few as approximately 1000 gate equivalences, where the encryption of one 64-bit plaintext requires 547 clock cycles. A fully pipelined implementation of PRESENT with 31 encryption stages achieves a throughput of 64 bit per clock cycle, which can be tranlsated into encryption throughputs of more than 50 Gbit/s.

Even though no attacks against PRESENT are known at the time of writing, it should be noted that it is a relatively new block cipher.

3.8 Discussion and Further Reading

DES History and Attacks Even though plain DES (i.e., non-3DES) is today mainly used in legacy applications, its history helps us understand the evolution of cryptography since the mid-1970s from an obscure discipline almost solely studied in government organizations towards an open discipline with many players in industry and academia. A summary of the DES history can be found in [165]. The two main analytical attacks developed against DES, differential and linear cryptanalysis, are today among the most powerful general methods for breaking block ciphers. Readers interested in the theory of block ciphers are encouraged to study these attacks. Good descriptions are given in [21, 114].

As we have seen in this chapter, DES should no longer be used since a brute-force attack can be accomplished at low cost in little time with cryptanalytical hardware. The two machines built outside governments, Deep Crack and COPACOBANA, are instructive examples of how to build low-cost "supercomputers" for very narrowly defined computational tasks. More information about Deep Crack can be found on the Internet [78] and about COPACOBANA in the articles [105, 88] and online at [47]. Readers interested in the fascinating area of cryptanalytical computers in general should take a look at the SHARCS (Special-purpose Hardware for Attacking Cryptographic Systems) workshop series, which started in 2005 and has information online [170].

DES Alternatives It should be noted that hundreds of block ciphers have been proposed over the last three decades, especially in the late 1980s and in the 1990s. DES has influenced the design of many other encryption algorithms. It is probably fair to say that the majority of today's successful block ciphers have borrowed ideas from DES. Some of the popular block ciphers are also based on Feistel networks as is DES. Examples of Feistel ciphers include Blowfish, CAST, KASUMI, Mars,

MISTY1, Twofish and RC6. One cipher which is well known and markedly different from DES is IDEA; it uses arithmetic in three different algebraic structures as atomic operations.

DES is a good example of a block cipher which is very efficient in hardware. The recent advent of pervasive computing has created a need for extremely small ciphers for applications such as RFID tags or low-cost smart cards, e.g., for high-volume public transportation payment tickets. Good references for PRESENT are [29, 147]. In addition to PRESENT, other recently proposed very small block ciphers include Clefia [48], HIGHT [93] and mCrypton [111]. A good overview of the new field of lightweight cryptography is given in the surveys [71, 98]. A more in-depth treatment of lightweight algorithms can be found in the Ph.D. dissertation [135].

Implementation With respect to software implementation of DES, an early reference is [20]. More advanced techniques are described in [106]. The powerful method of bit-slicing is applicable not only to DES but to most other ciphers.

Regarding DES hardware implementation, an early but still very interesting reference is [169]. There are many descriptions of high-performance implementations of DES on a variety of hardware platforms, including FPGAs [163], standard ASICs as well as more exotic semiconductor technology [67].

3.9 Lessons Learned

■ DES was the dominant symmetric encryption algorithm from the mid-1970s to the mid-1990s. Since 56-bit keys are no longer secure, the Advanced Encryption Standard (AES) was created.

■ Standard DES with 56-bit key length can be broken relatively easily nowadays through an exhaustive key search.

■ DES is quite robust against known analytical attacks: In practice it is very difficult to break the cipher with differential or linear cryptanalysis.

■ DES is reasonably efficient in software and very fast and small in hardware.

■ By encrypting with DES three times in a row, triple DES (3DES) is created, against which no practical attack is currently known.

■ The "default" symmetric cipher is nowadays often AES. In addition, the other four AES finalist ciphers all seem very secure and efficient.

■ Since about 2005 several proposals for lightweight ciphers have been made. They are suited for resource-constrained applications.

Problems

3.1. As stated in Sect. 3.5.2, one important property which makes DES secure is that the S-boxes are nonlinear. In this problem we verify this property by computing the output of S_1 for several pairs of inputs.

Show that $S_1(x_1) \oplus S_1(x_2) \neq S_1(x_1 \oplus x_2)$, where "$\oplus$" denotes bitwise XOR, for:

1. $x_1 = 000000$, $x_2 = 000001$
2. $x_1 = 111111$, $x_2 = 100000$
3. $x_1 = 101010$, $x_2 = 010101$

3.2. We want to verify that $IP(\cdot)$ and $IP^{-1}(\cdot)$ are truly inverse operations. We consider a vector $x = (x_1, x_2, \ldots, x_{64})$ of 64 bit. Show that $IP^{-1}(IP(x)) = x$ for the first five bits of x, i.e. for x_i, $i = 1, 2, 3, 4, 5$.

3.3. What is the output of the first round of the DES algorithm when the plaintext and the key are both all zeros?

3.4. What is the output of the first round of the DES algorithm when the plaintext and the key are both all ones?

3.5. Remember that it is desirable for good block ciphers that a change in one input bit affects many output bits, a property that is called diffusion or the avalanche effect. We try now to get a feeling for the avalanche property of DES. We apply an input word that has a "1" at bit position 57 and all other bits as well as the key are zero. (Note that the input word has to run through the initial permutation.)

1. How many S-boxes get different inputs compared to the case when an all-zero plaintext is provided?
2. What is the minimum number of output bits of the S-boxes that will change according to the S-box design criteria?
3. What is the output after the first round?
4. How many output bit after the first round have actually changed compared to the case when the plaintext is all zero? (Observe that we only consider a single round here. There will be more and more output differences after every new round. Hence the term *avalanche effect*.)

3.6. An avalanche effect is also desirable for the key: A one-bit change in a key should result in a dramatically different ciphertext if the plaintext is unchanged.

1. Assume an encryption with a given key. Now assume the key bit at position 1 (prior to $PC - 1$) is being flipped. Which S-boxes in which rounds are affected by the bit flip during DES encryption?
2. Which S-boxes in which DES rounds are affected by this bit flip during DES decryption?

3.7. A DES key K_w is called a *weak key* if encryption and decryption are identical operations:

$$DES_{K_w}(x) = DES_{K_w}^{-1}(x), \text{ for all } x \tag{3.1}$$

1. Describe the relationship of the subkeys in the encryption and decryption algorithm that is required so that Eq. (3.1) is fulfilled.
2. There are four weak DES keys. What are they?
3. What is the likelihood that a randomly selected key is weak?

3.8. DES has a somewhat surprising property related to bitwise complements of its inputs and outputs. We investigate the property in this problem.

We denote the bitwise complement of a number A (that is, all bits are flipped) by A'. Let $\oplus$ denote bitwise XOR. We want to show that if

$$y = \text{DES}_k(x)$$

then

$$y' = \text{DES}_{k'}(x'). \tag{3.2}$$

This states that if we complement the plaintext and the key, then the ciphertext output will also be the complement of the original ciphertext. Your task is to *prove* this property.

Try to prove this property using the following steps:

1. Show that for any bit strings A, B of equal length,

$$A' \oplus B' = A \oplus B$$

   and

$$A' \oplus B = (A \oplus B)'.$$

   (These two operations are needed for some of the following steps.)
2. Show that $PC - 1(k') = (PC - 1(k))'$.
3. Show that $LS_i(C'_{i-1}) = (LS_i(C_{i-1}))'$.
4. Using the two results from above, show that if k_i are the keys generated from k, then k'_i are the keys generated from k', where $i = 1, 2, \ldots, 16$.
5. Show that $IP(x') = (IP(x))'$.
6. Show that $E(R'_i) = (E(R_i))'$.
7. Using all previous results, show that if R_{i-1}, L_{i-1}, k_i generate R_i, then R'_{i-1}, L'_{i-1}, k'_i generate R'_i.
8. Show that Eq. (3.2) is true.

3.9. Assume we perform a known-plaintext attack against DES with one pair of plaintext and ciphertext. How many keys do we have to test in a worst-case scenario if we apply an exhaustive key search in a straightforward way? How many on average?

3.10. In this problem we want to study the clock frequency requirements for a hardware implementation of DES in real-world applications. The speed of a DES implementation is mainly determined by the time required to do one core iteration. This hardware kernel is then used 16 consecutive times in order to generate the encrypted output. (An alternative approach would be to build a hardware pipeline with 16 stages, resulting in 16-fold increased hardware costs.)

1. Let's assume that one core iteration can be performed in one clock cycle. Develop an expression for the required clock frequency for encrypting a stream of data with a data rate r [bit/sec]. Ignore the time needed for the initial and final permutation.
2. What clock frequency is required for encrypting a fast network link running at a speed of 1 Gb/sec? What is the clock frequency if we want to support a speed of 8 Gb/sec?

3.11. As the example of COPACOBANA [105] shows, key-search machines need not be prohibitive from a monetary point of view. We now consider a simple brute-force attack on DES which runs on COPACOBANA.

1. Compute the runtime of an average exhaustive key-search on DES assuming the following implementational details:

   - COPACOBANA platform with 20 FPGA modules
   - 6 FPGAs per FPGA module
   - 4 DES engines per FPGA
   - Each DES engine is fully pipelined and is capable of performing one encryption per clock cycle
   - 100 MHz clock frequency

2. How many COPACOBANA machines do we need in the case of an average search time of one hour?
3. Why does any design of a key-search machine constitute only an upper security threshold? By *upper security threshold* we mean a (complexity) measure which describes the maximum security that is provided by a given cryptographic algorithm.

3.12. We study a real-world case in this problem. A commercial file encryption program from the early 1990s used standard DES with 56 key bits. In those days, performing an exhaustive key search was considerably harder than nowadays, and thus the key length was sufficient for some applications. Unfortunately, the implementation of the key generation was flawed, which we are going to analyze. Assume that we can test 10^6 keys per second on a conventional PC.

The key is generated from a password consisting of 8 characters. The key is a simple concatenation of the 8 ASCII characters, yielding $64 = 8 \cdot 8$ key bits. With the permutation $PC - 1$ in the key schedule, the least significant bit (LSB) of each 8-bit character is ignored, yielding 56 key bits.

1. What is the size of the key space if all 8 characters are randomly chosen 8-bit ASCII characters? How long does an average key search take with a single PC?
2. How many key bits are used, if the 8 characters are randomly chosen 7-bit ASCII characters (i.e., the most significant bit is always zero)? How long does an average key search take with a single PC?
3. How large is the key space if, in addition to the restriction in Part 2, only letters are used as characters. Furthermore, unfortunately, all letters are converted

to capital letters before generating the key in the software. How long does an average key search take with a single PC?

3.13. This problem deals with the lightweight cipher PRESENT.

1. Calculate the state of PRESENT-80 after the execution of one round. You can use the following table to solve this problem with paper and pencil. Use the following values (in hexadecimal notation):
plaintext = 0000 0000 0000 0000,
key = BBBB 5555 5555 EEEE FFFF.

Plaintext	0000 0000 0000 0000
Round key	
State after KeyAdd	
State after S-Layer	
State after P-Layer	

2. Now calculate the round key for the second round using the following table.

Key	BBBB 5555 5555 EEEE FFFF
Key state after rotation	
Key state after S-box	
Key state after CounterAdd	
Round key for Round 2	

Chapter 4
The Advanced Encryption Standard (AES)

The *Advanced Encryption Standard (AES)* is the most widely used symmetric cipher today. Even though the term "Standard" in its name only refers to US government applications, the AES block cipher is also mandatory in several industry standards and is used in many commercial systems. Among the commercial standards that include AES are the Internet security standard IPsec, TLS, the Wi-Fi encryption standard IEEE 802.11i, the secure shell network protocol SSH (Secure Shell), the Internet phone Skype and numerous security products around the world. To date, there are no attacks better than brute-force known against AES.

In this chapter you will learn:

■ The design process of the US symmetric encryption standard, AES
■ The encryption and decryption function of AES
■ The internal structure of AES, namely:
　□ byte substitution layer
　□ diffusion layer
　□ key addition layer
　□ key schedule
■ Basic facts about Galois fields
■ Efficiency of AES implementations

4.1 Introduction

In 1999 the US National Institute of Standards and Technology (NIST) indicated
that DES should only be used for legacy systems and instead triple DES (3DES)
should be used. Even though 3DES resists brute-force attacks with today's technology, there are several problems with it. First, it is not very efficient with regard to
software implementations. DES is already not particularly well suited for software
and 3DES is three times slower than DES. Another disadvantage is the relatively
short block size of 64 bits, which is a drawback in certain applications, e.g., if one
wants to built a hash function from a block cipher (cf. Sect. 11.3.2). Finally, if one
is worried about attacks with quantum computers, which might become reality in a
few decades, key lengths on the order of 256 bits are desirable. All these consideration led NIST to the conclusion that an entirely new block cipher was needed as a
replacement for DES.

In 1997 NIST called for proposals for a new *Advanced Encryption Standard
(AES)*. Unlike the DES development, the selection of the algorithm for AES was
an open process administered by NIST. In three subsequent AES evaluation rounds,
NIST and the international scientific community discussed the advantages and disadvantages of the submitted ciphers and narrowed down the number of potential
candidates. In 2001, NIST declared the block cipher *Rijndael* as the new AES and
published it as a final standard (FIPS PUB 197). Rijndael was designed by two
young Belgian cryptographers.

Within the call for proposals, the following requirements for all AES candidate
submissions were mandatory:

- block cipher with 128 bit block size
- three key lengths must be supported: 128, 192 and 256 bit
- security relative to other submitted algorithms
- efficiency in software and hardware

The invitation for submitting suitable algorithms and the subsequent evaluation
of the successor of DES was a public process. A compact chronology of the AES
selection process is given here:

- The need for a new block cipher was announced on January 2, 1997, by NIST.
- A formal call for AES was announced on September 12, 1997.
- Fifteen candidate algorithms were submitted by researchers from several coun-
  tries by August 20, 1998.
- On August 9, 1999, five finalist algorithms were announced:

    - *Mars* by IBM Corporation
    - *RC6* by RSA Laboratories
    - *Rijndael*, by Joan Daemen and Vincent Rijmen
    - *Serpent*, by Ross Anderson, Eli Biham and Lars Knudsen
    - *Twofish*, by Bruce Schneier, John Kelsey, Doug Whiting, David Wagner, Chris
      Hall and Niels Ferguson

- On October 2, 2000, NIST announced that it had chosen Rijndael as the AES.
- On November 26, 2001, AES was formally approved as a US federal standard.

It is expected that AES will be the dominant symmetric-key algorithm for many commercial applications for the next few decades. It is also remarkable that in 2003 the US National Security Agency (NSA) announced that it allows AES to encrypt classified documents up to the level SECRET for all key lengths, and up to the TOP SECRET level for key lengths of either 192 or 256 bits. Prior to that date, only non-public algorithms had been used for the encryption of classified documents.

4.2 Overview of the AES Algorithm

The AES cipher is almost identical to the block cipher Rijndael. The Rijndael block and key size vary between 128, 192 and 256 bits. However, the AES standard only calls for a block size of 128 bits. Hence, only Rijndael with a block length of 128 bits is known as the AES algorithm. In the remainder of this chapter, we only discuss the standard version of Rijndael with a block length of 128 bits.

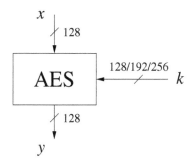

Fig. 4.1 AES input/output parameters

As mentioned previously, three key lengths must be supported by Rijndael as this was an NIST design requirement. The number of internal rounds of the cipher is a function of the key length according to Table 4.1.

Table 4.1 Key lengths and number of rounds for AES

key lengths	# rounds = n_r
128 bit	10
192 bit	12
256 bit	14

In contrast to DES, AES does not have a Feistel structure. Feistel networks do not encrypt an entire block per iteration, e.g., in DES, $64/2 = 32$ bits are encrypted in one round. AES, on the other hand, encrypts all 128 bits in one iteration. This is one reason why it has a comparably small number of rounds.

AES consists of so-called *layers*. Each layer manipulates all 128 bits of the data path. The data path is also referred to as the state of the algorithm. There are only three different types of layers. Each round, with the exception of the first, consists of all three layers as shown in Fig. 4.2: the plaintext is denoted as x, the ciphertext as y and the number of rounds as n_r. Moreover, the last round n_r does not make use of the MixColumn transformation, which makes the encryption and decryption scheme symmetric.

We continue with a brief description of the layers:

Key Addition layer A 128-bit round key, or subkey, which has been derived from the main key in the key schedule, is XORed to the state.

Byte Substitution layer (S-Box) Each element of the state is nonlinearly transformed using lookup tables with special mathematical properties. This introduces *confusion* to the data, i.e., it assures that changes in individual state bits propagate quickly across the data path.

Diffusion layer It provides *diffusion* over all state bits. It consists of two sublayers, both of which perform linear operations:

- The *ShiftRows* layer permutes the data on a byte level.
- The *MixColumn* layer is a matrix operation which combines (mixes) blocks of four bytes.

Similar to DES, the key schedule computes round keys, or subkeys, $(k_0, k_1, \ldots, k_{n_r})$ from the original AES key.

Before we describe the internal functions of the layers in Sect. 4.4, we have to introduce a new mathematical concept, namely *Galois fields*. Galois field computations are needed for all operations within the AES layers.

4.3 Some Mathematics: A Brief Introduction to Galois Fields

In AES, Galois field arithmetic is used in most layers, especially in the S-Box and the MixColumn layer. Hence, for a deeper understanding of the internals of AES, we provide an introduction to Galois fields as needed for this purpose before we continue with the algorithm in Sect. 4.4. A background on Galois fields is not required for a basic understanding of AES, and the reader can skip this section.

4.3.1 Existence of Finite Fields

A *finite field*, sometimes also called *Galois field*, is a set with a finite number of elements. Roughly speaking, a Galois field is a finite set of elements in which we

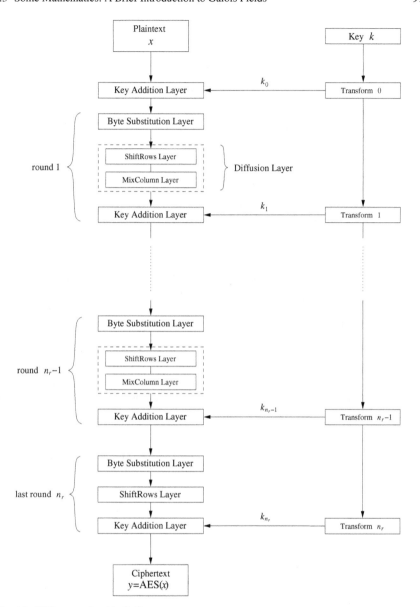

Fig. 4.2 AES encryption block diagram

can add, subtract, multiply and invert. Before we introduce the definition of a field, we first need the concept of a a simpler algebraic structure, a group.

Definition 4.3.1 Group

A group *is a set of elements G together with an operation ∘ which combines two elements of G. A group has the following properties:*

1. *The group operation ∘ is* closed. *That is, for all $a, b, \in G$, it holds that $a \circ b = c \in G$.*
2. *The group operation is* associative. *That is, $a \circ (b \circ c) = (a \circ b) \circ c$ for all $a, b, c \in G$.*
3. *There is an element $1 \in G$, called the* neutral element *(or* identity element*), such that $a \circ 1 = 1 \circ a = a$ for all $a \in G$.*
4. *For each $a \in G$ there exists an element $a^{-1} \in G$, called the* inverse *of a, such that $a \circ a^{-1} = a^{-1} \circ a = 1$.*
5. *A group G is* abelian (or commutative) *if, furthermore, $a \circ b = b \circ a$ for all $a, b \in G$.*

Roughly speaking, a group is set with one operation and the corresponding inverse operation. If the operation is called addition, the inverse operation is subtraction; if the operation is multiplication, the inverse operation is division (or multiplication with the inverse element).

Example 4.1. The set of integers $\mathbb{Z}_m = \{0, 1, \ldots, m-1\}$ and the operation addition modulo m form a group with the neutral element 0. Every element a has an inverse $-a$ such that $a + (-a) = 0 \mod m$. Note that this set does not form a group with the operation multiplication because most elements a do not have an inverse such that $a a^{-1} = 1 \mod m$.

◇

In order to have all four basic arithmetic operations (i.e., addition, subtraction, multiplication, division) in one structure, we need a set which contains an additive and a multiplicative group. This is what we call a field.

Definition 4.3.2 Field

A field *F is a set of elements with the following properties:*

- *All elements of F form an additive group with the group operation "+" and the neutral element 0.*
- *All elements of F except 0 form a multiplicative group with the group operation "×" and the neutral element 1.*
- *When the two group operations are mixed, the distributivity law holds, i.e., for all $a, b, c \in F$: $a(b + c) = (ab) + (ac)$.*

Example 4.2. The set $\mathbb{R}$ of real numbers is a field with the neutral element 0 for the additive group and the neutral element 1 for the multiplicative group. Every real number a has an additive inverse, namely $-a$, and every nonzero element a has a multiplicative inverse $1/a$.

◇

In cryptography, we are almost always interested in fields with a finite number of elements, which we call finite fields or Galois fields. The number of elements in the field is called the *order* or *cardinality* of the field. Of fundamental importance is the following theorem:

> **Theorem 4.3.1** *A field with order m only exists if m is a prime power, i.e., $m = p^n$, for some positive integer n and prime integer p. p is called the* characteristic *of the finite field.*

This theorem implies that there are, for instance, finite fields with 11 elements, or with 81 elements (since $81 = 3^4$) or with 256 elements (since $256 = 2^8$, and 2 is a prime). However, there is no finite field with 12 elements since $12 = 2^2 \cdot 3$, and 12 is thus not a prime power. In the remainder of this section we look at how finite fields can be built, and more importantly for our purpose, how we can do arithmetic in them.

4.3.2 Prime Fields

The most intuitive examples of finite fields are fields of prime order, i.e., fields with $n = 1$. Elements of the field $GF(p)$ can be represented by integers $0, 1, \ldots, p-1$. The two operations of the field are modular integer addition and integer multiplication *modulo p*.

> **Theorem 4.3.2** *Let p be a prime. The integer ring $\mathbb{Z}_p$ is denoted as $GF(p)$ and is referred to as a* prime field, *or as a* Galois field *with a prime number of elements. All nonzero elements of $GF(p)$ have an inverse. Arithmetic in $GF(p)$ is done modulo p.*

This means that if we consider the integer ring $\mathbb{Z}_m$ which was introduced in Sect. 1.4.2, i.e., integers with modular addition and multiplication, and m happens to be a prime, $\mathbb{Z}_m$ is not only a ring but also a finite field.

In order to do arithmetic in a prime field, we have to follow the rules for integer rings: Addition and multiplication are done modulo p, the additive inverse of any element a is given by $a + (-a) = 0 \bmod p$, and the multiplicative inverse of any nonzero element a is defined as $a \cdot a^{-1} = 1$. Let's have a look at an example of a prime field.

Example 4.3. We consider the finite field $GF(5) = \{0, 1, 2, 3, 4\}$. The tables below describe how to add and multiply any two elements, as well as the additive and

addition   **additive inverse**

```
+ 0 1 2 3 4
0 0 1 2 3 4                                                  −0 = 0
1 1 2 3 4 0                                                  −1 = 4
2 2 3 4 0 1                                                  −2 = 3
3 3 4 0 1 2                                                  −3 = 2
4 4 0 1 2 3                                                  −4 = 1
```

multiplication   **multiplicative inverse**

```
× 0 1 2 3 4
0 0 0 0 0 0                                                  0⁻¹ does not exist
1 0 1 2 3 4                                                  1⁻¹ = 1
2 0 2 4 1 3                                                  2⁻¹ = 3
3 0 3 1 4 2                                                  3⁻¹ = 2
4 0 4 3 2 1                                                  4⁻¹ = 4
```

multiplicative inverse of the field elements. Using these tables, we can perform all calculations in this field without using modular reduction explicitly.

◇

A very important prime field is $GF(2)$, which is the smallest finite field that exists. Let's have a look at the multiplication and addition tables for the field.

Example 4.4. Let's consider the small finite field $GF(2) = \{0,1\}$. Arithmetic is simply done modulo 2, yielding the following arithmetic tables:

addition   **multiplication**

```
+ 0 1                           × 0 1
0 0 1                           0 0 0
1 1 0                           1 0 1
```

◇

As we saw in Chap. 2 on stream ciphers, $GF(2)$ addition, i.e., modulo 2 addition, is equivalent to an XOR gate. What we learn from the example above is that $GF(2)$ multiplication is equivalent to the logical AND gate. The field $GF(2)$ is important for AES.

4.3.3 Extension Fields $GF(2^m)$

In AES the finite field contains 256 elements and is denoted as $GF(2^8)$. This field was chosen because each of the field elements can be represented by one byte. For the S-Box and MixColumn transforms, AES treats every byte of the internal data path as an element of the field $GF(2^8)$ and manipulates the data by performing arithmetic in this finite field.

However, if the order of a finite field is not prime, and 2^8 is clearly not a prime, the addition and multiplication operation cannot be represented by addition and multiplication of integers modulo 2^8. Such fields with $m > 1$ are called *extension fields*. In order to deal with extension fields we need (1) a different notation for field elements and (2) different rules for performing arithmetic with the elements. We will see in the following that elements of extension fields can be represented as *polynomials*, and that computation in the extension field is achieved by performing a certain type of *polynomial arithmetic*.

In extension fields $GF(2^m)$ elements are not represented as integers but as polynomials with coefficients in $GF(2)$. The polynomials have a maximum degree of $m-1$, so that there are m coefficients in total for every element. In the field $GF(2^8)$, which is used in AES, each element $A \in GF(2^8)$ is thus represented as:

$$A(x) = a_7x^7 + \cdots + a_1x + a_0, \quad a_i \in GF(2) = \{0,1\}.$$

Note that there are exactly $256 = 2^8$ such polynomials. The set of these 256 polynomials is the finite field $GF(2^8)$. It is also important to observe that every polynomial can simply be stored in digital form as an 8-bit vector

$$A = (a_7, a_6, a_5, a_4, a_3, a_2, a_1, a_0).$$

In particular, we do *not* have to store the factors x^7, x^6, etc. It is clear from the bit positions to which power x^i each coefficient belongs.

4.3.4 Addition and Subtraction in $GF(2^m)$

Let's now look at addition and subtraction in extension fields. The key addition layer of AES uses addition. It turns out that these operations are straightforward. They are simply achieved by performing standard polynomial addition and subtraction: We merely add or subtract coefficients with equal powers of x. The coefficient additions or subtractions are done in the underlying field $GF(2)$.

Definition 4.3.3 Extension field addition and subtraction
Let $A(x), B(x) \in GF(2^m)$. The sum of the two elements is then computed according to:

$$C(x) = A(x) + B(x) = \sum_{i=0}^{m-1} c_i x^i, \quad c_i \equiv a_i + b_i \bmod 2$$

and the difference is computed according to:

$$C(x) = A(x) - B(x) = \sum_{i=0}^{m-1} c_i x^i, \quad c_i \equiv a_i - b_i \equiv a_i + b_i \bmod 2.$$

Note that we perform modulo 2 addition (or subtraction) with the coefficients. As we saw in Chap. 2, addition and subtraction modulo 2 are the same operation. Moreover, addition modulo 2 is equal to bitwise XOR. Let's have a look at an example in the field $GF(2^8)$ which is used in AES:

Example 4.5. Here is how the sum $C(x) = A(x) + B(x)$ of two elements from $GF(2^8)$ is computed:

$$
\begin{array}{r}
A(x) = x^7 + x^6 + x^4 + \quad\quad 1 \\
B(x) = \quad\quad\quad\quad x^4 + x^2 + 1 \\
\hline
C(x) = x^7 + x^6 + \quad\quad x^2 \quad\quad
\end{array}
$$

◇

Note that if we computed the difference of the two polynomials $A(x) - B(x)$ from the example above, we would get the same result as for the sum.

4.3.5 Multiplication in $GF(2^m)$

Multiplication in $GF(2^8)$ is the core operation of the MixColumn transformation of AES. In a first step, two elements (represented by their polynomials) of a finite field $GF(2^m)$ are multiplied using the standard polynomial multiplication rule:

$$A(x) \cdot B(x) = (a_{m-1}x^{m-1} + \cdots + a_0) \cdot (b_{m-1}x^{m-1} + \cdots + b_0)$$
$$C'(x) = c'_{2m-2}x^{2m-2} + \cdots + c'_0,$$

where:

$$c'_0 = a_0 b_0 \bmod 2$$
$$c'_1 = a_0 b_1 + a_1 b_0 \bmod 2$$
$$\vdots$$
$$c'_{2m-2} = a_{m-1} b_{m-1} \bmod 2.$$

Note that all coefficients a_i, b_i and c_i are elements of $GF(2)$, and that coefficient arithmetic is performed in $GF(2)$. In general, the product polynomial $C(x)$ will have a degree higher than $m - 1$ and has to be reduced. The basic idea is an approach similar to the case of multiplication in prime fields: in $GF(p)$, we multiply the two integers, divide the result by a prime, and consider only the remainder. Here is what we are doing in extension fields: The product of the multiplication is divided by a certain polynomial, and we consider only the remainder after the polynomial division. We need irreducible polynomials for the module reduction. We recall from Sect. 2.3.1 that irreducible polynomials are roughly comparable to prime numbers, i.e., their only factors are 1 and the polynomial itself.

Definition 4.3.4 Extension field multiplication
Let $A(x), B(x) \in GF(2^m)$ and let

$$P(x) \equiv \sum_{i=0}^{m} p_i x^i, \quad p_i \in GF(2)$$

be an irreducible polynomial. Multiplication of the two elements $A(x), B(x)$ is performed as

$$C(x) \equiv A(x) \cdot B(x) \bmod P(x).$$

Thus, every field $GF(2^m)$ requires an irreducible polynomial $P(x)$ of degree m with coefficients from $GF(2)$. Note that not all polynomials are irreducible. For example, the polynomial $x^4 + x^3 + x + 1$ is reducible since

$$x^4 + x^3 + x + 1 = (x^2 + x + 1)(x^2 + 1)$$

and hence cannot be used to construct the extension field $GF(2^4)$. Since primitive polynomials are a special type of irreducible polynomial, the polynomials in Table 2.3 can be used for constructing fields $GF(2^m)$. For AES, the irreducible polynomial

$$P(x) = x^8 + x^4 + x^3 + x + 1$$

is used. It is part of the AES specification.

Example 4.6. We want to multiply the two polynomials $A(x) = x^3 + x^2 + 1$ and $B(x) = x^2 + x$ in the field $GF(2^4)$. The irreducible polynomial of this Galois field is given as

$$P(x) = x^4 + x + 1.$$

The plain polynomial product is computed as:

$$C'(x) = A(x) \cdot B(x) = x^5 + x^3 + x^2 + x.$$

We can now reduce $C'(x)$ using the polynomial division method we learned in school. However, sometimes it is easier to reduce each of the leading terms x^4 and

x^5 individually:

$$x^4 = 1 \cdot P(x) + (x+1)$$
$$x^4 \equiv x+1 \bmod P(x)$$
$$x^5 \equiv x^2 + x \bmod P(x).$$

Now, we only have to insert the reduced expression for x^5 into the intermediate result $C'(x)$:

$$C(x) \equiv x^5 + x^3 + x^2 + x \bmod P(x)$$
$$C(x) \equiv (x^2 + x) + (x^3 + x^2 + x) = x^3$$
$$A(x) \cdot B(x) \equiv x^3.$$

◇

It is important not to confuse multiplication in $GF(2^m)$ with integer multiplication, especially if we are concerned with software implementations of Galois fields. Recall that the polynomials, i.e., the field elements, are normally stored as bit vectors in the computers. If we look at the multiplication from the previous example, the following very atypical operation is being performed on the bit level:

$$
\begin{array}{ccccc}
A & \cdot & B & = & C \\
(x^3 + x^2 + 1) & \cdot & (x^2 + x) & = & x^3 \\
(1\,1\,0\,1) & \cdot & (0\,1\,1\,0) & = & (1\,0\,0\,0).
\end{array}
$$

This computation is **not** identical to integer arithmetic. If the polynomials are interpreted as integers, i.e., $(1101)_2 = 13_{10}$ and $(0110)_2 = 6_{10}$, the result would have been $(1001110)_2 = 78_{10}$, which is clearly not the same as the Galois field multiplication product. Hence, even though we can represent field elements as integers data types, we cannot make use of the integer arithmetic provided

4.3.6 Inversion in $GF(2^m)$

Inversion in $GF(2^8)$ is the core operation of the Byte Substitution transformation, which contains the AES S-Boxes. For a given finite field $GF(2^m)$ and the corresponding irreducible reduction polynomial $P(x)$, the inverse A^{-1} of a nonzero element $A \in GF(2^m)$ is defined as:

$$A^{-1}(x) \cdot A(x) = 1 \bmod P(x).$$

For small fields — in practice this often means fields with 2^{16} or fewer elements — lookup tables which contain the precomputed inverses of all field elements are often used. Table 4.2 shows the values which are used within the S-Box of AES. The table contains all inverses in $GF(2^8)$ modulo $P(x) = x^8 + x^4 + x^3 + x + 1$ in hexadecimal notation. A special case is the entry for the field element 0, for which an inverse does not exist. However, for the AES S-Box, a substitution table is needed that is defined for every possible input value. Hence, the designers defined the S-Box such that the input value 0 is mapped to the output value 0.

Table 4.2 Multiplicative inverse table in $GF(2^8)$ for bytes xy used within the AES S-Box

		Y														
		0	1	2	3	4	5	6	7	8	9	A	B	C	D	E	F
	0	00	01	8D	F6	CB	52	7B	D1	E8	4F	29	C0	B0	E1	E5	C7
	1	74	B4	AA	4B	99	2B	60	5F	58	3F	FD	CC	FF	40	EE	B2
	2	3A	6E	5A	F1	55	4D	A8	C9	C1	0A	98	15	30	44	A2	C2
	3	2C	45	92	6C	F3	39	66	42	F2	35	20	6F	77	BB	59	19
	4	1D	FE	37	67	2D	31	F5	69	A7	64	AB	13	54	25	E9	09
	5	ED	5C	05	CA	4C	24	87	BF	18	3E	22	F0	51	EC	61	17
	6	16	5E	AF	D3	49	A6	36	43	F4	47	91	DF	33	93	21	3B
	7	79	B7	97	85	10	B5	BA	3C	B6	70	D0	06	A1	FA	81	82
X	8	83	7E	7F	80	96	73	BE	56	9B	9E	95	D9	F7	02	B9	A4
	9	DE	6A	32	6D	D8	8A	84	72	2A	14	9F	88	F9	DC	89	9A
	A	FB	7C	2E	C3	8F	B8	65	48	26	C8	12	4A	CE	E7	D2	62
	B	0C	E0	1F	EF	11	75	78	71	A5	8E	76	3D	BD	BC	86	57
	C	0B	28	2F	A3	DA	D4	E4	0F	A9	27	53	04	1B	FC	AC	E6
	D	7A	07	AE	63	C5	DB	E2	EA	94	8B	C4	D5	9D	F8	90	6B
	E	B1	0D	D6	EB	C6	0E	CF	AD	08	4E	D7	E3	5D	50	1E	B3
	F	5B	23	38	34	68	46	03	8C	DD	9C	7D	A0	CD	1A	41	1C

Example 4.7. From Table 4.2 the inverse of

$$x^7 + x^6 + x = (11000010)_2 = (C2)_{hex} = (xy)$$

is given by the element in row C, column 2:

$$(2F)_{hex} = (00101111)_2 = x^5 + x^3 + x^2 + x + 1.$$

This can be verified by multiplication:

$$(x^7 + x^6 + x) \cdot (x^5 + x^3 + x^2 + x + 1) \equiv 1 \mod P(x).$$

◇

Note that the table above does not contain the S-Box itself, which is a bit more complex and will be described in Sect. 4.4.1.

As an alternative to using lookup tables, one can also explicitly compute inverses. The main algorithm for computing multiplicative inverses is the extended Euclidean algorithm, which is introduced in Sect. 6.3.1.

4.4 Internal Structure of AES

In the following, we examine the internal structure of AES. Figure 4.3 shows the graph of a single AES round. The 16-byte input $A_0, \ldots, A_{15}$ is fed byte-wise into the

S-Box. The 16-byte output $B_0, \ldots, B_{15}$ is permuted byte-wise in the ShiftRows layer and mixed by the MixColumn transformation $c(x)$. Finally, the 128-bit subkey k_i is XORed with the intermediate result. We note that AES is a byte-oriented cipher.

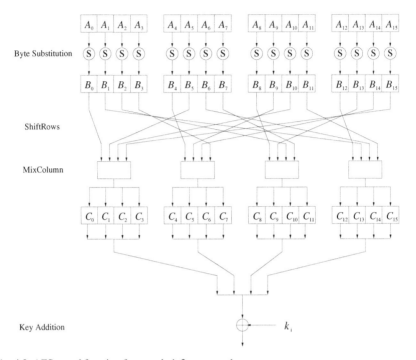

Fig. 4.3 AES round function for rounds $1, 2, \ldots, n_r - 1$

This is in contrast to DES, which makes heavy use of bit permutation and can thus be considered to have a bit-oriented structure.

In order to understand how the data moves through AES, we first imagine that the state A (i.e., the 128-bit data path) consisting of 16 bytes $A_0, A_1, \ldots, A_{15}$ is arranged in a four-by-four byte matrix:

A_0	A_4	A_8	A_{12}
A_1	A_5	A_9	A_{13}
A_2	A_6	A_{10}	A_{14}
A_3	A_7	A_{11}	A_{15}

As we will see in the following, AES operates on elements, columns or rows of the current state matrix. Similarly, the key bytes are arranged into a matrix with four rows and four (128-bit key), six (192-bit key) or eight (256-bit key) columns. Here is, as an example, the state matrix of a 192-bit key:

K_0	K_4	K_8	K_{12}	K_{16}	K_{20}
K_1	K_5	K_9	K_{13}	K_{17}	K_{21}
K_2	K_6	K_{10}	K_{14}	K_{18}	K_{22}
K_3	K_7	K_{11}	K_{15}	K_{19}	K_{23}

We discuss now what happens in each of the layers.

4.4.1 Byte Substitution Layer

As shown in Fig. 4.3, the first layer in each round is the *Byte Substitution layer*. The Byte Substitution layer can be viewed as a row of 16 parallel S-Boxes, each with 8 input and output bits. Note that all 16 S-Boxes are identical, unlike DES where eight different S-Boxes are used. In the layer, each state byte A_i is replaced, i.e., substituted, by another byte B_i:

$$S(A_i) = B_i.$$

The S-Box is the only nonlinear element of AES, i.e., it holds that $\text{ByteSub}(A) + \text{ByteSub}(B) \neq \text{ByteSub}(A + B)$ for two states A and B. The S-Box substitution is a bijective mapping, i.e., each of the $2^8 = 256$ possible input elements is one-to-one mapped to one output element. This allows us to uniquely reverse the S-Box, which is needed for decryption. In software implementations the S-Box is usually realized as a 256-by-8 bit lookup table with fixed entries, as given in Table 4.3.

Table 4.3 AES S-Box: Substitution values in hexadecimal notation for input byte (xy)

		0	1	2	3	4	5	6	7	8	9	A	B	C	D	E	F
	0	63	7C	77	7B	F2	6B	6F	C5	30	01	67	2B	FE	D7	AB	76
	1	CA	82	C9	7D	FA	59	47	F0	AD	D4	A2	AF	9C	A4	72	C0
	2	B7	FD	93	26	36	3F	F7	CC	34	A5	E5	F1	71	D8	31	15
	3	04	C7	23	C3	18	96	05	9A	07	12	80	E2	EB	27	B2	75
	4	09	83	2C	1A	1B	6E	5A	A0	52	3B	D6	B3	29	E3	2F	84
	5	53	D1	00	ED	20	FC	B1	5B	6A	CB	BE	39	4A	4C	58	CF
	6	D0	EF	AA	FB	43	4D	33	85	45	F9	02	7F	50	3C	9F	A8
	7	51	A3	40	8F	92	9D	38	F5	BC	B6	DA	21	10	FF	F3	D2
x	8	CD	0C	13	EC	5F	97	44	17	C4	A7	7E	3D	64	5D	19	73
	9	60	81	4F	DC	22	2A	90	88	46	EE	B8	14	DE	5E	0B	DB
	A	E0	32	3A	0A	49	06	24	5C	C2	D3	AC	62	91	95	E4	79
	B	E7	C8	37	6D	8D	D5	4E	A9	6C	56	F4	EA	65	7A	AE	08
	C	BA	78	25	2E	1C	A6	B4	C6	E8	DD	74	1F	4B	BD	8B	8A
	D	70	3E	B5	66	48	03	F6	0E	61	35	57	B9	86	C1	1D	9E
	E	E1	F8	98	11	69	D9	8E	94	9B	1E	87	E9	CE	55	28	DF
	F	8C	A1	89	0D	BF	E6	42	68	41	99	2D	0F	B0	54	BB	16

Example 4.8. Let's assume the input byte to the S-Box is $A_i = (C2)_{hex}$, then the substituted value is

$$S((C2)_{hex}) = (25)_{hex}.$$

On a bit level — and remember, the only thing that is ultimate of interest in encryption is the manipulation of bits — this substitution can be described as:

$$S(11000010) = (00100101).$$

◇

Even though the S-Box is bijective, it does not have any fixed points, i.e., there aren't any input values A_i such that $S(A_i) = A_i$. Even the zero-input is not a fixed point: $S(00000000) = (01100011)$.

Example 4.9. Let's assume the input to the Byte Substitution layer is

$$(C2, C2, \ldots, C2)$$

in hexadecimal notation. The output state is then

$$(25, 25, \ldots, 25).$$

◇

Mathematical description of the S-Box For readers who are interested in how the S-Box entries are constructed, a more detailed description now follows. This description, however, is not necessary for a basic understanding of AES, and the remainder of this subsection can be skipped without problem. Unlike the DES S-Boxes, which are essentially random tables that fulfill certain properties, the AES S-Boxes have a strong algebraic structure. An AES S-Box can be viewed as a two-step mathematical transformation (Fig. 4.4).

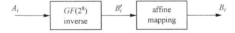

Fig. 4.4 The two operations within the AES S-Box which computes the function $B_i = S(A_i)$

The first part of the substitution is a Galois field inversion, the mathematics of which were introduced in Sect. 4.3.2. For each input element A_i, the inverse is computed: $B_i' = A_i^{-1}$, where both A_i and B_i' are considered elements in the field $GF(2^8)$ with the fixed irreducible polynomial $P(x) = x^8 + x^4 + x^3 + x + 1$. A lookup table with all inverses is shown in Table 4.2. Note that the inverse of the zero element does not exist. However, for AES it is defined that the zero element $A_i = 0$ is mapped to itself.

In the second part of the substitution, each byte B_i' is multiplied by a constant bit-matrix followed by the addition of a constant 8-bit vector. The operation is described by:

$$\begin{pmatrix} b_0 \\ b_1 \\ b_2 \\ b_3 \\ b_4 \\ b_5 \\ b_6 \\ b_7 \end{pmatrix} \equiv \begin{pmatrix} 1 & 0 & 0 & 0 & 1 & 1 & 1 & 1 \\ 1 & 1 & 0 & 0 & 0 & 1 & 1 & 1 \\ 1 & 1 & 1 & 0 & 0 & 0 & 1 & 1 \\ 1 & 1 & 1 & 1 & 0 & 0 & 0 & 1 \\ 1 & 1 & 1 & 1 & 1 & 0 & 0 & 0 \\ 0 & 1 & 1 & 1 & 1 & 1 & 0 & 0 \\ 0 & 0 & 1 & 1 & 1 & 1 & 1 & 0 \\ 0 & 0 & 0 & 1 & 1 & 1 & 1 & 1 \end{pmatrix} \begin{pmatrix} b'_0 \\ b'_1 \\ b'_2 \\ b'_3 \\ b'_4 \\ b'_5 \\ b'_6 \\ b'_7 \end{pmatrix} + \begin{pmatrix} 1 \\ 1 \\ 0 \\ 0 \\ 0 \\ 1 \\ 1 \\ 0 \end{pmatrix} \mod 2.$$

Note that $B' = (b'_7, \ldots, b'_0)$ is the bitwise vector representation of $B'_i(x) = A_i^{-1}(x)$. This second step is referred to as *affine mapping*. Let's look at an example of how the S-Box computations work.

Example 4.10. We assume the S-Box input $A_i = (11000010)_2 = (C2)_{hex}$. From Table 4.2 we can see that the inverse is:

$$A_i^{-1} = B'_i = (2F)_{hex} = (00101111)_2.$$

We now apply the B'_i bit vector as input to the affine transformation. Note that the least significant bit (lsb) b'_0 of B'_i is at the rightmost position.

$$B_i = (00100101) = (25)_{hex}$$

Thus, $S((C2)_{hex}) = (25)_{hex}$, which is exactly the result that is also given in the S-Box Table 4.3.

◇

If one computes both steps for all 256 possible input elements of the S-Box and stores the results, one obtains Table 4.3. In most AES implementations, in particular in virtually all software realizations of AES, the S-Box outputs are *not explicitly computed* as shown here, but rather lookup tables like Table 4.3 are used. However, for hardware implementations it is sometimes advantageous to realize the S-Boxes as digital circuits which actually compute the inverse followed by the affine mapping.

The advantage of using inversion in $GF(2^8)$ as the core function of the Byte Substitution layer is that it provides a high degree of nonlinearity, which in turn provides optimum protection against some of the strongest known analytical attacks. The affine step "destroys" the algebraic structure of the Galois field, which in turn is needed to prevent attacks that would exploit the finite field inversion.

4.4.2 Diffusion Layer

In AES, the Diffusion layer consists of two sublayers, the *ShiftRows* transformation and the *MixColumn* transformation. We recall that diffusion is the spreading of the influence of individual bits over the entire state. Unlike the nonlinear S-Box, the diffusion layer performs a linear operation on state matrices A, B, i.e., $\text{DIFF}(A) + \text{DIFF}(B) = \text{DIFF}(A + B)$.

ShiftRows Sublayer

The ShiftRows transformation cyclically shifts the second row of the state matrix by three bytes to the right, the third row by two bytes to the right and the fourth row by one byte to the right. The first row is not changed by the ShiftRows transformation. The purpose of the ShiftRows transformation is to increase the diffusion properties of AES. If the input of the ShiftRows sublayer is given as a state matrix $B = (B_0, B_1, \ldots, B_{15})$:

B_0	B_4	B_8	B_{12}
B_1	B_5	B_9	B_{13}
B_2	B_6	B_{10}	B_{14}
B_3	B_7	B_{11}	B_{15}

the output is the new state:

$$
\begin{array}{|c|c|c|c|}
\hline
B_0 & B_4 & B_8 & B_{12} \\\hline
B_5 & B_9 & B_{13} & B_1 \\\hline
B_{10} & B_{14} & B_2 & B_6 \\\hline
B_{15} & B_3 & B_7 & B_{11} \\\hline
\end{array}
\begin{array}{l}
\text{no shift} \\
\longrightarrow \text{three positions right shift} \\
\longrightarrow \text{two positions right shift} \\
\longrightarrow \text{one position right shift}
\end{array}
\tag{4.1}
$$

MixColumn Sublayer

The *MixColumn* step is a linear transformation which mixes each column of the state matrix. Since every input byte influences four output bytes, the MixColumn operation is the major diffusion element in AES. The combination of the ShiftRows and MixColumn layer makes it possible that after only three rounds every byte of the state matrix depends on all 16 plaintext bytes.

In the following, we denote the 16-byte input state by B and the 16-byte output state by C:

$$\text{MixColumn}(B) = C,$$

where B is the state after the ShiftRows operation as given in Expression (4.1).

Now, each 4-byte column is considered as a vector and multiplied by a fixed 4×4 matrix. The matrix contains *constant* entries. Multiplication and addition of the coefficients is done in $GF(2^8)$. As an example, we show how the first four output bytes are computed:

$$\begin{pmatrix} C_0 \\ C_1 \\ C_2 \\ C_3 \end{pmatrix} = \begin{pmatrix} 02 & 03 & 01 & 01 \\ 01 & 02 & 03 & 01 \\ 01 & 01 & 02 & 03 \\ 03 & 01 & 01 & 02 \end{pmatrix} \begin{pmatrix} B_0 \\ B_5 \\ B_{10} \\ B_{15} \end{pmatrix}.$$

The second column of output bytes (C_4, C_5, C_6, C_7) is computed by multiplying the four input bytes (B_4, B_9, B_{14}, B_3) by the same constant matrix, and so on. Figure 4.3 shows which input bytes are used in each of the four MixColumn operations.

We discuss now the details of the vector–matrix multiplication which forms the MixColum operations. We recall that each state byte C_i and B_i is an 8-bit value representing an element from $GF(2^8)$. All arithmetic involving the coefficients is done in this Galois field. For the constants in the matrix a hexadecimal notation is used: "01" refers to the $GF(2^8)$ polynomial with the coefficients (00000001), i.e., it is the element 1 of the Galois field; "02" refers to the polynomial with the bit vector (00000010), i.e., to the polynomial x; and "03" refers to the polynomial with the bit vector (00000011), i.e., the Galois field element $x+1$.

The additions in the vector–matrix multiplication are $GF(2^8)$ additions, that is simple bitwise XORs of the respective bytes. For the multiplication of the constants, we have to realize multiplications with the constants 01, 02 and 03. These are quite efficient, and in fact, the three constants were chosen such that software implementation is easy. Multiplication by 01 is multiplication by the identity and does not involve any explicit operation. Multiplication by 02 and 03 can be done through table look-up in two 256-by-8 tables. As an alternative, multiplication by 02 can also be implemented as a multiplication by x, which is a left shift by one bit, and a modular reduction with $P(x) = x^8 + x^4 + x^3 + x + 1$. Similarly, multiplication by 03, which represents the polynomial $(x+1)$, can be implemented by a left shift by one bit and addition of the original value followed by a modular reduction with $P(x)$.

Example 4.11. We continue with our example from Sect. 4.4.1 and assume that the input state to the MixColumn layer is

$$B = (25, 25, \dots, 25).$$

In this special case, only two multiplications in $GF(2^8)$ have to be done. These are $02 \cdot 25$ and $03 \cdot 25$, which can be computed in polynomial notation:

$$02 \cdot 25 = x \cdot (x^5 + x^2 + 1)$$
$$= x^6 + x^3 + x,$$
$$03 \cdot 25 = (x+1) \cdot (x^5 + x^2 + 1)$$
$$= (x^6 + x^3 + x) + (x^5 + x^2 + 1)$$
$$= x^6 + x^5 + x^3 + x^2 + x + 1.$$

Since both intermediate values have a degree smaller than 8, no modular reduction with $P(x)$ is necessary.

The output bytes of C result from the following addition in $GF(2^8)$:

$$
\begin{array}{rcccc}
01 \cdot 25 = & & x^5+ & x^2+ & 1 \\
01 \cdot 25 = & & x^5+ & x^2+ & 1 \\
02 \cdot 25 = & x^6+ & x^3+ & x & \\
03 \cdot 25 = & x^6+ x^5+ & x^3+ x^2+ & x+ & 1 \\
\hline
C_i = & & x^5+ & x^2+ & 1,
\end{array}
$$

where $i = 0, \ldots, 15$. This leads to the output state $C = (25, 25, \ldots, 25)$.

◇

4.4.3 Key Addition Layer

The two inputs to the *Key Addition layer* are the current 16-byte state matrix and a subkey which also consists of 16 bytes (128 bits). The two inputs are combined through a bitwise XOR operation. Note that the XOR operation is equal to addition in the Galois field $GF(2)$. The subkeys are derived in the key schedule that is described below in Sect. 4.4.4.

4.4.4 Key Schedule

The *key schedule* takes the original input key (of length 128, 192 or 256 bit) and derives the subkeys used in AES. Note that an XOR addition of a subkey is used both at the input and output of AES. This process is sometimes referred to as key whitening. The number of subkeys is equal to the number of rounds plus one, due to the key needed for key whitening in the first key addition layer, cf. Fig. 4.2. Thus, for the key length of 128 bits, the number of rounds is $n_r = 10$, and there are 11 subkeys, each of 128 bits. The AES with a 192-bit key requires 13 subkeys of length 128 bits, and AES with a 256-bit key has 15 subkeys. The AES subkeys are computed recursively, i.e., in order to derive subkey k_i, subkey k_{i-1} must be known, etc.

The AES key schedule is word-oriented, where 1 word = 32 bits. Subkeys are stored in a key expansion array W that consists of words. There are different key schedules for the three different AES key sizes of 128, 192 and 256 bit, which are all fairly similar. We introduce the three key schedules in the following.

Key Schedule for 128-Bit Key AES

The 11 subkeys are stored in a key expansion array with the elements $W[0], \ldots, W[43]$. The subkeys are computed as depicted in Fig. 4.5. The elements $K_0, \ldots, K_{15}$ denote the bytes of the original AES key.

First, we note that the first subkey k_0 is the original AES key, i.e., the key is copied into the first four elements of the key array W. The other array elements are

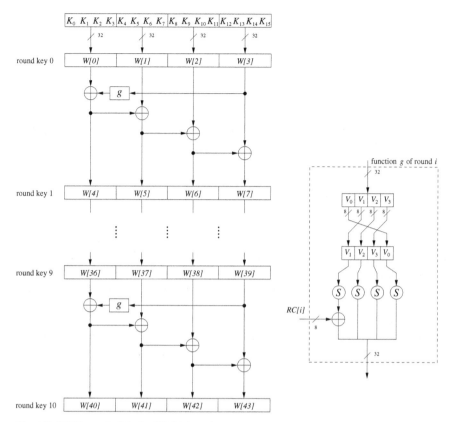

Fig. 4.5 AES key schedule for 128-bit key size

computed as follows. As can be seen in the figure, the leftmost word of a subkey $W[4i]$, where $i = 1, \ldots, 10$, is computed as:

$$W[4i] = W[4(i-1)] + g(W[4i-1]).$$

Here $g()$ is a nonlinear function with a four-byte input and output. The remaining three words of a subkey are computed recursively as:

$$W[4i+j] = W[4i+j-1] + W[4(i-1)+j],$$

where $i = 1, \ldots, 10$ and $j = 1, 2, 3$. The function $g()$ rotates its four input bytes, performs a byte-wise S-Box substitution, and adds a *round coefficient RC* to it. The round coefficient is an element of the Galois field $GF(2^8)$, i.e, an 8-bit value. It is only added to the leftmost byte in the function $g()$. The round coefficients vary from round to round according to the following rule:

$$RC[1] = x^0 = (0000\,0001)_2,$$
$$RC[2] = x^1 = (0000\,0010)_2,$$
$$RC[3] = x^2 = (0000\,0100)_2,$$
$$\vdots$$
$$RC[10] = x^9 = (0011\,0110)_2.$$

The function $g()$ has two purposes. First, it adds nonlinearity to the key schedule. Second, it removes symmetry in AES. Both properties are necessary to thwart certain block cipher attacks.

Key Schedule for 192-Bit Key AES

AES with 192-bit key has 12 rounds and, thus, 13 subkeys of 128 bit each. The subkeys require 52 words, which are stored in the array elements $W[0], \ldots, W[51]$. The computation of the array elements is quite similar to the 128-bit key case and is shown in Fig. 4.6. There are eight iterations of the key schedule. (Note that these key schedule iterations do *not* correspond to the 12 AES rounds.) Each iteration computes six new words of the subkey array W. The subkey for the first AES round is formed by the array elements $(W[0], W[1], W[2], W[3])$, the second subkey by the elements $(W[4], W[5], W[6], W[7])$, and so on. Eight round coefficients $RC[i]$ are needed within the function $g()$. They are computed as in the 128-bit case and range from $RC[1], \ldots, RC[8]$.

Key Schedule for 256-Bit Key AES

AES with 256-bit key needs 15 subkeys. The subkeys are stored in the 60 words $W[0], \ldots, W[59]$. The computation of the array elements is quite similar to the 128-bit key case and is shown in Fig. 4.7. The key schedule has seven iterations, where each iteration computes eight words for the subkeys. (Again, note that these key schedule iterations do not correspond to the 14 AES rounds.) The subkey for the first AES round is formed by the array elements $(W[0], W[1], W[2], W[3])$, the second subkey by the elements $(W[4], W[5], W[6], W[7])$, and so on. There are seven round coefficients $RC[1], \ldots, RC[7]$ within the function $g()$ needed, that are computed as in the 128-bit case. This key schedule also has a function $h()$ with 4-byte input and output. The function applies the S-Box to all four input bytes.

In general, when implementing any of the key schedules, two different approaches exist:

1. Precomputation All subkeys are expanded first into the array W. The encryption (decryption) of a plaintext (ciphertext) is executed afterwards. This approach is often taken in PC and server implementations of AES, where large pieces of data are encrypted under one key. Please note that this approach requires $(n_r + 1) \cdot 16$ bytes of memory, e.g., $11 \cdot 16 = 176$ bytes if the key size is 128 bits. This is the reason

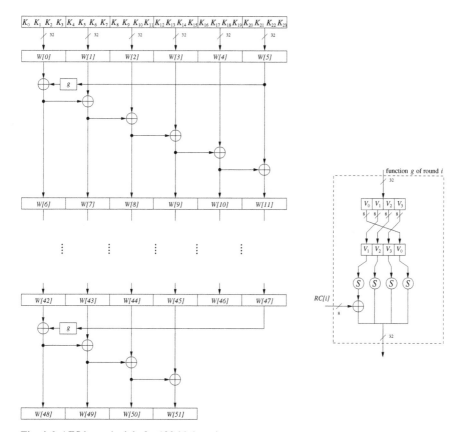

Fig. 4.6 AES key schedule for 192-bit key sizes

why such an implementation on a device with limited memory resources, such as a smart card, is sometimes not desireable.

2. On-the-fly A new subkey is derived for every new round during the encryption (decryption) of a plaintext (ciphertext). Please note that when decrypting ciphertexts, the last subkey is XORed first with the ciphertext. Therefore, it is required to recursively derive all subkeys first and then start with the decryption of a ciphertext and the on-the-fly generation of subkeys. As a result of this overhead, the decryption of a ciphertext is always slightly slower than the encryption of a plaintext when the on-the-fly generation of subkeys is used.

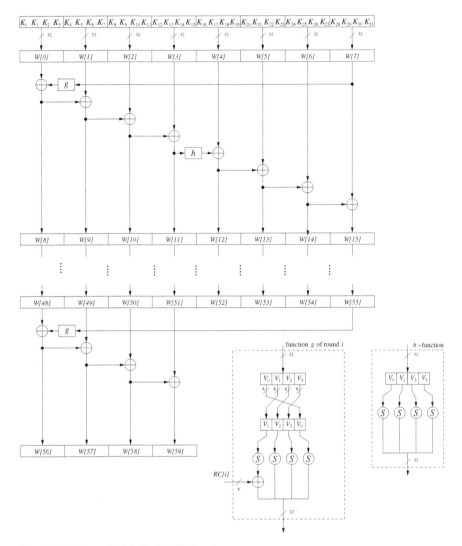

Fig. 4.7 AES key schedule for 256-bit key size

4.5 Decryption

Because AES is not based on a Feistel network, all layers must actually be inverted, i.e., the Byte Substitution layer becomes the Inv Byte Substitution layer, the ShiftRows layer becomes the Inv ShiftRows layer, and the MixColumn layer becomes Inv MixColumn layer. However, as we will see, it turns out that the inverse layer operations are fairly similar to the layer operations used for encryption. In ad-

dition, the order of the subkeys is reversed, i.e., we need a reversed key schedule. A
block diagram of the decryption function is shown in Fig. 4.8.

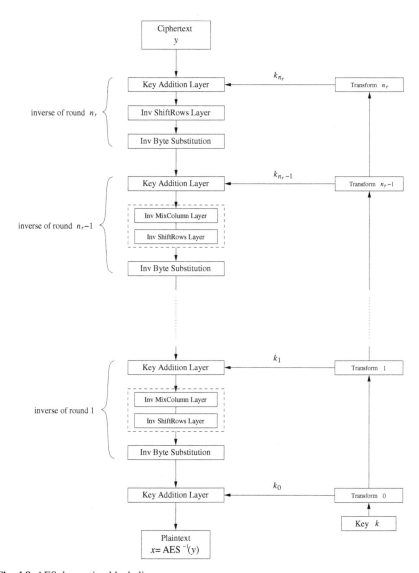

Fig. 4.8 AES decryption block diagram

Since the last encryption round does not perform the MixColum operation, the
first decryption round also does not contain the corresponding inverse layer. All
other decryption rounds, however, contain all AES layers. In the following, we discuss the inverse layers of the general AES decryption round (Fig. 4.9). Since the

XOR operation is its own inverse, the key addition layer in the decryption mode is
the same as in the encryption mode: it consists of a row of plain XOR gates.

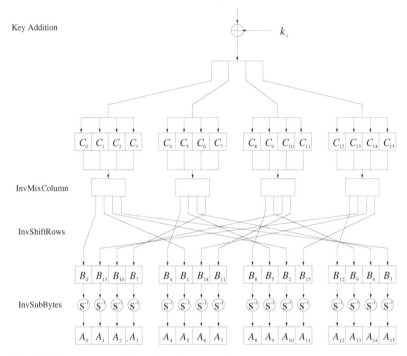

Fig. 4.9 AES decryption round function $1, 2, \ldots, n_r - 1$

Inverse MixColumn Sublayer

After the addition of the subkey, the inverse MixColumn step is applied to the state
(again, the exception is the first decryption round). In order to reverse the MixColumn operation, the inverse of its matrix must be used. The input is a 4-byte column
of the State C which is multiplied by the inverse 4×4 matrix. The matrix contains
constant entries. Multiplication and addition of the coefficients is done in $GF(2^8)$.

$$\begin{pmatrix} B_0 \\ B_1 \\ B_2 \\ B_3 \end{pmatrix} = \begin{pmatrix} 0E & 0B & 0D & 09 \\ 09 & 0E & 0B & 0D \\ 0D & 09 & 0E & 0B \\ 0B & 0D & 09 & 0E \end{pmatrix} \begin{pmatrix} C_0 \\ C_1 \\ C_2 \\ C_3 \end{pmatrix}$$

The second column of output bytes (B_4, B_5, B_6, B_7) is computed by multiplying the
four input bytes (C_4, C_5, C_6, C_7) by the same constant matrix, and so on. Each value

B_i and C_i is an element from $GF(2^8)$. Also, the constants are elements from $GF(2^8)$. The notation for the constants is hexadecimal and is the same as was used for the MixColumn layer, for example:

$$0B = (0B)_{hex} = (0000\,1011)_2 = x^3 + x + 1.$$

Additions in the vector–matrix multiplication are bitwise XORs.

Inverse ShiftRows Sublayer

In order to reverse the ShiftRows operation of the encryption algorithm, we must shift the rows of the state matrix in the opposite direction. The first row is not changed by the inverse ShiftRows transformation. If the input of the ShiftRows sublayer is given as a state matrix $B = (B_0, B_1, \ldots, B_{15})$:

B_0	B_4	B_8	B_{12}
B_1	B_5	B_9	B_{13}
B_2	B_6	B_{10}	B_{14}
B_3	B_7	B_{11}	B_{15}

the inverse ShiftRows sublayer yields the output:

B_0	B_4	B_8	B_{12}	no shift
B_{13}	B_1	B_5	B_9	⟵ three positions left shift
B_{10}	B_{14}	B_2	B_6	⟵ two positions left shift
B_7	B_{11}	B_{15}	B_3	⟵ one position left shift

Inverse Byte Substitution Layer

The inverse S-Box is used when decrypting a ciphertext. Since the AES S-Box is a bijective, i.e., a one-to-one mapping, it is possible to construct an inverse S-Box such that:

$$A_i = S^{-1}(B_i) = S^{-1}(S(A_i)),$$

where A_i and B_i are elements of the state matrix. The entries of the inverse S-Box are given in Table 4.4.

For readers who are interested in the details of how the entries of inverse S-Box are constructed, we provide a derivation. However, for a functional understanding of AES, the remainder of this section can be skipped. In order to reverse the S-Box substitution, we first have to compute the inverse of the affine transformation. For this, each input byte B_i is considered an element of $GF(2^8)$. The inverse affine transformation on each byte B_i is defined by:

Table 4.4 Inverse AES S-Box: Substitution values in hexadecimal notation for input byte (xy)

x \ y	0	1	2	3	4	5	6	7	8	9	A	B	C	D	E	F
0	52	09	6A	D5	30	36	A5	38	BF	40	A3	9E	81	F3	D7	FB
1	7C	E3	39	82	9B	2F	FF	87	34	8E	43	44	C4	DE	E9	CB
2	54	7B	94	32	A6	C2	23	3D	EE	4C	95	0B	42	FA	C3	4E
3	08	2E	A1	66	28	D9	24	B2	76	5B	A2	49	6D	8B	D1	25
4	72	F8	F6	64	86	68	98	16	D4	A4	5C	CC	5D	65	B6	92
5	6C	70	48	50	FD	ED	B9	DA	5E	15	46	57	A7	8D	9D	84
6	90	D8	AB	00	8C	BC	D3	0A	F7	E4	58	05	B8	B3	45	06
7	D0	2C	1E	8F	CA	3F	0F	02	C1	AF	BD	03	01	13	8A	6B
8	3A	91	11	41	4F	67	DC	EA	97	F2	CF	CE	F0	B4	E6	73
9	96	AC	74	22	E7	AD	35	85	E2	F9	37	E8	1C	75	DF	6E
A	47	F1	1A	71	1D	29	C5	89	6F	B7	62	0E	AA	18	BE	1B
B	FC	56	3E	4B	C6	D2	79	20	9A	DB	C0	FE	78	CD	5A	F4
C	1F	DD	A8	33	88	07	C7	31	B1	12	10	59	27	80	EC	5F
D	60	51	7F	A9	19	B5	4A	0D	2D	E5	7A	9F	93	C9	9C	EF
E	A0	E0	3B	4D	AE	2A	F5	B0	C8	EB	BB	3C	83	53	99	61
F	17	2B	04	7E	BA	77	D6	26	E1	69	14	63	55	21	0C	7D

$$\begin{pmatrix} b'_0 \\ b'_1 \\ b'_2 \\ b'_3 \\ b'_4 \\ b'_5 \\ b'_6 \\ b'_7 \end{pmatrix} \equiv \begin{pmatrix} 0&1&0&1&0&0&1&0 \\ 0&0&1&0&1&0&0&1 \\ 1&0&0&1&0&1&0&0 \\ 0&1&0&0&1&0&1&0 \\ 0&0&1&0&0&1&0&1 \\ 1&0&0&1&0&0&1&0 \\ 0&1&0&0&1&0&0&1 \\ 1&0&1&0&0&1&0&0 \end{pmatrix} \begin{pmatrix} b_0 \\ b_1 \\ b_2 \\ b_3 \\ b_4 \\ b_5 \\ b_6 \\ b_7 \end{pmatrix} + \begin{pmatrix} 0 \\ 0 \\ 0 \\ 0 \\ 0 \\ 1 \\ 0 \\ 1 \end{pmatrix} \bmod 2,$$

where $(b_7, \ldots, b_0)$ is the bitwise vector representation of $B_i(x)$, and $(b'_7, \ldots, b'_0)$ the result after the inverse affine transformation.

In the second step of the inverse S-Box operation, the Galois field inverse has to be reversed. For this, note that $A_i = (A_i^{-1})^{-1}$. This means that the inverse operation is reversed by computing the inverse again. In our notation we thus have to compute

$$A_i = (B'_i)^{-1} \in GF(2^8)$$

with the fixed reduction polynomial $P(x) = x^8 + x^4 + x^3 + x + 1$. Again, the zero element is mapped to itself. The vector $A_i = (a_7, \ldots, a_0)$ (representing the field element $a_7 x^7 + \cdots + a_1 x + a_0$) is the result of the substitution:

$$A_i = S^{-1}(B_i).$$

Decryption Key Schedule

Since decryption round one needs the last subkey, the second decryption round needs the second-to-last subkey and so on, we need the subkey in reversed order as shown in Fig. 4.8. In practice this is mainly achieved by computing the entire key schedule first and storing all 11, 13 or 15 subkeys, depending on the number or rounds AES is using (which in turn depends on the three key lengths supported by AES). This precomputation adds usually a small latency to the decryption operation relative to encryption.

4.6 Implementation in Software and Hardware

We briefly comment on the efficiency of the AES cipher with respect to software and hardware implementation.

Software

Unlike DES, AES was designed such that an efficient software implementation is possible. A straightforward implementation of AES which directly follows the data path description, such as the description given in this chapter, is well suited for 8-bit processors such as those found on smart cards, but is not particularly efficient on 32-bit or 64-bit machines, which are common in today's PCs. In a naïve implementation, all time-critical functions (Byte Substitution, ShiftRows, MixColumn) operate on individual bytes. Processing 1 byte per instruction is inefficient on modern 32-bit or 64-bit processors.

However, the Rijndael designers proposed a method which results in fast software implementations. The core idea is to merge all round functions (except the rather trivial key addition) into one table look-up. This results in four tables, each of which consists of 256 entries, where each entry is 32 bits wide. These tables are named a *T-Box*. Four table accesses yield 32 output bits of one round. Hence, one round can be computed with 16 table look-ups. On a 1.2-GHz Intel processor, a throughput of 400 Mbit/s (or 50 MByte/s) is possible. The fastest known implementation on a 64-bit Athlon CPU achieves a theoretical throughput of more than 1.6 Gbit/s. However, conventional hard disc encryption tools with AES or an open-source implementation of AES reach a perfomance of a few hundred Mbit/s on similar platforms.

Hardware

Compared to DES, AES requires more hardware resources for an implementation. However, due to the high integration density of modern integrated circuits, AES can be implemented with very high throughputs in modern ASIC or FPGA (field programmable gate array — these are programmable hardware devices) technology. Commercial AES ASICs can exceed throughputs of 10Gbit/sec. Through parallelization of AES encryption units on one chip, the speed can be further increased. It can be said that symmetric encryption with today's ciphers is extremely fast, not only compared to asymmetric cryptosystems but also compared to other algorithms needed in modern communication systems, such as data compression or signal processing schemes.

4.7 Discussion and Further Reading

AES Algorithm and Security A detailed description of the design principles of AES can be found in [52]. This book by the Rijndael inventors describes the design of the block cipher. Recent research in context to AES can be found online in the *AES Lounge* [68]. This website is a dissemination effort within ECRYPT, the Network of Excellence in Cryptology, and is a rich resource of activities around AES. It gives many links to further information and papers regarding implementation and theoretical aspects of AES.

There is currently no analytical attack against AES known which has a complexity less than a brute-force attack. An elegant algebraic description was found [122], which in turn triggered speculations that this could lead to attacks. Subsequent research showed that an attack is, in fact, not feasible. By now, the common assumption is that the approach will not threaten AES. A good summary on algebraic attacks can be found in [43]. In addition, there have been proposals for many other attacks, including square attack, impossible differential attack or related key attack. Again, a good source for further references is the *AES Lounge*.

The standard reference for the mathematics of finite fields is [110]. A very accessible but brief introduction is also given in [19]. The International Workshop on the Arithmetic of Finite Fields (WAIFI), a relatively new workshop series, is concerned with both the applications and the theory of Galois fields [171].

Implementation As mentioned in Sect. 4.6, in most software implementations on modern CPUs special lookup tables are being used (T-Boxes). An early detailed description of the construction of T-Boxes can be found in [51, Sect. 5]. A description of a high-speed software implementation on modern 32-bit and 64-bit CPUs is given in [116, 115]. The bit slicing technique which was developed in the context of DES is also applicable to AES and can lead to very fast code as shown in [117].

A strong indication for the importance of AES was the recent introduction of special AES instructions by Intel in CPUs starting in 2008. The instructions allow these machines to compute the round operation particularly quickly.

There is wealth of literature dealing with hardware implementation of AES. A good introduction to the area of AES hardware architectures is given in [104, Chap. 10]. As an example of the variety of AES implementations, reference [86] describes a very small FPGA implementation with 2.2Mbit/s and a very fast pipelined FPGA implementation with 25Gbit/s. It is also possible to use the DSP blocks (i.e., fast arithmetic units) available on modern FPGAs for AES, which can also yield throughputs beyond 50Mbit/s [63]. The basic idea in all high-speed architectures is to process several plaintext blocks in parallel by means of pipelining. On the other end of the performance spectrum are lightweight architectures which are optimized for applications such as RFID. The basic idea here is to serialize the data path, i.e., one round is processed in several time steps. Good references are [75, 42].

4.8 Lessons Learned

- AES is a modern block cipher which supports three key lengths of 128, 192 and 256 bit. It provides excellent long-term security against brute-force attacks.
- AES has been studied intensively since the late 1990s and no attacks have been found that are better than brute-force.
- AES is not based on Feistel networks. Its basic operations use Galois field arithmetic and provide strong diffusion and confusion.
- AES is part of numerous open standards such as IPsec or TLS, in addition to being the mandatory encryption algorithm for US government applications. It seems likely that the cipher will be the dominant encryption algorithm for many years to come.
- AES is efficient in software and hardware.

Problems

4.1. Since May 26, 2002, the AES (Advanced Encryption Standard) describes the official standard of the US government.

1. The evolutionary history of AES differs from that of DES. Briefly describe the differences of the AES history in comparison to DES.
2. Outline the fundamental events of the developing process.
3. What is the name of the algorithm that is known as AES?
4. Who developed this algorithm?
5. Which block sizes and key lengths are supported by this algorithm?

4.2. For the AES algorithm, some computations are done by Galois Fields (GF). With the following problems, we practice some basic computations.

Compute the multiplication and addition table for the prime field $GF(7)$. A multiplication table is a square (here: 7×7) table which has as its rows and columns all field elements. Its entries are the products of the field element at the corresponding row and column. Note that the table is symmetric along the diagonal. The addition table is completely analogous but contains the sums of field elements as entries.

4.3. Generate the multiplication table for the extension field $GF(2^3)$ for the case that the irreducible polynomial is $P(x) = x^3 + x + 1$. The multiplication table is in this case a 8×8 table. (Remark: You can do this manually or write a program for it.)

4.4. Addition in $GF(2^4)$: Compute $A(x) + B(x) \mod P(x)$ in $GF(2^4)$ using the irreducible polynomial $P(x) = x^4 + x + 1$. What is the influence of the choice of the reduction polynomial on the computation?

1. $A(x) = x^2 + 1$, $B(x) = x^3 + x^2 + 1$
2. $A(x) = x^2 + 1$, $B(x) = x + 1$

4.5. Multiplication in $GF(2^4)$: Compute $A(x) \cdot B(x) \mod P(x)$ in $GF(2^4)$ using the irreducible polynomial $P(x) = x^4 + x + 1$. What is the influence of the choice of the reduction polynomial on the computation?

1. $A(x) = x^2 + 1$, $B(x) = x^3 + x^2 + 1$
2. $A(x) = x^2 + 1$, $B(x) = x + 1$

4.6. Compute in $GF(2^8)$:

$$(x^4 + x + 1)/(x^7 + x^6 + x^3 + x^2),$$

where the irreducible polynomial is the one used by AES, $P(x) = x^8 + x^4 + x^3 + x + 1$. Note that Table 4.2 contains a list of all multiplicative inverses for this field.

4.7. We consider the field $GF(2^4)$, with $P(x) = x^4 + x + 1$ being the irreducible polynomial. Find the inverses of $A(x) = x$ and $B(x) = x^2 + x$. You can find the inverses either by trial and error, i.e., brute-force search, or by applying the Euclidean algorithm for polynomials. (However, the Euclidean algorithm is only sketched in this chapter.) Verify your answer by multiplying the inverses you determined by A and B, respectively.

4.8. Find all irreducible polynomials

1. of degree 3 over $GF(2)$,
2. of degree 4 over $GF(2)$.

The best approach for doing this is to consider all polynomials of lower degree and check whether they are factors. Please note that we only consider monic irreducible polynomials, i.e., polynomials with the highest coefficient equal to one.

4.9. We consider AES with 128-bit block length and 128-bit key length. What is the output of the first round of AES if the plaintext consists of 128 ones, and the first subkey (i.e., the first subkey) also consists of 128 ones? You can write your final results in a rectangular array format if you wish.

4.10. In the following, we check the *diffusion properties* of AES after a single round. Let $W = (w_0, w_1, w_2, w_3) = (0x01000000, 0x00000000, 0x00000000, 0x00000000)$ be the input in 32-bit chunks to a 128-bit AES. The subkeys for the computation of the result of the first round of AES are $W_0, \ldots, W_7$ with 32 bits each are given by

$$W_0 = (0x2B7E1516),$$
$$W_1 = (0x28AED2A6),$$
$$W_2 = (0xABF71588),$$
$$W_3 = (0x09CF4F3C),$$
$$W_4 = (0xA0FAFE17),$$
$$W_5 = (0x88542CB1),$$
$$W_6 = (0x23A33939),$$
$$W_7 = (0x2A6C7605).$$

Use this book to figure out how the input is processed in the first round (e.g., S-Boxes). For the solution, you might also want to write a short computer program or use an existing one. In any case, indicate all intermediate steps for the computation of *ShiftRows*, *SubBytes* and *MixColumns*!

1. Compute the output of the first round of AES to the input W and the subkeys $W_0, \ldots, W_7$.
2. Compute the output of the first round of AES for the case that *all* input bits are zero.
3. How many output bits have changed? Remark that we only consider a single round — after every further round, more output bits will be affected (*avalanche effect*).

4.11. The MixColumn transformation of AES consists of a matrix–vector multiplication in the field $GF(2^8)$ with $P(x) = x^8 + x^4 + x^3 + x + 1$. Let $b = (b_7 x^7 + \ldots + b_0)$ be one of the (four) input bytes to the vector–matrix multiplication. Each input byte is multiplied with the constants 01, 02 and 03. Your task is to provide exact equations for computing those three constant multiplications. We denote the result by $d = (d_7 x^7 + \ldots + d_0)$.

1. Equations for computing the 8 bits of $d = 01 \cdot b$.
2. Equations for computing the 8 bits of $d = 02 \cdot b$.
3. Equations for computing the 8 bits of $d = 03 \cdot b$.

Note: The AES specification uses "01" to represent the polynomial 1, "02" to represent the polynomial x, and "03" to represent $x + 1$.

4.12. We now look at the gate (or bit) complexity of the MixColumn function, using the results from problem 4.11. We recall from the discussion of stream ciphers that a 2-input XOR gate performs a $GF(2)$ addition.

1. How many 2-input XOR gates are required to perform one constant multiplication by 01, 02 and 03, respectively, in $GF(2^8)$.
2. What is the overall gate complexity of a hardware implementation of one matrix–vector multiplication?
3. What is the overall gate complexity of a hardware implementation of the entire Diffusion layer? We assume permutations require no gates.

4.13. We consider the first part of the ByteSub operation, i.e, the Galois field inversion.

1. Using Table 4.2, what is the inverse of the bytes 29, F3 and 01, where each byte is given in hexadecimal notation?
2. Verify your answer by performing a $GF(2^8)$ multiplication with your answer and the input byte. Note that you have to represent each byte first as polynomials in $GF(2^8)$. The MSB of each byte represents the x^7 coefficient.

4.14. Your task is to compute the S-Box, i.e., the ByteSub, values for the input bytes 29, F3 and 01, where each byte is given in hexadecimal notation.

1. First, look up the inverses using Table 4.2 to obtain values B'. Now, perform the affine mapping by computing the matrix–vector multiplication and addition.
2. Verify your result using the S-Box Table 4.3.
3. What is the value of $S(0)$?

4.15. *Derive* the bit representation for the following round constants within the key schedule:

- $RC[8]$
- $RC[9]$
- $RC[10]$

4.16. For the following, we assume AES with 192-bit key length. Furthermore, let us assume an ASIC which can check $3 \cdot 10^7$ keys per second.

1. If we use 100,000 such ICs in parallel, how long does an average key search take? Compare this period of time with the age of the universe (approx. 10^{10} years).
2. Assume Moore's Law will still be valid for the next few years, how many years do we have to wait until we can build a key search machine to perform an average key search of AES-192 in 24 hours? Again, assume that we use 100,000 ICs in parallel.

Chapter 5
More About Block Ciphers

A block cipher is much more than just an encryption algorithm. It can be used as a versatile building block with which a diverse set of cryptographic mechanisms can be realized. For instance, we can use them for building different types of block-based encryption schemes, and we can even use block ciphers for realizing stream ciphers. The different ways of encryption are called *modes of operation* and are discussed in this chapter. Block ciphers can also be used for constructing hash functions, message authentication codes which are also knowns as MACs, or key establishment protocols, all of which will be described in later chapters. There are also other uses for block ciphers, e.g., as pseudo-random generators. In addition to modes of operation, this chapter also discusses two very useful techniques for increasing the security of block ciphers, namely key whitening and multiple encryption.

In this chapter you will learn

- the most important modes of operation for block ciphers in practice
- security pitfalls when using modes of operations
- the principles of key whitening
- why double encryption is not a good idea, and the meet-in-the-middle attack
- triple encryption

5.1 Encryption with Block Ciphers: Modes of Operation

In the previous chapters we introduced how DES, 3DES and AES encrypt a block
of data. Of course, in practice one wants typically to encrypt more than one single
8-byte or 16-byte block of plaintext, e.g., when encrypting an e-mail or a computer
file. There are several ways of encrypting long plaintexts with a block cipher. We
introduce several popular modes of operation in this chapter, including

- Electronic Code Book mode (ECB),
- Cipher Block Chaining mode (CBC),
- Cipher Feedback mode (CFB),
- Output Feedback mode (OFB),
- Counter mode (CTR).

The latter three modes use the block cipher as a building block for a stream cipher.

All of the five modes have one goal: They encrypt data and thus provide confidentiality for a message sent from Alice to Bob. In practice, we often not only want
to keep data confidential, but Bob also wants to know whether the message is really coming from Alice. This is called authentication and the Galois Counter mode
(GCM), which we will also introduce, is a mode of operation that lets the receiver
(Bob) determine whether the message was really sent by the person he shares a key
with (Alice). Moreover, authentication also allows Bob to detect whether the ciphertext was altered during transmission. More on authentication is found in Chap. 10.

The ECB and CFB modes require that the length of the plaintext be an exact
multiple of the block size of the cipher used, e.g., a multiple of 16 bytes in the
case of AES. If the plaintext does not have this length, it must be padded. There
are several ways of doing this padding in practice. One possible padding method
is to append a single "1" bit to the plaintext and then to append as many "0" bits
as necessary to reach a multiple of the block length. Should the plaintext be an
exact multiple of the block length, an extra block consisting only of padding bits is
appended.

5.1.1 Electronic Codebook Mode (ECB)

The *Electronic Code Book (ECB)* mode is the most straightforward way of encrypting a message. In the following, let $e_k(x_i)$ denote the encryption of plaintext block
x_i with key k using some arbitrary block cipher. Let $e_k^{-1}(y_i)$ denote the decryption
of ciphertext block y_i with key k. Let us assume that the block cipher encrypts (decrypts) blocks of size b bits. Messages which exceed b bits are partitioned into b-bit
blocks. If the length of the message is not a multiple of b bits, it must be padded to
a multiple of b bits prior to encryption. As shown in Fig. 5.1, in ECB mode each
block is encrypted separately. The block cipher can, for instance, be AES or 3DES.

Encryption and decryption in the ECB mode is formally described as follows:

Fig. 5.1 Encryption and decryption in ECB mode

Definition 5.1.1 Electronic Codebook Mode (ECB)
Let $e()$ be a block cipher of block size b, and let x_i and y_i be bit strings of length b.
Encryption: $y_i = e_k(x_i), \quad i \geq 1$
Decryption: $x_i = e_k^{-1}(y_i) = e_k^{-1}(e_k(x_i)), \quad i \geq 1$

It is straightforward to verify the correctness of the ECB mode:

$$e_k^{-1}(y_i) = e_k^{-1}(e_k(x_i)) = x_i.$$

The ECB mode has advantages. Block synchronization between the encryption and decryption parties Alice and Bob is not necessary, i.e., if the receiver does not receive all encrypted blocks due to transmission problems, it is still possible to decrypt the received blocks. Similarly, bit errors, e.g., caused by noisy transmission lines, only affect the corresponding block but not succeeding blocks. Also, block ciphers operating in ECB mode can be parallelized, e.g., one encryption unit encrypts (or decrypts) block 1, the next one block 2, and so on. This is an advantage for high-speed implementations, but many other modes such as the CFB do not allow parallelization.

However, as is often the case in cryptography, there are some unexpected weaknesses associated with the ECB mode which we will discuss in the following. The main problem of the ECB mode is that it encrypts highly deterministically. This means that identical plaintext blocks result in identical ciphertext blocks, as long as the key does not change. The ECB mode can be viewed as a gigantic code book — hence the mode's name — which maps every input to a certain output. Of course, if the key is changed the entire code book changes, but as long as the key is static the book is fixed. This has several undesirable consequences. First, an attacker recognizes if the same message has been sent twice simply by looking at the ciphertext. Deducing information from the ciphertext in this way is called *traffic analysis*. For instance, if there is a fixed header that always precedes a message, the header always results in the same ciphertext. From this, an attacker can, for instance, learn when a new message has been sent. Second, plaintext blocks are encrypted independently of previous blocks. If an attacker reorders the ciphertext blocks, this might result in valid plaintext and the reordering might not be detected. We demonstrate two simple attacks which exploit these weaknesses of the ECB mode.

The ECB mode is susceptible to *substitution attacks*, because once a particular plaintext to ciphertext block mapping $x_i \rightarrow y_i$ is known, a sequence of ciphertext blocks can easily be manipulated. We demonstrate how a substitution attack could work in the real world. Imagine the following example of an electronic wire transfer betweens banks.

Example 5.1. Substitution attack against electronic bank transfer
Let's assume a protocol for wire transfers between banks (Fig. 5.2). There are five fields which specify a transfer: the sending bank's ID and account number, the receiving bank's ID and account number, and the amount. We assume now (and this is a major simplification) that each of the fields has exactly the size of the block cipher width, e.g., 16 bytes in the case of AES. Furthermore, the encryption key between the two banks does not change too frequently. Due to the nature of the ECB, an attacker can exploit the deterministic nature of this mode of operation by simple substitution of the blocks. The attack details are as follows:

Block #	1	2	3	4	5
	Sending Bank A	Sending Account #	Receiving Bank B	Receiving Account #	Amount $

Fig. 5.2 Example for a substitution attack against ECB encryption

1. The attacker, Oscar, opens one account at bank A and one at bank B.
2. Oscar taps the encrypted line of the banking communication network.
3. He sends \$1.00 transfers from his account at bank A to his account at bank B repeatedly. He observes the ciphertexts going through the communication network. Even though he cannot decipher the random-looking ciphertext blocks, he can check for ciphertext blocks that repeat. After a while he can recognize the five blocks of his own transfer. He now stores blocks 1, 3 and 4 of these transfers. These are the encrypted versions of the ID numbers of both banks as well as the encrypted version of his account at bank B.
4. Recall that the two banks do not change the key too frequently. This means that the same key is used for several other transfers between bank A and B. By comparing blocks 1 and 3 of *all* subsequent messages with the ones he has stored, Oscar recognizes all transfers that are made from some account at bank A to some account at bank B. He now simply replaces block 4 — which contains the receiving account number — with the block 4 that he stored before. This block contains Oscar's account number in encrypted form. As a consequence, *all transfers* from some account of bank A to some account of bank B are redirected to go into Oscar's B account! Note that bank B now has means of detecting that the block 4 has been replaced in some of the transfers it receives.
5. Withdraw money from bank B quickly and fly to a country that has a relaxed attitude about the extradition of white-collar criminals.
   ◇

What's interesting about this attack is that it works completely without attacking the block cipher itself. So even if we would use AES with a 256-bit key and if we would encrypt each block, say, 1000 times, this would not prevent the attack. It should be stressed, however, that this is not an attack that breaks the block cipher itself. Messages that are unknown to Oscar still remain confidential. He simply replaced parts of the ciphertext with some other (previous) ciphertexts. This is called a violation of the *integrity* of the message. There are available techniques for preserving the integrity of a message, namely message authentication codes (MACs) and digital signatures. Both are widely used in practice to prevent such an attack, and are introduced in Chaps. 10 and 12. Also, the Galois Counter mode, which is described below, is an encryption mode with a built-in integrity check. Note that this attack only works if the key between bank A and B is not changed too frequently. This is another reason why frequent key freshness is a good idea.

We now look at another problem posed by the ECB mode.

Example 5.2. Encryption of bitmaps in ECB mode
Figure 5.3 clearly shows a major disadvantage of the ECB mode: Identical plaintexts are mapped to identical ciphertexts. In case of a simple bitmap, the information (text in the picture) can still be read out from the encrypted picture even though we used AES with a 256-bit key for encryption. This is because the background consists of only a few different plaintext blocks which yields a fairly uniformly looking background in the ciphertext. On the other hand, all plaintext blocks which contain part of the letters result in random-looking ciphertexts. These random-looking ciphertexts are clearly distinguishable from the uniform background by the human eye.

CRYPTOGRAPHY
AND
DATA SECURITY

Fig. 5.3 Image and encrypted image using AES with 256-bit key in ECB mode

◇

This weakness is similar to the attack of the substitution cipher that was introduced in the first example. In both cases, statistical properties in the plaintext are
preserved in the ciphertext. Note that unlike an attack against the substitution cipher
or the above banking transfer attack, an attacker does not have to do anything in the
case here. The human eye automatically makes use of the statistical information.

Both attacks above were examples of the weakness of a deterministic encryption
scheme. Thus, it is usually preferable that different ciphertexts are produced every
time we encrypt the same plaintext. This behavior is called *probabilistic encryption*. This can be achieved by introducing some form of randomness, typically in
form of an initialization vector (IV). The following modes of operation all encrypt
probabilistically by means of an IV.

5.1.2 Cipher Block Chaining Mode (CBC)

There are two main ideas behind the *Cipher Block Chaining (CBC)* mode. First, the
encryption of all blocks are "chained together" such that ciphertext y_i depends not
only on block x_i but on all previous plaintext blocks as well. Second, the encryption
is randomized by using an initialization vector (IV). Here are the details of the CBC
mode.

The ciphertext y_i, which is the result of the encryption of plaintext block x_i, is
fed back to the cipher input and XORed with the succeeding plaintext block x_{i+1}.
This XOR sum is then encrypted, yielding the next ciphertext y_{i+1}, which can then
be used for encrypting x_{i+2}, and so on. This process is shown on the left-hand side
of Fig. 5.4. For the first plaintext block x_1 there is no previous ciphertext. For this an
IV is added to the first plaintext, which also allows us to make each CBC encryption
nondeterministic. Note that the first ciphertext y_1 depends on plaintext x_1 (and the
IV). The second ciphertext depends on the IV, x_1 and x_2. The third ciphertext y_3
depends on the IV and x_1, x_2, x_3, and so on. The last ciphertext is a function of all
plaintext blocks and the IV.

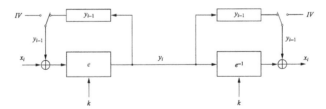

Fig. 5.4 Encryption and decryption in CBC mode

When decrypting a ciphertext block y_i in CBC mode, we have to reverse the two
operations we have done on the encryption side. First, we have to reverse the block
cipher encryption by applying the decryption function $e^{-1}()$. After this we have to

undo the XOR operation by again XORing the correct ciphertext block. This can be expressed for general blocks y_i as $e_k^{-1}(y_i) = x_i \oplus y_{i-1}$. The right-hand side of Fig. 5.4 shows this process. Again, if the first ciphertext block y_1 is decrypted, the result must be XORed with the initialization vector IV to determine the plaintext block x_1, i.e., $x_1 = IV \oplus e_k^{-1}(y_1)$. The entire process of encryption and decryption can be described as:

Definition 5.1.2 Cipher block chaining mode (CBC)
Let $e()$ be a block cipher of block size b; let x_i and y_i be bit strings of length b; and IV be a nonce of length b.
Encryption (first block): $y_1 = e_k(x_1 \oplus IV)$
Encryption (general block): $y_i = e_k(x_i \oplus y_{i-1}), \quad i \geq 2$
Decryption (first block): $x_1 = e_k^{-1}(y_1) \oplus IV$
Decryption (general block): $x_i = e_k^{-1}(y_i) \oplus y_{i-1}, \quad i \geq 2$

We now verify the mode, i.e., we show that the decryption actually reverses the encryption. For the decryption of the first block y_1, we obtain:

$$d(y_1) = e_k^{-1}(y_1) \oplus IV = e_k^{-1}(e_k(x_1 \oplus IV)) \oplus IV = (x_1 \oplus IV) \oplus IV = x_1$$

For the decryption of all subsequent blocks y_i, $i \geq 2$, we obtain:

$$d(y_i) = e_k^{-1}(y_i) \oplus y_{i-1} = e_k^{-1}(e_k(x_i \oplus y_{i-1})) \oplus y_{i-1} = (x_i \oplus y_{i-1}) \oplus y_{i-1} = x_i$$

If we choose a new IV every time we encrypt, the CBC mode becomes a probabilistic encryption scheme. If we encrypt a string of blocks $x_1, \ldots, x_t$ once with a first IV and a second time with a different IV, the two resulting ciphertext sequences look completely unrelated to each other for an attacker. Note that we do *not* have to keep the IV secret. However, in most cases, we want the IV to be a nonce, i.e., a number used only once. There are many different ways of generating and agreeing on initialization values. In the simplest case, a randomly chosen number is transmitted in the clear between the two communication parties prior to the encrypted session. Alternatively it is a counter value that is known to Alice and Bob, and it is incremented every time a new session starts (which requires that the counter value must be stored between sessions). It could be derived from values such as Alice's and Bob's ID number, e.g., their IP addresses, together with the current time. Also, in order to strengthen any of these methods, we can take a value as described above and ECB-encrypt it once using the block cipher with the key known to Alice and Bob, and use the resulting ciphertext as the IV. There are some advanced attacks which also require that the IV is nonpredictable.

It is instructive to discuss whether the substitution attack against the bank transfer that worked for the ECB mode is applicable to the CBC mode. If the IV is properly chosen for every wire transfer, the attack will not work at all since Oscar will not recognize any patterns in the ciphertext. If the IV is kept the same for several transfers, he would recognize the transfers from his account at bank A to his account at bank B. However, if he substitutes ciphertext block 4, which is his encrypted account number, in other wire transfers going from bank A to B, bank B would decrypt block 4 and 5 to some random value. Even though money would not be redirected into Oscar's account, it might be redirected to some other random account. The amount would be a random value too. This is obviously also highly undesirable for banks. This example shows that even though Oscar cannot perform specific manipulations, ciphertext alterations by him can cause random changes to the plaintext, which can have major negative consequences. Hence in many, if not in most, real-world systems, encryption itself is not sufficient: we also have to protect the integrity of the message. This can be achieved by message authentication codes (MACs) or digital signatures, which are introduced in Chap. 12. The Galois Counter mode described below provides encryption and integrity check simultaneously.

5.1.3 Output Feedback Mode (OFB)

In the *Output Feedback (OFB)* mode a block cipher is used to build a stream cipher encryption scheme. This scheme is shown in Fig. 5.5. Note that in OFB mode the key stream is not generated bitwise but instead in a blockwise fashion. The output of the cipher gives us b key stream bits, where b is the width of the block cipher used, with which we can encrypt b plaintext bits using the XOR operation.

The idea behind the OFB mode is quite simple. We start with encrypting an IV with a block cipher. The cipher output gives us the first set of b key stream bits. The next block of key stream bits is computed by feeding the previous cipher output back into the block cipher and encrypting it. This process is repeated as shown in Fig. 5.5.

The OFB mode forms a synchronous stream cipher (cf. Fig. 2.3) as the key stream does not depend on the plain or ciphertext. In fact, using the OFB mode is quite similar to using a standard stream cipher such as RC4 or Trivium. Since the OFB mode forms a stream cipher, encryption and decryption are exactly the same operation. As can be seen in the right-hand part of Fig. 5.5, the receiver does not use the block cipher in decryption mode $e^{-1}()$ to decrypt the ciphertext. This is because the actual encryption is performed by the XOR function, and in order to reverse it, i.e., to decrypt it, we simply have to perform another XOR function on the receiver side. This is in contrast to ECB and CBC mode, where the data is actually being encrypted and decrypted by the block cipher.

Encryption and decryption using the OFB scheme is as follows:

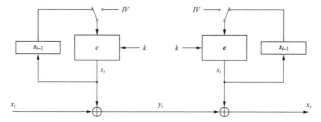

Fig. 5.5 Encryption and decryption in OFB mode

> **Definition 5.1.3** Output feedback mode (OFB)
> *Let $e()$ be a block cipher of block size b; let x_i, y_i and s_i be bit strings of length b; and IV be a nonce of length b.*
> **Encryption (first block):** $s_1 = e_k(IV)$ *and* $y_1 = s_1 \oplus x_1$
> **Encryption (general block):** $s_i = e_k(s_{i-1})$ *and* $y_i = s_i \oplus x_i, \quad i \geq 2$
> **Decryption (first block):** $s_1 = e_k(IV)$ *and* $x_1 = s_1 \oplus y_1$
> **Decryption (general block):** $s_i = e_k(s_{i-1})$ *and* $x_i = s_i \oplus y_i, \quad i \geq 2$

As a result of the use of an IV, the OFB encryption is also nondeterministic, hence, encrypting the same plaintext twice results in different ciphertexts. As in the case for the CBC mode, the IV should be a nonce. One advantage of the OFB mode is that the block cipher computations are independent of the plaintext. Hence, one can precompute one or several blocks s_i of key stream material.

5.1.4 Cipher Feedback Mode (CFB)

The *Cipher Feedback (CFB)* mode also uses a block cipher as a building block for a stream cipher. It is similar to the OFB mode but instead of feeding back the output of the block cipher, the ciphertext is fed back. (Hence, a somewhat more accurate term for this mode would have been "Ciphertext Feedback mode".) As in the OFB mode, the key stream is not generated bitwise but instead in a blockwise fashion. The idea behind the CFB mode is as follows: To generate the first key stream block s_1, we encrypt an IV. For all subsequent key stream blocks $s_2, s_3, \ldots$, we encrypt the previous ciphertext. This scheme is shown in Fig. 5.6.

Since the CFB mode forms a stream cipher, encryption and decryption are exactly the same operation. The CFB mode is an example of an asynchronous stream cipher (cf. Fig. 2.3) since the stream cipher output is also a function of the ciphertext.

The formal description of the CFB mode follows:

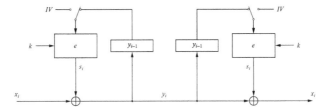

Fig. 5.6 Encryption and decryption in CFB mode

Definition 5.1.4 Cipher feedback mode (CFB)
Let $e()$ be a block cipher of block size b; let x_i and y_i be bit strings of length b; and IV be a nonce of length b.
Encryption (first block): $y_1 = e_k(IV) \oplus x_1$
Encryption (general block): $y_i = e_k(y_{i-1}) \oplus x_i, \quad i \geq 2$
Decryption (first block): $x_1 = e_k(IV) \oplus y_1$
Decryption (general block): $x_i = e_k(y_{i-1}) \oplus y_i, \quad i \geq 2$

As a result of the use of an IV, the CFB encryption is also nondeterministic, hence, encrypting the same plaintext twice results in different ciphertexts. As in the case for the CBC and OFB modes, the IV should be a nonce.

A variant of the CFB mode can be used in situations where short plaintext blocks are to be encrypted. Let's use the encryption of the link between a (remote) keyboard and a computer as an example. The plaintexts generated by the keyboard are typically only 1 byte long, e.g., an ASCII character. In this case, only 8 bits of the key stream are used for encryption (it does not matter which ones we choose as they are all secure), and the ciphertext also only consists of 1 byte. The feedback of the ciphertext as input to the block cipher is a bit tricky. The previous block cipher input is shifted by 8 bit positions to the left, and the 8 least significant positions of the input register are filled with the ciphertext byte. This process repeats. Of course, this approach works not only for plaintext blocks of length 8, but for any lengths shorter than the cipher output.

5.1.5 Counter Mode (CTR)

Another mode which uses a block cipher as a stream cipher is the Counter (CTR) mode. As in the OFB and CFB modes, the key stream is computed in a blockwise fashion. The input to the block cipher is a counter which assumes a different value every time the block cipher computes a new key stream block. Figure 5.7 shows the principle.

We have to be careful how to initialize the input to the block cipher. We must prevent using the same input value twice. Otherwise, if an attacker knows one of

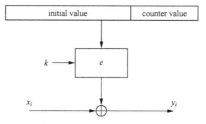

Fig. 5.7 Encryption and decryption in counter mode

the two plaintexts that were encrypted with the same input, he can compute the key stream block and thus immediately decrypt the other ciphertext. In order to achieve this uniqueness, often the following approach is taken in practice. Let's assume a block cipher with an input width of 128 bits, such as an AES. First we choose an IV that is a nonce with a length smaller than the block length, e.g., 96 bits. The remaining 32 bits are then used by a counter with the value CTR which is initialized to zero. For every block that is encrypted during the session, the counter is incremented but the IV stays the same. In this example, the number of blocks we can encrypt without choosing a new IV is 2^{32}. Since every block consists of 8 bytes, a maximum of $8 \times 2^{32} = 2^{35}$ bytes, or about 32 Gigabytes, can be encrypted before a new IV must be generated. Here is a formal description of the Counter mode with a cipher input construction as just introduced:

Definition 5.1.5 Counter mode (CTR)
Let $e()$ be a block cipher of block size b, and let x_i and y_i be bit strings of length b. The concatenation of the initialization value IV and the counter CTR_i is denoted by $(IV\|CTR_i)$ and is a bit string of length b.
Encryption: $y_i = e_k(IV\|CTR_i) \oplus x_i, \quad i \geq 1$
Decryption: $x_i = e_k(IV\|CTR_i) \oplus y_i, \quad i \geq 1$

Please note that the string $(IV\|CTR_1)$ does not have to be kept secret. It can, for instance, be generated by Alice and sent to Bob together with the first ciphertext block. The counter CTR can either be a regular integer counter or a slightly more complex function such as a maximum-length LFSR.

One might wonder why so many modes are needed. One attractive feature of the Counter mode is that it can be parallelized because, unlike the OFB or CFB mode, it does not require any feedback. For instance, we can have two block cipher engines running in parallel, where the first block cipher encrypts the counter value CTR_1 and the other CTR_2 at the same time. When the two block cipher engines are finished, the first engine encrypts the value CTR_3 and the other one CTR_4, and so on. This scheme would allow us to encrypt at twice the data rate of a single implementation. Of course, we can have more than two block ciphers running in parallel, increasing the speed-up proportionally. For applications with high throughput demands, e.g., in networks with data rates in the range of Gigabits per second, encryption modes that can be parallelized are very desirable.

5.1.6 Galois Counter Mode (GCM)

The *Galois Counter Mode (GCM)* is an encryption mode which also computes a message authentication code (MAC) [160]. A MAC provides a cryptographic checksum that is computed by the sender, Alice, and appended to the message. Bob also computes a MAC from the message and checks whether his MAC is the same as the one computed by Alice. This way, Bob can make sure that (1) the message was really created by Alice and (2) that nobody tampered with the ciphertext during transmission. These two properties are called message authentication and integrity, respectively. Much more about MACs is found in Chap. 12. We presented a slightly simplified version of the GCM mode in the following.

GCM protects the confidentiality of the plaintext x by using an encryption in counter mode. Additionally, GCM protects not only the authenticity of the plaintext x but also the authenticity of a string *AAD* called *additional authenticated data*. This authenticated data is, in contrast to the plaintext, left in clear in this mode of operation. In practice, the string *AAD* might include addresses and parameters in a network protocol.

The GCM consists of an underlying block cipher and a Galois field multiplier with which the two GCM functions *authenticated encryption* and *authenticated decryption* are realized. The cipher needs to have a block size of 128 bits such as AES. On the sender side, GCM encrypts data using the Counter Mode (CTR) followed by the computation of a MAC value. For encryption, first an initial counter is derived from an IV and a serial number. Then the initial counter value is incremented, and this value is encrypted and XORed with the first plaintext block. For subsequent plaintexts, the counter is incremented and then encrypted. Note that the underlying block cipher is only used in encryption mode. GCM allows for precomputation of the block cipher function if the initialization vector is known ahead of time.

For authentication, GCM performs a chained Galois field multiplication. For every plaintext x_i an intermediate authentication parameter g_i is derived. g_i is computed as the XOR sum of the current ciphertext y_i and g_i, and multiplied by the constant H. The value H is a hash subkey which is generated by encryption of the all-zero input with the block cipher. All multiplications are in the 128-bit Galois field $GF(2^{128})$ with the irreducible polynomial $P(x) = x^{128} + x^7 + x^2 + x + 1$. Since only one multiplication is required per block cipher encryption, the GCM mode adds very little computational overhead to the encryption.

Definition 5.1.6 Basic Galois Counter mode (GCM)
Let $e()$ be a block cipher of block size 128 bit; let x be the plaintext consisting of the blocks $x_1, \ldots, x_n$; and let AAD be the additional authenticated data.

1. **Encryption**

    a. *Derive a counter value CTR_0 from the IV and compute $CTR_1 = CTR_0 + 1$.*
    b. *Compute ciphertext: $y_i = e_k(CTR_i) \oplus x_i, \quad i \geq 1$*

2. **Authentication**

    a. *Generate authentication subkey $H = e_k(0)$*
    b. *Compute $g_0 = AAD \times H$ (Galois field multiplication)*
    c. *Compute $g_i = (g_{i-1} \oplus y_i) \times H, \quad 1 \leq i \leq n$ (Galois field multiplication)*
    d. *Final authentication tag: $T = (g_n \times H) \oplus e_k(CTR_0)$*

Figure 5.8 shows a diagram of the GCM.

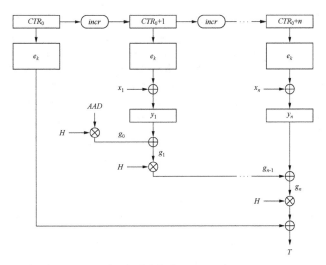

Fig. 5.8 Basic authenticated encryption in Galois Counter mode

The receiver of the packet $[(y_1, \ldots, y_n), T, ADD]$ decrypts the ciphertext by also applying the Counter mode. To check the authenticity of the data, the receiver also computes an authentication tag T' using the received ciphertext and ADD as input. He employs exactly the same steps as the sender. If T and T' match, the receiver is

assured that the ciphertext (and *ADD*) were not manipulated in transit and that only
the sender could have generated the message.

5.2 Exhaustive Key Search Revisited

In Sect. 3.5.1 we saw that given a plaintext–ciphertext pair (x_1, y_1) a DES key can
be exhaustively searched using the simple algorithm:

$$DES_{k_i}(x_1) \stackrel{?}{=} y_1, \quad i = 0, 1, \dots, 2^{56} - 1. \tag{5.1}$$

For most other block ciphers, however, a key search is somewhat more complicated.
Somewhat surprisingly, a brute-force attack can produce *false positive* results, i.e.,
keys k_i are found that are not the one used for the encryption, yet they perform a
correct encryption in Eq. (5.1). The likelihood of this occurring is related to the
relative size of the key space and the plaintext space.

A brute-force attack is still possible, but several pairs of plaintext–ciphertext are
needed. The length of the respective plaintext required to break the cipher with a
brute-force attack is referred to as *unicity distance*. After trying every possible key,
there should be just one plaintext that makes sense.

Let's first look why one pair (x_1, y_1) might not be sufficient to identify the correct
key. For illustration purposes we assume a cipher with a block width of 64 bit and a
key size of 80 bit. If we encrypt x_1 under all possible 2^{80} keys, we obtain 2^{80} ciphertexts. However, there exist only 2^{64} different ones, and thus some keys must map x_1
to the same ciphertext. If we run through all keys for a given plaintext–ciphertext
pair, we find on average $2^{80}/2^{64} = 2^{16}$ keys that perform the mapping $e_k(x_1) = y_1$.
This estimation is valid since the encryption of a plaintext for a given key can be
viewed as a random selection of a 64-bit ciphertext string. The phenomenon of multiple "paths" between a given plaintext and ciphertext is depicted in Fig. 5.9, in
which $k^{(i)}$ denote the keys that map x_1 to y_1. These keys can be considered key
candidates.

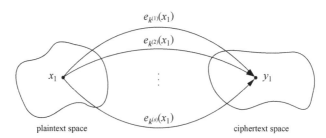

Fig. 5.9 Multiple keys map between one plaintext and one ciphertext

Among the approximately 2^{16} key candidates $k^{(i)}$ is the correct one that was used by to perform the encryption. Let's call this one the target key. In order to identify the target key we need a second plaintext–ciphertext pair (x_2, y_2). Again, there are about 2^{16} key candidates that map x_2 to y_2. One of them is the target key. The other keys can be viewed as randomly drawn from the 2^{80} possible ones. It is crucial to note that the target key must be present in *both* sets of key candidates. To determine the effectiveness of a brute-force attack, the crucial question is now: What is the likelihood that another (false!) key is contained in both sets? The answer is given by the following theorem:

Theorem 5.2.1 *Given a block cipher with a key length of κ bits and block size of n bits, as well as t plaintext–ciphertext pairs $(x_1, y_1), \ldots, (x_t, y_t)$, the expected number of false keys which encrypt all plaintexts to the corresponding ciphertexts is:*

$$2^{\kappa - tn}$$

Returning to our example and assuming two plaintext–ciphertext pairs, the likelihood of a false key k_f that performs both encryptions $e_{k_f}(x_1) = y_1$ and $e_{k_f}(x_2) = y_2$ is:

$$2^{80 - 2 \cdot 64} = 2^{-48}$$

This value is so small that for almost all practical purposes it is sufficient to test two plaintext–ciphertext pairs. If the attacker chooses to test three pairs, the likelihood of a false key decreases to $2^{80 - 3 \cdot 64} = 2^{-112}$. As we saw from this example, the likelihood of a false alarm decreases rapidly with the number t of plaintext–ciphertext pairs. In practice, typically we only need a few pairs.

The theorem above is not only important if we consider an individual block cipher but also if we perform multiple encryptions with a cipher. This issue is addressed in the following section.

5.3 Increasing the Security of Block Ciphers

In some situations we wish to increase the security of block ciphers, e.g., if a cipher such as DES is available in hardware or software for legacy reasons in a given application. We discuss two general approaches to strengthen a cipher, multiple encryption and key whitening. Multiple encryption, i.e., encrypting a plaintext more than once, is already a fundamental design principle of block ciphers, since the round function is applied many times to the cipher. Our intuition tells us that the security of a block cipher against both brute-force and analytical attacks increases by performing multiple encryptions in a row. Even though this is true in principle, there are a few surprising facts. For instance, doing double encryption does very little to increase the brute-force resistance over a single encryption. We study this counterintuitive fact in the next section. Another very simple yet effective approach to increase the brute-force resistance of block ciphers is called key whitening; it is also discussed below.

We note here that when using AES, we already have three different security levels given by the key lengths of 128, 192 and 256 bits. Given that there are no realistic attacks known against AES with any of those key lengths, there appears no reason to perform multiple encryption with AES for practical systems. However, for some selected older ciphers, especially for DES, multiple encryption can be a useful tool.

5.3.1 Double Encryption and Meet-in-the-Middle Attack

Let's assume a block cipher with a key length of κ bits. For *double encryption*, a plaintext x is first encrypted with a key k_L, and the resulting ciphertext is encrypted again using a second key k_R. This scheme is shown in Fig. 5.10.

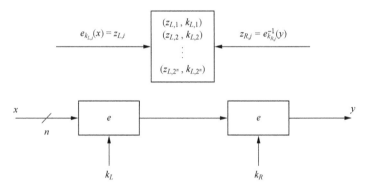

Fig. 5.10 Double encryption and meet-in-the-middle attack

A naïve brute-force attack would require us to search through all possible combinations of both keys, i.e., the effective key lengths would be 2κ and an exhaustive key search would require $2^\kappa \cdot 2^\kappa = 2^{2\kappa}$ encryptions (or decryptions). However, using the *meet-in-the-middle* attack, the key space is drastically reduced. This is a divide-and-conquer attack in which Oscar first brute-force-attacks the encryption on the left-hand side, which requires 2^κ cipher operations, and then the right encryption, which again requires 2^κ operations. If he succeeds with this attack, the total complexity is $2^\kappa + 2^\kappa = 2 \cdot 2^\kappa = 2^{\kappa+1}$. This is barely more complex than a key search of a single encryption and of course is much less complex than performing $2^{2\kappa}$ search operations.

The attack has two phases. In the first one, the left encryption is brute-forced and a lookup table is computed. In the second phase the attacker tries to find a match in the table which reveals both encryption keys. Here are the details of this approach.

Phase I: Table Computation For a given plaintext x_1, compute a lookup table for all pairs $(k_{L,i}, z_{L,i})$, where $e_{k_{L,i}}(x_1) = z_{L,i}$ and $i = 1, 2, \ldots, 2^\kappa$. These computations are symbolized by the left arrow in the figure. The $z_{L,i}$ are the intermediate values that occur in between the two encryptions. This list should be ordered by the values of the $z_{L,i}$. The number of entries in the table is 2^κ, with each entry being $n + \kappa$ bits wide. Note that one of the keys we used for encryption must be the correct target key, but we still do not know which one it is.

Phase II: Key Matching In order to find the key, we now decrypt y_1, i.e., we perform the computations symbolized by the right arrow in the figure. We select the first possible key $k_{R,1}$, e.g., the all-zero key, and compute:

$$e_{k_{R,1}}^{-1}(x_1) = z_{R,1}.$$

We now check whether $z_{R,1}$ is equal to any of the $z_{L,i}$ values in the table which we computed in the first phase. If it is *not* in the table, we increment the key to $k_{R,1}$, decrypt y_1 again, and check whether this value is in the table. We continue until we have a match.

We now have what is called a *collision* of two values, i.e., $z_{L,i} = z_{R,j}$. This gives us two keys: The value $z_{L,i}$ is associated with the key $k_{L,i}$ from the left encryption, and $k_{R,j}$ is the key we just tested from the right encryption. This means there exists a key pair $(k_{L,i}, k_{R,j})$ which performs the double encryption:

$$e_{k_{R,j}}(e_{k_{L,i}}(x_1)) = y_1 \qquad (5.2)$$

As discussed in Sect. 5.2, there is a chance that this is not the target key pair we are looking for since there are most likely several possible key pairs that perform the mapping $x_1 \rightarrow y_1$. Hence, we have to verify additional key candidates by encrypting several plaintext–ciphertext pairs according to Eq. (5.2). If the verification fails for any of the pairs $(x_1, y_1), (x_2, y_2), \ldots$, we go back to beginning of Phase II and increment the key k_R again and continue with the search.

Let's briefly discuss how many plaintext–ciphertext pairs we will need to rule out faulty keys with a high likelihood. With respect to multiple mappings between a plaintext and a ciphertext as depicted in Fig. 5.9, double encryption can be modeled as a cipher with 2κ key bits and n block bits. In practice, one often has $2\kappa > n$, in which case we need several plaintext–ciphertext pairs. The theorem in Sect. 5.2 can easily be adopted to the case of multiple encryption, which gives us a useful guideline about how many (x, y) pairs should be available:

Theorem 5.3.1 *Given are l subsequent encryptions with a block cipher with a key length of κ bits and block size of n bits, as well as t plaintext–ciphertext pairs $(x_1, y_1), \ldots, (x_t, y_t)$. The expected number of false keys which encrypt all plaintexts to the corresponding ciphertexts is given by:*
$$2^{l\kappa - tn}$$

Let's look at an example.

Example 5.3. As an example, if we double-encrypt with DES and choose to test three plaintext–ciphertext pairs, the likelihood of a faulty key pair surviving all three key tests is:

$$2^{2 \cdot 56 - 3 \cdot 64} = 2^{-80}.$$

◊

Let us examine the computational complexity of the meet-in-the-middle attack. In the first phase of the attack, corresponding to the left arrow in the figure, we perform 2^κ encryptions and store them in 2^κ memory locations. In the second stage, corresponding to the right arrow in the figure, we perform a maximum of 2^κ decryptions and table look-ups. We ignore multiple key tests at this stage. The total cost for the meet-in-the-middle attack is:

$$\text{number of encryptions and decryptions} = 2^\kappa + 2^\kappa = 2^{\kappa+1}$$
$$\text{number of storage locations} = 2^\kappa$$

This compares to 2^κ encryptions or decryptions and essentially no storage cost in the case of a brute-force attack against a single encryption. Even though the storage requirements go up quite a bit, the costs in computation and memory are still only proportional to 2^κ. Thus, it is widely believed that double encryption is not worth the effort. Instead, triple encryption should be used; this method is described in the following section.

Note that for a more exact complexity analysis of the meet-in-the-middle attack, we would also need take the cost of sorting the table entries in Phase I into account as well as the table look-ups in Phase II. For our purposes, however, we can ignore these additional costs.

5.3.2 Triple Encryption

Compared to double encryption, a much more secure approach is the encryption of a block of data three times in a row:

$$y = e_{k_3}(e_{k_2}(e_{k_1}(x))).$$

In practice, often a variant of the triple encryption from above is used:

$$y = e_{k_1}(e_{k_2}^{-1}(e_{k_3}(x))).$$

This type of triple encryption is sometimes referred to as encryption–decryption–encryption (EDE). The reason for this has nothing to do with security. If $k_1 = k_2$, the operation effectively performed is

$$y = e_{k_3}(x),$$

which is single encryption. Since it is sometimes desirable that one implementation can perform both triple encryption and single encryption, i.e., in order to interoperate with legacy systems, EDE is a popular choice for triple encryption. Moreover, for a 112-bit security, it is sufficient to choose two different keys k_1 and k_2 and set $k_3 = k_1$ in case of 3DES.

Of course, we can still perform a meet-in-the-middle attack as shown in Fig. 5.11.

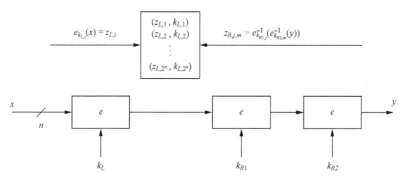

Fig. 5.11 Triple encryption and sketch of a meet-in-the-middle attack

Again, we assume κ bits per key. The problem for an attacker is that she has to compute a lookup table either after the first or after the second encryption. In both cases, the attacker has to compute two encryptions or decryptions in a row in order to reach the lookup table. Here lies the cryptographic strength of triple encryption: There are 2^{2k} possibilities to run through all possible keys of two encryptions or decryptions. In the case of 3DES, this forces an attacker to perform 2^{112} key tests, which is entirely infeasible with current technology. In summary, the meet-in-the-middle attack reduces the *effective key length* of triple encryption from 3κ to 2κ. Because of this, it is often said that the *effective key length* of triple DES is 112 bits as opposed to $3 \cdot 56 = 168$ bits which are actually used as input to the cipher.

5.3.3 Key Whitening

Using an extremely simple technique called *key whitening*, it is possible to make block ciphers such as DES much more resistant against brute-force attacks. The basic scheme is shown in Fig. 5.12.

In addition to the regular cipher key k, two whitening keys k_1 and k_2 are used to XOR-mask the plaintext and ciphertext. This process can be expressed as:

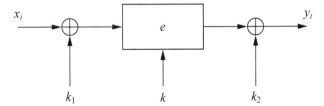

Fig. 5.12 Key whitening of a block cipher

Definition 5.3.1 Key whitening for block ciphers
Encryption: $y = e_{k,k_1,k_2}(x) = e_k(x \oplus k_1) \oplus k_2$
Decryption: $x = e_{k,k_1,k_2}^{-1}(x) = e_k^{-1}(y \oplus k_2) \oplus k_1$

It is important to stress that key whitening does not strengthen block ciphers against most analytical attacks such as linear and differential cryptanalysis. This is in contrast to multiple encryption, which often also increases the resistance to analytical attacks. Hence, key whitening is not a "cure" for inherently weak ciphers. Its main application is ciphers that are relatively strong against analytical attacks but possess too short a key space. The prime example of such a cipher is DES. A variant of DES which uses key whitening is *DESX*. In the case of DESX, the key k_2 is derived from k and k_1. Please note that most modern block ciphers such as AES already apply key whitening internally by adding a subkey prior to the first round and after the last round.

Let's now discuss the security of key whitening. A naïve brute-force attack against the scheme requires $2^{\kappa+2n}$ search steps, where κ is the bit length of the key and n the block size. Using the meet-in-the-middle attack introduced in Sect. 5.3, the computational load can be reduced to approximately $2^{\kappa+n}$ steps, plus storage of 2^n data sets. However, if the adversary Oscar can collect 2^m plaintext–ciphertext pairs, a more advanced attack exists with a computational complexity of

$$2^{\kappa+n-m}$$

cipher operations. Even though we do not introduce the attack here, we'll briefly discuss its consequences if we apply key whitening to DES. We assume that the attacker knows 2^m plaintext–ciphertext pairs. Note that the designer of a security system can often control how many plaintext–ciphertext are generated before a new key is established. Thus, the parameter m cannot be arbitrarily increased by the attacker. Also, since the number of known plaintexts grows exponentially with m, values beyond, say, $m = 40$, seem quite unrealistic. As a practical example, let's assume key whitening of DES, and that Oscar can collect a maximum of 2^{32} plaintexts. He now has to perform

$$2^{56+64-32} = 2^{88}$$

DES computations. Given that with today's technology even 2^{56} DES operations require several days with special hardware, performing 2^{88} encryptions is completely out of reach. Note that the number of plaintexts (which Oscar is not supposed to know in most circumstances) corresponds to 32 GByte of data, the collection of which is also a formidable task in most real-world situations.

A particular attractive feature of key whitening is that the additional computational load is negligible. A typical block cipher implementation in software requires several hundred instructions for encrypting one input block. In contrast, a 64-bit XOR operation only takes 2 instructions on a 32-bit machine, so that the performance impact due to key whitening is in the range of 1% or less in most cases.

5.4 Discussion and Further Reading

Modes of Operation After the AES selection process, the US National Institute of Standards and Technology (NIST) supported the process of evaluating new modes of operations in a series of special publications and workshops [124]. Currently, there are eight approved block cipher modes: five for confidentiality (ECB, CBC, CFB, OFB, CTR), one for authentication (CMAC) and two combined modes for confidentiality and authentication (CCM, GCM). The modes are widely used in practice and are part of many standards, e.g., for computer networks or banking.

Other Applications for Block Ciphers The most important application of block ciphers in practice, in addition to data encryption, is *Message Authentication Codes (MACs)*, which are discussed in Chap. 12. The schemes CBC-MAC, OMAC and PMAC are constructed with a block cipher. *Authenticated Encryption (AE)* uses block ciphers to both encrypt and generate a MAC in order to provide confidentiality and authentication, respectively. In addition to the GCM introduced in this chapter, other AE modes include the *EAX* mode, *OCB* mode, and *GC* mode.

Another application is the *Cryptographically Secure Pseudo Random Number Generators (CSPRNG)* built from block ciphers. In fact, the stream cipher modes introduced in this chapter, OFB, CFB and CTR mode, form CSPRNGs. There are also standards such as [4, Appendix A.2.4] which explicitly specify random number generators from block ciphers.

Block ciphers can also be used to build *cryptographic hash functions*, as discussed in Chap. 11.

Extending Brute-Force Attacks Even though there are no algorithmic shortcuts to brute-force attacks, there are methods which are efficient if several exhaustive key searches have to be performed. Those methods are called time–memory tradeoff attacks (TMTO). The general idea is to encrypt a fixed plaintext under a large number of keys and to store certain intermediate results. This is the precomputation phase, which is typically at least as complex as a single brute-force attack and which results in large lookup tables. In the online phase, a search through the tables takes place which is considerably faster than a brute-force attack. Thus, after the precomputation phase, individual keys can be found much more quickly. TMTO attacks were originally proposed by Hellman [91] and were improved with the introduction of distinguished points by Rivest [145]. More recently rainbow tables were proposed to further improve TMTO attacks [131]. A limiting factor of TMTO attacks in practice is that for each individual attack it is required that the same piece of known plaintext was encrypted, e.g., a file header.

Block Ciphers and Quantum Computers With the potential rise of quantum computers in the future, the security of currently used crypto algorithms has to be reevaluated. (It should be noted that the possible existence of quantum computers in a few decades from now is hotly debated.) Whereas all popular existing asymmetric algorithms such as RSA are vulnerable to attacks using quantum computers [153], symmetric algorithms are much more resilient. A potential quantum computer using Grover's algorithm [87] would require only $2^{(n/2)}$ steps in order to perform a complete key search on a cipher with a keyspace of 2^n elements. Hence, key lengths of more than 128 bit are required if resistance against quantum computer attacks is desired. This observation was also the motivation for requiring the 192-bit and 256-bit key lengths for AES. Interestingly, it can be shown that there can be no quantum algorithm which performs such an attack more efficiently than Grover's algorithm [16].

5.5 Lessons Learned

- There are many different ways to encrypt with a block cipher. Each mode of operation has some advantages and disadvantages.
- Several modes turn a block cipher into a stream cipher.
- There are modes that perform encryption together together with authentication, i.e., a cryptographic checksum protects against message manipulation.
- The straightforward ECB mode has security weaknesses, independent of the underlying block cipher.
- The counter mode allows parallelization of encryption and is thus suited for high-speed implementations.
- Double encryption with a given block cipher only marginally improves the resistance against brute-force attacks.
- Triple encryption with a given block cipher roughly *doubles* the key length. Triple DES (3DES) has an effective key length of 112 bits.
- Key whitening enlarges the DES key length without much computational overhead.

Problems

5.1. Consider the storage of data in encrypted form in a large database using AES. One record has a size of 16 bytes. Assume that the records are not related to one another. Which mode would be best suited and why?

5.2. We consider known-plaintext attacks on block ciphers by means of an exhaustive key search where the key is k bits long. The block length counts n bits with $n > k$.

1. How many plaintexts and ciphertexts are needed to successfully break a block cipher running in ECB mode? How many steps are done in the worst case?
2. Assume that the initialization vector IV for running the considered block cipher in CBC mode is known. How many plaintexts and ciphertexts are now needed to break the cipher by performing an exhaustive key search? How many steps need now maximally be done? Briefly describe the attack.
3. How many plaintexts and ciphertexts are necessary, if you do *not* know the IV?
4. Is breaking a block cipher in CBC mode by means of an exhaustive key search considerably more difficult than breaking an ECB mode block cipher?

5.3. In a company, all files which are sent on the network are automatically encrypted by using AES-128 in CBC mode. A fixed key is used, and the IV is changed once per day. The network encryption is file-based, so that the IV is used at the beginning of every file.

You managed to spy out the fixed AES-128 key, but do not know the recent IV. Today, you were able to eavesdrop two different files, one with unidentified content and one which is known to be an automatically generated temporary file and only contains the value 0xFF. Briefly describe how it is possible to obtain the unknown initialization vector and how you are able to determine the content of the unknown file.

5.4. Keeping the IV secret in OFB mode does not make an exhaustive key search more complex. Describe how we can perform a brute-force attack with unknown IV. What are the requirements regarding plaintext and ciphertext?

5.5. Describe how the OFB mode can be attacked if the IV is *not* different for each execution of the encryption operation.

5.6. Propose an OFB mode scheme which encrypts one byte of plaintext at a time, e.g., for encrypting key strokes from a remote keyboard. The block cipher used is AES. Perform one block cipher operation for every new plaintext byte. Draw a block diagram of your scheme and pay particular attention to the bit lengths used in your diagram (cf. the descripton of a byte mode at the end of Sect. 5.1.4).

5.7. As is so often true in cryptography, it is easy to weaken a seemingly strong scheme by small modifications. Assume a variant of the OFB mode by which we only feed back the 8 most significant bits of the cipher output. We use AES and fill the remaining 120 input bits to the cipher with 0s.

1. Draw a block diagram of the scheme.
2. Why is this scheme weak if we encrypt moderately large blocks of plaintext, say 100 kByte? What is the maximum number of known plaintexts an attacker needs to completely break the scheme?
3. Let the feedback byte be denoted by FB. Does the scheme become cryptographically stronger if we feedback the 128-bit value $FB, FB, \ldots, FB$ to the input (i.e., we copy the feedback byte 16 times and use it as AES input)?

5.8. In the text, a variant of the CFB mode is proposed which encrypts individual bytes. Draw a block diagram for this mode when using AES as block cipher. Indicate the width (in bit) of each line in your diagram.

5.9. We are using AES in counter mode for encrypting a hard disk with 1 TB of capacity. What is the maximum length of the IV?

5.10. Sometimes error propagation is an issue when choosing a mode of operation in practice. In order to analyze the propagation of errors, let us assume a bit error (i.e., a substitution of a "0" bit by a "1" bit or vice versa) in a ciphertext block y_i.

1. Assume an error occurs during the transmission in one block of ciphertext, let's say y_i. Which cleartext blocks are affected on Bob's side when using the ECB mode?
2. Again, assume block y_i contains an error introduced during transmission. Which cleartext blocks are affected on Bob's side when using the CBC mode?
3. Suppose there is an error in the cleartext x_i on Alice's side. Which cleartext blocks are affected on Bob's side when using the CBC mode?
4. Assume a single bit error occurs in the transmission of a ciphertext character in 8-bit CFB mode. How far does the error propagate? Describe exactly *how* each block is affected.
5. Prepare an overview of the effect of bit errors in a ciphertext block for the modes ECB, CBC, CFB, OFB and CTR. Differentiate between random bit errors and specific bit errors when decrypting y_i.

5.11. Besides simple bit errors, the deletion or insertion of a bit yields even more severe effects since the synchronization of blocks is disrupted. In most cases, the decryption of subsequent blocks will be incorrect. A special case is the CFB mode with a feedback width of 1 bit. Show that the synchronization is automatically restored after $\kappa + 1$ steps, where κ is the block size of the block cipher.

5.12. We now analyze the security of DES double encryption (2DES) by doing a cost-estimate:
$$2DES(x) = DES_{K_2}(DES_{K_1}(x))$$

1. First, let us assume a pure key search without any memory usage. For this purpose, the whole key space spanned by K_1 and K_2 has to be searched. How much does a key-search machine for breaking 2DES (worst case) in 1 week cost?
   In this case, assume ASICs which can perform 10^7 keys per second at a cost of \$5 per IC. Furthermore, assume an overhead of 50% for building the key search machine.

2. Let us now consider the meet-in-the-middle (or time-memory tradeoff) attack, in which we can use memory. Answer the following questions:

   ■ How many entries have to be stored?
   ■ How many bytes (not bits!) have to be stored for each entry?
   ■ How costly is a key search in one week? Please note that the key space has to be searched before filling up the memory completely. Then we can begin to search the key space of the second key. Assume the same hardware for both key spaces.

   For a rough cost estimate, assume the following costs for hard disk space: $8/10 GByte, where 1 GByte = 10^9 Byte.
3. Assuming Moore's Law, when do the costs move below $1 million?

5.13. Imagine that aliens — rather than abducting earthlings and performing strange experiments on them — drop a computer on planet Earth that is particularly suited for AES key searches. In fact, it is so powerful that we can search through 128, 192 and 256 key bits in a matter of days. Provide guidelines for the number of plaintext–ciphertext pairs the aliens need so that they can rule out false keys with a reasonable likelihood. (**Remark**: Since the existence of both aliens and human-built computers for such key lengths seem extremely unlikely at the time of writing, this problem is pure science fiction.)

5.14. Given multiple plaintext–ciphertext pairs, your objective is to attack an encryption scheme based upon multiple encryptions.

1. You want to break an encryption system E, which makes use of triple AES-192 encryption (e.g. block length $n = 128$ bit, key size of $k = 192$ bit). How many tuples (x_i, y_i) with $y_i = e_K(x_i)$ do you need to level down the probability of finding a key K, which matches the condition $y_i = e_K(x_i)$ for one particular i, but fails for most other values of i (a so called *false positive*), to $Pr(K' \neq K) = 2^{-20}$?
2. What is the maximum key size of a block cipher that you could still effectively attack with an error probability of at most $Pr(K' \neq K) = 2^{-10} = 1/1024$, if this cipher always uses double encryption ($l = 2$) and has a block length of $n = 80$ bit?
3. Estimate the success probability, if you are provided with four plaintext–ciphertext blocks which are double encrypted using AES-256 ($n = 128$ bits, $k = 256$ bits). Please justify your results.

Note that this is a purely theoretical problem. Key spaces of size 2^{128} and beyond can not be brute-forced.

5.15. 3DES with three different keys can be broken with about 2^{2k} encryptions and 2^k memory cells, $k = 56$. Design the corresponding attack. How many pairs (x, y) should be available so that the probability to determine an incorrect key triple (k_1, k_2, k_3) is sufficiently low?

5.16. This is your chance to break a cryptosystem. As we know by now, cryptography is a tricky business. The following problem illustrates how easy it is to turn a strong scheme into a weak one with minor modifications.

We saw in this chapter that key whitening is a good technique for strengthening block ciphers against brute-force attacks. We now look at the following variant of key whitening against DES, which we'll call DESA:

$$DESA_{k,k_1}(x) = DES_k(x) \oplus k_1.$$

Even though the method looks similar to key whitening, it hardly adds to the security. Your task is to show that breaking the scheme is roughly as difficult as a brute-force attack against single DES. Assume you have a few pairs of plaintext–ciphertext.

Chapter 6
Introduction to Public-Key Cryptography

Before we learn about the basics of public-key cryptography, let us recall that the term *public-key cryptography* is used interchangeably with *asymmetric cryptography*; they both denote exactly the same thing and are used synonymously.

As stated in Chap. 1, symmetric cryptography has been used for at least 4000 years. Public-key cryptography, on the other hand, is quite new. It was publicly introduced by Whitfield Diffie, Martin Hellman and Ralph Merkle in 1976. Much more recently, in 1997 British documents which were declassified revealed that the researchers James Ellis, Clifford Cocks and Graham Williamson from the UK's Government Communications Headquarters (GCHQ) discovered and realized the principle of public-key cryptography a few years earlier, in 1972. However, it is still being debated whether the government office fully recognized the far-reaching consequences of public-key cryptography for commercial security applications.

In this chapter you will learn:

- A brief history of public-key cryptography
- The pros and cons of public-key cryptography
- Some number theoretical topics that are needed for understanding public-key algorithms, most importantly the extended Euclidean algorithm

6.1 Symmetric vs. Asymmetric Cryptography

In this chapter we will see that asymmetric, i.e., public-key, algorithms are very different from symmetric algorithms such as AES or DES. Most public-key algorithms are based on number-theoretic functions. This is quite different from symmetric ciphers, where the goal is usually *not* to have a compact mathematical description between input and output. Even though mathematical structures are often used for small blocks *within* symmetric ciphers, for instance, in the AES S-Box, this does not mean that the entire cipher forms a compact mathematical description.

Symmetric Cryptography Revisited

In order to understand the principle of asymmetric cryptography, let us first recall the basic symmetric encryption scheme in Fig. 6.1.

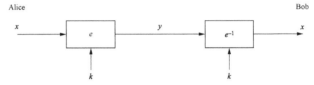

Fig. 6.1 Principle of symmetric-key encryption

Such a system is symmetric with respect to two properties:

1. The *same secret key* is used for encryption and decryption.
2. The encryption and decryption *function* are very similar (in the case of DES they are essentially identical).

There is a simple analogy for symmetric cryptography, as shown in Fig. 6.2. Assume there is a safe with a strong lock. Only Alice and Bob have a copy of the key for the lock. The action of encrypting of a message can be viewed as putting the message in the safe. In order to read, i.e., decrypt, the message, Bob uses his key and opens the safe.

Modern symmetric algorithms such as AES or 3DES are very secure, fast and are in widespread use. However, there are several shortcomings associated with symmetric-key schemes, as discussed below.

Key Distribution Problem The key must be established between Alice and Bob using a secure channel. Remember that the communication link for the message is not secure, so sending the key over the channel directly — which would be the most convenient way of transporting it — can't be done.

Number of Keys Even if we solve the key distribution problem, we must potentially deal with a very large number of keys. If each pair of users needs a separate pair of keys in a network with n users, there are

Fig. 6.2 Analogy for symmetric encryption: a safe with one lock

$$\frac{n \cdot (n-1)}{2}$$

key pairs, and every user has to store $n-1$ keys securely. Even for mid-size networks, say, a corporation with 2000 people, this requires more than 4 million key pairs that must be generated and transported via secure channels. More about this problem is found in Sect. 13.1.3. (There are smarter ways of dealing with keys in symmetric cryptography networks as detailed in Sect. 13.2; however, those approaches have other problems such as a single point of failure.)

No Protection Against Cheating by Alice or Bob Alice and Bob have the same capabilities, since they possess the same key. As a consequence, symmetric cryptography cannot be used for applications where we would like to prevent cheating by either Alice or Bob as opposed to cheating by an outsider like Oscar. For instance, in e-commerce applications it is often important to prove that Alice actually sent a certain message, say, an online order for a flat screen TV. If we only use symmetric cryptography and Alice changes her mind later, she can always claim that Bob, the vendor, has falsely generated the electronic purchase order. Preventing this is called *nonrepudiation* and can be achieved with asymmetric cryptography, as discussed in Sect. 10.1.1. Digital signatures, which are introduced in Chap. 10, provide nonrepudiation.

Fig. 6.3 Analogy for public-key encryption: a safe with public lock for depositing a message and a secret lock for retrieving a message

Principles of Asymmetric Cryptography

In order to overcome these drawbacks, Diffie, Hellman and Merkle had a revolutionary proposal based on the following idea: It is not necessary that the key possessed by the person who *encrypts* the message (that's Alice in our example) is secret. The crucial part is that Bob, the receiver, can only *decrypt* using a secret key. In order to realize such a system, Bob publishes a public encryption key which is known to everyone. Bob also has a matching secret key, which is used for decryption. Thus, Bob's key k consists of two parts, a public part, k_{pub}, and a private one, k_{pr}.

A simple analogy of such a system is shown in Fig. 6.3. This systems works quite similarly to the good old mailbox on the corner of a street: Everyone can put a letter in the box, i.e., encrypt, but only a person with a private (secret) key can retrieve letters, i.e., decrypt. If we assume we have cryptosystems with such a functionality, a basic protocol for public-key encryption looks as shown in Fig. 6.4.

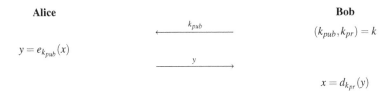

Fig. 6.4 Basic protocol for public-key encryption

By looking at that protocol you might argue that even though we can encrypt a message without a secret channel for key establishment, we still cannot exchange a key if we want to encrypt with, say, AES. However, the protocol can easily be modified for this use. What we have to do is to *encrypt a symmetric key*, e.g., an AES key, using the public-key algorithm. Once the symmetric key has been decrypted by Bob, both parties can use it to encrypt and decrypt messages using symmetric ciphers. Figure 6.5 shows a basic key transport protocol where we use AES as the symmetric cipher for illustration purposes (of course, one can use any other symmetric algorithm in such a protocol). The main advantage of the protocol in Fig. 6.5 over the protocol in Fig. 6.4 is that the payload is encrypted with a symmetric cipher, which tends to be much faster than an asymmetric algorithm.

From the discussion so far, it looks as though asymmetric cryptography is a desirable tool for security applications. The question remains how one can build public-key algorithms. In Chaps. 7, 8 and 9 we introduce most asymmetric schemes of practical relevance. They are all built from one common principle, the one-way function. The informal definition of it is as follows:

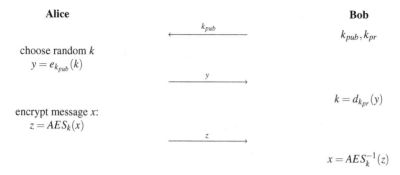

Fig. 6.5 Basic key transport protocol with AES as an example of a symmetric cipher

---

Definition 6.1.1 One-way function
A function $f()$ is a one-way function if:

1. $y = f(x)$ is computationally easy, and
2. $x = f^{-1}(y)$ is computationally infeasible.

Obviously, the adjectives "easy" and "infeasible" are not particularly exact. In mathematical terms, a function is easy to compute if it can be evaluated in polynomial time, i.e., its running time is a polynomial expression. In order to be useful in practical crypto schemes, the computation $y = f(x)$ should be sufficiently fast that it does not lead to unacceptably slow execution times in an application. The inverse computation $x = f^{-1}(y)$ should be so computationally intensive that it is not feasible to evaluate it in any reasonable time period, say, 10,000 years, when using the best known algorithm.

There are two popular one-way functions which are used in practical public-key schemes. The first is the integer factorization problem, on which RSA is based. Given two large primes, it is easy to compute the product. However, it is very difficult to factor the resulting product. In fact, if each of the primes has 150 or more decimal digits, the resulting product cannot be factored, even with thousands of PCs running for many years. The other one-way function that is used widely is the discrete logarithm problem. This is not quite as intuitive and is introduced in Chap. 8.

6.2 Practical Aspects of Public-Key Cryptography

Actual public-key algorithms will be introduced in the next chapters, since there is some mathematics we must study first. However, it is very interesting to look at the principal security functions of public-key cryptography which we address in this section.

6.2.1 Security Mechanisms

As shown in the previous section, public-key schemes can be used for encryption of data. It turns out that we can do many other, previously unimaginable, things with public-key cryptography. The main functions that they can provide are listed below:

Main Security Mechanisms of Public-Key Algorithms:

Key Establishment There are protocols for establishing secret keys over an insecure channel. Examples for such protocols include the Diffie–Hellman key exchange (DHKE) or RSA key transport protocols.

Nonrepudiation Providing nonrepudiation and message integrity can be realized with digital signature algorithms, e.g., RSA, DSA or ECDSA.

Identification We can identify entities using challenge-and-response protocols together with digital signatures, e.g., in applications such as smart cards for banking or for mobile phones.

Encryption We can encrypt messages using algorithms such as RSA or Elgamal.

We note that identification and encryption can also be achieved with symmetric ciphers, but they typically require much more effort with key management. It looks as though public-key schemes can provide all functions required by modern security protocols. Even though this is true, the major drawback in practice is that encryption of data is very computationally intensive — or more colloquially: extremely slow — with public-key algorithms. Many block and stream ciphers can encrypt about one hundred to one thousand times faster than public-key algorithms. Thus, somewhat ironically, public-key cryptography is rarely used for the actual encryption of data. On the other hand, symmetric algorithms are poor at providing nonrepudiation and key establishment functionality. In order to use the best of both worlds, most practical protocols are *hybrid protocols* which incorporate both symmetric and public-key algorithms. Examples include the SSL/TLS potocol that is commonly used for secure Web connections, or IPsec, the security part of the Internet communication protocol.

6.2.2 The Remaining Problem: Authenticity of Public Keys

From the discussion so far we've seen that a major advantage of asymmetric schemes is that we can freely distribute public keys, as shown in the protocols in Figs. 6.4 and 6.5. However, in practice, things are a bit more tricky because we still have to assure the *authenticity* of public keys. In other words: Do we really know that a certain public key belongs to a certain person? In practice, this issue is often solved with what is called *certificates*. Roughly speaking, certificates bind a public key to a certain identity. This is a major issue in many security application, e.g., when doing e-commerce transactions on the Internet. We discuss this topic in more detail in Sect. 13.3.2.

Another problem, which is not as fundamental, is that public-key algorithms require very long keys, resulting in slow execution times. The issue of key lengths and security is discussed below.

6.2.3 Important Public-Key Algorithms

In the previous chapters, we learned about some block ciphers, DES and AES. However, there exist many other symmetric algorithms. Several hundred algorithms have been proposed over the years and even though a lot were found not to be secure, there exist many cryptographically strong ones as discussed in Sect. 3.7. The situation is quite different for asymmetric algorithms. There are only three major families of public-key algorithms which are of practical relevance. They can be classified based on their underlying computational problem.

Public-Key Algorithm Families of Practical Relevance

Integer-Factorization Schemes Several public-key schemes are based on the fact that it is difficult to factor large integers. The most prominent representative of this algorithm family is RSA.

Discrete Logarithm Schemes There are several algorithms which are based on what is known as the discrete logarithm problem in finite fields. The most prominent examples include the Diffie–Hellman key exchange, Elgamal encryption or the Digital Signature Algorithm (DSA).

Elliptic Curve (EC) Schemes A generalization of the discrete logarithm algorithm are elliptic curve public-key schemes. The most popular examples include Elliptic Curve Diffie–Hellman key exchange (ECDH) and the Elliptic Curve Digital Signature Algorithm (ECDSA).

The first two families were proposed in the mid-1970s, and elliptic curves were proposed in the mid-1980s. There are no known attacks against any of the schemes if the parameters, especially the operand and key lengths, are chosen carefully. Algorithms belonging to each of the families will be introduced in Chaps. 7, 8 and 9. It is important to note that each of the three families can be used to provide the main public-key mechanisms of key establishment, nonrepudiation through digital signatures and encryption of data.

In addition to the three families above, there have been proposals for several other public-key schemes. They often lack cryptographic maturity, i.e., it is not known how robust they are against mathematical attacks. Multivariate quadratic

(MQ) or some lattice-based schemes are examples of this. Another common problem is that they have poor implementation characteristics, like key lengths in the range of megabytes, e.g., the McEliece cryptosystems. However, there are also some other schemes, for instance, hyperelliptic curve cryptosystems, which are both as efficient and secure as the three established families shown above, but which simply have not gained widespread adoption. For most applications it is recommended to use public-key schemes from the three established algorithm families.

6.2.4 Key Lengths and Security Levels

All three of the established public-key algorithm families are based on number-theoretic functions. One distinguishing feature of them is that they require arithmetic with very long operands and keys. Not surprisingly, the longer the operands and keys, the more secure the algorithms become. In order to compare different algorithms, one often considers the *security level*. An algorithm is said to have a "security level of n bit" if the best known attack requires 2^n steps. This is a quite natural definition because symmetric algorithms with a security level of n have a key of length n bit. The relationship between cryptographic strength and security is not as straightforward in the asymmetric case, though. Table 6.1 shows recommended bit lengths for public-key algorithms for the four security levels 80, 128, 192 and 256 bit. We see from the table that RSA-like schemes and discrete-logarithm schemes require very long operands and keys. The key length of elliptic curve schemes is significantly smaller, yet still twice as long as symmetric ciphers with the same cryptographic strength.

Table 6.1 Bit lengths of public-key algorithms for different security levels

Algorithm Family	Cryptosystems	Security Level (bit)			
		80	128	192	256
Integer factorization	RSA	1024 bit	3072 bit	7680 bit	15360 bit
Discrete logarithm	DH, DSA, Elgamal	1024 bit	3072 bit	7680 bit	15360 bit
Elliptic curves	ECDH, ECDSA	160 bit	256 bit	384 bit	512 bit
Symmetric-key	AES, 3DES	80 bit	128 bit	192 bit	256 bit

You may want to compare this table with the one given in Sect. 1.3.2, which provides information about the security estimations of symmetric-key algorithms. In order to provide long-term security, i.e., security for a timespan of several decades, a security level of 128 bit should be chosen, which requires fairly long keys for all three algorithm families.

An undesired consequence of the long operands is that public-key schemes are extremely arithmetically intensive. As mentioned earlier, it is not uncommon that one public-operation, say a digital signature, is by 2–3 orders of magnitude slower than the encryption of one block using AES or 3DES. Moreover, the computational complexity of the three algorithm families grows roughly with the cube bit length. As an example, increasing the bit length from 1024 to 3076 in a given RSA signature generation software results in an execution that is $3^3 = 27$ times slower! On modern PCs, execution times in the range of several 10 msec to a few 100 msec are common, which does not pose a problem for many applications. However, public-key performance can be a more serious bottleneck in constrained devices where small CPUs are prevalent, e.g., mobile phones or smart cards, or on network servers that have to compute many public-key operations per second. Chaps. 7, 8 and 9 introduce several techniques for implementing public-key algorithms reasonably efficiently.

6.3 Essential Number Theory for Public-Key Algorithms

We will now study a few techniques from number theory which are essential for public-key cryptography. We introduce the Euclidean algorithm, Euler's phi function as well as Fermat's Little Theorem and Euler's theorem. All are important for asymmetric algorithms, especially for understanding the RSA crypto scheme.

6.3.1 Euclidean Algorithm

We start with the problem of computing the *greatest common divisor (gcd)*. The gcd of two positive integers r_0 and r_1 is denoted by

$$\gcd(r_0, r_1)$$

and is the largest positive number that divides both r_0 and r_1. For instance $\gcd(21,9) = 3$. For small numbers, the gcd is easy to calculate by factoring both numbers and finding the highest common factor.

Example 6.1. Let $r_0 = 84$ and $r_1 = 30$. Factoring yields

$$r_0 = 84 = 2 \cdot 2 \cdot 3 \cdot 7$$
$$r_1 = 30 = 2 \cdot 3 \cdot 5$$

The gcd is the product of all common prime factors:

$$2 \cdot 3 = 6 = \gcd(30, 84)$$

◇

For the large numbers which occur in public-key schemes, however, factoring often is not possible, and a more efficient algorithm is used for gcd computations, the Euclidean algorithm. The algorithm, which is also referred to as Euclid's algorithm, is based on the simple observation that

$$\gcd(r_0, r_1) = \gcd(r_0 - r_1, r_1),$$

where we assume that $r_0 > r_1$, and that both numbers are positive integers. This property can easily be proven: Let $\gcd(r_0, r_1) = g$. Since g divides both r_0 and r_1, we can write $r_0 = g \cdot x$ and $r_1 = g \cdot y$, where $x > y$, and x and y are coprime integers, i.e., they do not have common factors. Moreover, it is easy to show that $(x - y)$ and y are also coprime. It follows from here that:

$$\gcd(r_0 - r_1, r_1) = \gcd(g \cdot (x - y), g \cdot y) = g.$$

Let's verify this property with the numbers from the previous example:

Example 6.2. Again, let $r_0 = 84$ and $r_1 = 30$. We now look at the gcd of $(r_0 - r_1)$ and r_1:

$$r_0 - r_1 = 54 = 2 \cdot 3 \cdot 3 \cdot 3$$
$$r_1 = 30 = 2 \cdot 3 \cdot 5$$

The largest common factor still is $2 \cdot 3 = 6 = \gcd(30, 54) = \gcd(30, 84)$.

◇

It also follows immediately that we can apply the process iteratively:

$$\gcd(r_0, r_1) = \gcd(r_0 - r_1, r_1) = \gcd(r_0 - 2r_1, r_1) = \cdots = \gcd(r_0 - m \, r_1, r_1)$$

as long as $(r_0 - m \, r_1) > 0$. The algorithm uses the fewest number of steps if we choose the maximum value for m. This is the case if we compute:

$$\gcd(r_0, r_1) = \gcd(r_0 \bmod r_1, r_1).$$

Since the first term $(r_0 \bmod r_1)$ is smaller than the second term r_1, we usually swap them:

$$\gcd(r_0, r_1) = \gcd(r_1, r_0 \bmod r_1).$$

The core observation from this process is that we can **reduce the problem of finding the gcd of two given numbers to that of the gcd of two smaller numbers.** This process can be applied recursively until we obtain finally $\gcd(r_l, 0) = r_l$. Since each iteration preserves the gcd of the previous iteration step, it turns out that this final gcd is the gcd of the original problem, i.e.,

$$\gcd(r_0, r_1) = \cdots = \gcd(r_l, 0) = r_l.$$

We first show some examples for finding the gcd using the Euclidean algorithm and then discuss the algorithm a bit more formally.

Example 6.3. Let $r_0 = 27$ and $r_1 = 21$. Fig. 6.6 gives us some feeling for the algorithm by showing how the lengths of the parameters shrink in every iteration. The shaded parts in the iteration are the new remainders $r_2 = 6$ (first iteration), and $r_3 = 3$ (second iteration) which form the input terms for the next iterations. Note that in the last iteration the remainder is $r_4 = 0$, which indicates the termination of the algorithm. ◇

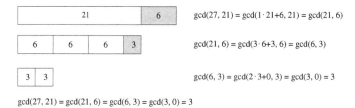

gcd(27, 21) = gcd(1·21+6, 21) = gcd(21, 6)

gcd(21, 6) = gcd(3·6+3, 6) = gcd(6, 3)

gcd(6, 3) = gcd(2·3+0, 3) = gcd(3, 0) = 3

gcd(27, 21) = gcd(21, 6) = gcd(6, 3) = gcd(3, 0) = 3

Fig. 6.6 Example of the Euclidean algorithm for the input values $r_0 = 27$ and $r_1 = 21$

It is also helpful to look at the Euclidean algorithm with slightly larger numbers, as happens in Example 6.4.

Example 6.4. Let $r_0 = 973$ and $r_1 = 301$. The gcd is then computed as

$973 = 3 \cdot 301 + 70$	$\gcd(973, 301) = \gcd(301, 70)$
$301 = 4 \cdot 70 + 21$	$\gcd(301, 70) = \gcd(70, 21)$
$70 = 3 \cdot 21 + 7$	$\gcd(70, 21) = \gcd(21, 7)$
$21 = 3 \cdot 7 + 0$	$\gcd(21, 7) = \gcd(7, 0) = 7$

◇

By now we should have an idea of Euclid's algorithm, and we can give a more formal description of the algorithm.

Euclidean Algorithm
Input: positive integers r_0 and r_1 with $r_0 > r_1$
Output: $\gcd(r_0, r_1)$
Initialization: $i = 1$
Algorithm:

1   DO
1.1 $i = i + 1$
1.2 $r_i = r_{i-2} \bmod r_{i-1}$
      WHILE $r_i \neq 0$
2   RETURN
      $\gcd(r_0, r_1) = r_{i-1}$

Note that the algorithm terminates if a remainder with the value $r_i = 0$ is computed. The remainder computed in the previous iteration, denoted by r_{l-1}, is the gcd of the original problem.

The Euclidean algorithm is very efficient, even with the very long numbers typically used in public-key cryptography. The number of iterations is close to the number of digits of the input operands. That means, for instance, that the number of iterations of a gcd involving 1024-bit numbers is 1024 times a constant. Of course, algorithms with a few thousand iterations can easily be executed on today's PCs, making the algorithms very efficient in practice.

6.3.2 Extended Euclidean Algorithm

So far, we have seen that finding the gcd of two integers r_0 and r_1 can be done by recursively reducing the operands. However, it turns out that finding the gcd is not the main application of the Euclidean algorithm. An extension of the algorithm allows us to compute modular inverses, which is of major importance in public-key cryptography. In addition to computing the gcd, the *extended Euclidean algorithm* (EEA) computes a linear combination of the form:

$$\gcd(r_0, r_1) = s \cdot r_0 + t \cdot r_1$$

where s and t are integer coefficients. This equation is often referred to as *Diophantine equation*.

The question now is: how do we compute the two coefficients s and t? The idea behind the algorithm is that we execute the standard Euclidean algorithm, but we express the current remainder r_i in every iteration as a linear combination of the form

$$r_i = s_i r_0 + t_i r_1. \tag{6.1}$$

If we succeed with this, we end up in the last iteration with the equation:

$$r_l = \gcd(r_0, r_1) = s_l r_0 + t_l r_1 = s r_0 + t r_1.$$

This means that the last coefficient s_l is the coefficient s in Eq. (6.1) we are looking for, and also $t_l = t$. Let's look at an example.

Example 6.5. We consider the extended Euclidean algorithm with the same values as in the previous example, $r_0 = 973$ and $r_1 = 301$. On the left-hand side, we compute the standard Euclidean algorithm, i.e., we compute new remainders $r_2, r_3, \ldots$. Also, we have to compute the integer quotient q_{i-1} in every iteration. On the right-hand side we compute the coefficients s_i and t_i such that $r_i = s_i r_0 + t_i r_1$. The coefficients are always shown in brackets.

i	$r_{i-2} = q_{i-1} \cdot r_{i-1} + r_i$	$r_i = [s_i]r_0 + [t_i]r_1$
2	$973 = 3 \cdot 301 + 70$	$70 = [1]r_0 + [-3]r_1$
3	$301 = 4 \cdot 70 + 21$	$21 = 301 - 4 \cdot 70$
		$= r_1 - 4(1r_0 - 3\,r_1)$
		$= [-4]r_0 + [13]r_1$
4	$70\ = 3 \cdot 21 + 7$	$7\ = 70 - 3 \cdot 21$
		$= (1r_0 - 3r_1) - 3(-4r_0 + 13r_1)$
		$= [13]r_0 + [-42]r_1$
	$21\ = 3 \cdot 7 + 0$	

The algorithm computed the three parameters $\gcd(973, 301) = 7$, $s = 13$ and $t = -42$. The correctness can be verified by:

$$\gcd(973, 301) = 7 = [13]973 + [-42]301 = 12649 - 12642.$$

◇

You should carefully watch the algebraic steps taking place in the right column of the example above. In particular, observe that the linear combination on the right-hand side is always constructed with the help of the *previous* linear combinations. We will now derive recursive formulae for computing s_i and r_i in every iteration. Assume we are in iteration with index i. In the two previous iterations we computed the values

$$r_{i-2} = [s_{i-2}]r_0 + [t_{i-2}]r_1 \tag{6.2}$$
$$r_{i-1} = [s_{i-1}]r_0 + [t_{i-1}]r_1 \tag{6.3}$$

In the current iteration i we first compute the quotient q_{i-1} and the new remainder r_i from r_{i-1} and r_{i-2}:

$$r_{i-2} = q_{i-1} \cdot r_{i-1} + r_i.$$

This equation can be rewritten as:

$$r_i = r_{i-2} - q_{i-1} \cdot r_{i-1}. \tag{6.4}$$

Recall that our goal is to represent the new remainder r_i as a linear combination of r_0 and r_1 as shown in Eq. (6.1). The core step for achieving this happens now: in Eq. (6.4) we simply substitute r_{i-2} by Eq. (6.2) and r_{i-1} by Eq. (6.3):

$$r_i = (s_{i-2}r_0 + t_{i-2}r_1) - q_{i-1}(s_{i-1}r_0 + t_{i-1}r_1)$$

If we rearrange the terms we obtain the desired result:

$$r_i = [s_{i-2} - q_{i-1}s_{i-1}]r_0 + [t_{i-2} - q_{i-1}t_{i-1}]r_1 \tag{6.5}$$
$$r_i = [s_i]r_0 + [t_i]r_1$$

Eq. (6.5) also gives us immediately the recursive formulae for computing s_i and t_i, namely $s_i = s_{i-2} - q_{i-1}s_{i-1}$ and $t_i = t_{i-2} - q_{i-1}t_{i-1}$. These recursions are valid for index values $i \geq 2$. Like any recursion, we need starting values for s_0, s_1, t_0, t_1. These initial values (which we derive in Problem 6.13) can be shown to be $s_0 = 1, s_1 = 0, t_0 = 0, t_1 = 1$.

Extended Euclidean Algorithm (EEA)
Input: positive integers r_0 and r_1 with $r_0 > r_1$
Output: $\gcd(r_0, r_1)$, as well as s and t such that $\gcd(r_0, r_1) = s \cdot r_0 + t \cdot r_1$.
Initialization:
$s_0 = 1 \qquad t_0 = 0$
$s_1 = 0 \qquad t_1 = 1$
$i \;\; = 1$
Algorithm:

1   DO
1.1 $i \;\;\; = i + 1$
1.2 $r_i \;\;\; = r_{i-2} \bmod r_{i-1}$
1.3 $q_{i-1} = (r_{i-2} - r_i)/r_{i-1}$
1.4 $s_i \;\;\; = s_{i-2} - q_{i-1} \cdot s_{i-1}$
1.5 $t_i \;\;\; = t_{i-2} - q_{i-1} \cdot t_{i-1}$
       WHILE $r_i \neq 0$
2   RETURN
       $\gcd(r_0, r_1) = r_{i-1}$
       $s = s_{i-1}$
       $t = t_{i-1}$

As mentioned above, the main application of the EEA in asymmetric cryptography is to compute the inverse modulo of an integer. We already encountered this problem in the context of the affine cipher in Chap. 1. For the affine cipher, we were required to find the inverse of the key value a modulo 26. With the Euclidean algorithm, this is straightforward. Let's assume we want to compute the inverse of $r_1 \bmod r_0$ where $r_1 < r_0$. Recall from Sect. 1.4.2 that the inverse only exists if $\gcd(r_0, r_1) = 1$. Hence, if we apply the EEA, we obtain $s \cdot r_0 + t \cdot r_1 = 1 = \gcd(r_0, r_1)$. Taking this equation modulo r_0 we obtain:

$$s \cdot r_0 + t \cdot r_1 = 1$$
$$s \cdot 0 + t \cdot r_1 \equiv 1 \bmod r_0$$
$$r_1 \cdot t \equiv 1 \bmod r_0 \qquad\qquad (6.6)$$

Equation (6.6) is exactly the definition of the inverse of r_1. That means, that t itself is the inverse of r_1:

$$t = r_1^{-1} \bmod r_0.$$

Thus, if we need to compute an inverse $a^{-1} \bmod m$, we apply the EEA with the input parameters m and a. The output value t that is computed is the inverse. Let's look at an example.

Example 6.6. Our goal is to compute $12^{-1} \bmod 67$. The values 12 and 67 are relatively prime, i.e., $\gcd(67, 12) = 1$. If we apply the EEA, we obtain the coefficients s and t in $\gcd(67, 12) = 1 = s \cdot 67 + t \cdot 12$. Starting with the values $r_0 = 67$ and $r_1 = 12$, the algorithm proceeds as follows:

i	q_{i-1}	r_i	s_i	t_i
2	5	7	1	-5
3	1	5	-1	6
4	1	2	2	-11
5	2	1	-5	**28**

This gives us the linear combination

$$-5 \cdot 67 + 28 \cdot 12 = 1$$

As shown above, the inverse of 12 follows from here as

$$12^{-1} \equiv 28 \bmod 67.$$

This result can easily be verified

$$28 \cdot 12 = 336 \equiv 1 \bmod 67.$$

◇

Note that the s coefficient is not needed and is in practice often not computed. Please note also that the result of the algorithm can be a negative value for t. The result is still correct, however. We have to compute $t = t + r_0$, which is a valid operation since $t \equiv t + r_0 \bmod r_0$.

For completeness, we show how the EEA can also be used for computing multiplicative inverses in Galois fields. In modern cryptography this is mainly relevant for the derivation of the AES S-Boxes and for elliptic curve public-key algorithms. The EEA can be used completely analogously with polynomials instead of integers. If we want to compute an inverse in a finite field $GF(2^m)$, the inputs to the algorithm are the field element $A(x)$ and the irreducible polynomial $P(x)$. The EEA computes the auxiliary polynomials $s(x)$ and $t(x)$, as well as the greatest common divisor $\gcd(P(x), A(x))$ such that:

$$s(x)P(x) + t(x)A(x) = \gcd(P(x), A(x)) = 1$$

Note that since $P(x)$ is irreducible, the gcd is always equal to 1. If we take the equation above and reduce both sides modulo $P(x)$, it is straightforward to see that the auxiliary polynomial $t(x)$ is equal to the inverse of $A(x)$:

$$s(x)0 + t(x)A(x) \equiv 1 \bmod P(x)$$
$$t(x) \equiv A^{-1}(x) \bmod P(x)$$

We give at this point an example of the algorithm for the small field $GF(2^3)$.

Example 6.7. We are looking for the inverse of $A(x) = x^2$ in the finite field $GF(2^3)$ with $P(x) = x^3 + x + 1$. The initial values for the $t(x)$ polynomial are: $t_0(x) = 0$, $t_1(x) = 1$.

Iteration	$r_{i-2}(x)$	$= [q_{i-1}(x)] r_{i-1}(x) + [r_i(x)]$	$t_i(x)$
2	$x^3 + x + 1$	$= [x] x^2 + [x+1]$	$t_2 = t_0 - q_1 t_1 = 0 - x1 \equiv x$
3	x^2	$= [x](x+1) + [x]$	$t_3 = t_1 - q_2 t_2 = 1 - x(x) \equiv 1 + x^2$
4	$x + 1$	$= [1]x + [1]$	$t_4 = t_2 - q_3 t_3 = x - 1(1+x^2)$
			$t_4 \equiv 1 + x + x^2$
5	x	$= [x]1 + [0]$	Termination since $r_5 = 0$

Note that polynomial coefficients are computed in $GF(2)$, and since addition and multiplication are the same operations, we can always replace a negative coefficient (such as $-x$) by a positive one. The new quotient and the new remainder that are computed in every iteration are shown in brackets above. The polynomials $t_i(x)$ are computed according to the recursive formula that was used for computing the integers t_i earlier in this section. The EEA terminates if the remainder is 0, which is the case in the iteration with index 5. The inverse is now given as the last $t_i(x)$ value that was computed, i.e., $t_4(x)$:

$$A^{-1}(x) = t(x) = t_4(x) = x^2 + x + 1.$$

Here is the check that $t(x)$ is in fact the inverse of x^2, where we use the properties that $x^3 \equiv x+1 \bmod P(x)$ and $x^4 \equiv x^2 + x \bmod P(x)$:

$$t_4(x) \cdot x^2 = x^4 + x^3 + x^2$$
$$\equiv (x^2 + x) + (x+1) + x^2 \bmod P(x)$$
$$\equiv 1 \bmod P(x)$$

◇

Note that in every iteration of the EEA, one uses long division (not shown above) to determine the new quotient $q_{i-1}(x)$ and the new remainder $r_i(x)$.

The inverse Table 4.2 in Chap. 4 was computed using the extended Euclidean algorithm.

6.3.3 Euler's Phi Function

We now look at another tool that is useful for public-key cryptosystems, especially for RSA. We consider the ring $\mathbb{Z}_m$, i.e., the set of integers $\{0, 1, \ldots, m-1\}$. We are interested in the (at the moment seemingly odd) problem of knowing *how many* numbers in this set are relatively prime to *m*. This quantity is given by *Euler's phi function*, which is defined as follows:

> **Definition 6.3.1** Euler's Phi Function
> *The number of integers in $\mathbb{Z}_m$ relatively prime to m is denoted by $\Phi(m)$.*

We first look at some examples and calculate Euler's phi function by actually counting all the integers in $\mathbb{Z}_m$ which are relatively prime.

Example 6.8. Let $m = 6$. The associated set is $\mathbb{Z}_6 = \{0,1,2,3,4,5\}$.
$\gcd(0,6) = 6$
$\gcd(1,6) = 1 \star$
$\gcd(2,6) = 2$
$\gcd(3,6) = 3$
$\gcd(4,6) = 2$
$\gcd(5,6) = 1 \star$
Since there are two numbers in the set which are relatively prime to 6, namely 1 and 5, the phi function takes the value 2, i.e., $\Phi(6) = 2$.
◇

Here is another example:

Example 6.9. Let $m = 5$. The associated set is $\mathbb{Z}_5 = \{0,1,2,3,4\}$.
$\gcd(0,5) = 5$
$\gcd(1,5) = 1 \star$
$\gcd(2,5) = 1 \star$
$\gcd(3,5) = 1 \star$
$\gcd(4,5) = 1 \star$
This time we have four numbers which are relatively prime to 5, hence, $\Phi(5) = 4$.
◇

From the examples above we can guess that calculating Euler's phi function by running through all elements and computing the gcd is extremely slow if the numbers are large. In fact, computing Euler's phi function in this naïve way is completely out of reach for the large numbers occurring in public-key cryptography. Fortunately, there exists a relation to calculate it much more easily if we know the factorization of *m*, which is given in following theorem.

Theorem 6.3.1 *Let m have the following canonical factorization*

$$m = p_1^{e_1} \cdot p_2^{e_2} \cdot \ldots \cdot p_n^{e_n},$$

where the p_i are distinct prime numbers and e_i are positive integers,
then

$$\Phi(m) = \prod_{i=1}^{n} (p_i^{e_i} - p_i^{e_i-1}).$$

Since the value of n, i.e., the number of distinct prime factors, is always quite small even for large numbers m, evaluating the product symbol $\prod$ is computationally easy. Let's look at an example where we calculate Euler's phi function using the relation:

Example 6.10. Let $m = 240$. The factorization of 240 in the canonical factorization form is

$$m = 240 = 16 \cdot 15 = 2^4 \cdot 3 \cdot 5 = p_1^{e_1} \cdot p_2^{e_2} \cdot p_3^{e_3}$$

There are three distinct prime factors, i.e., $n = 3$. The value for Euler's phi functions follows then as:

$$\Phi(m) = (2^4 - 2^3)(3^1 - 3^0)(5^1 - 5^0) = 8 \cdot 2 \cdot 4 = 64.$$

That means that 64 integers in the range $\{0, 1, \ldots, 239\}$ are coprime to $m = 240$. The alternative method, which would have required to evaluate the gcd 240 times, would have been much slower even for this small number.

◇

It is important to stress that we need to know the factorization of m in order to calculate Euler's phi function quickly in this manner. As we will see in the next chapter, this property is at the heart of the RSA public-key scheme: Conversely, if we know the factorization of a certain number, we can compute Euler's phi function and decrypt the ciphertext. If we do not know the factorization, we cannot compute the phi function and, hence, cannot decrypt.

6.3.4 Fermat's Little Theorem and Euler's Theorem

We describe next two theorems which are quite useful in public-key crpytography. We start with *Fermat's Little Theorem*.[1] The theorem is helpful for primality testing and in many other aspects of public-key cryptography. The theorem gives a seemingly surprising result if we do exponentiations modulo an integer.

---

[1] You should not confuse this with Fermat's Last Theorem, one of the most famous number-theoretical problems, which was proved in the 1990s after 350 years.

> **Theorem 6.3.2** Fermat's Little Theorem
> *Let a be an integer and p be a prime, then:*
>
> $$a^p \equiv a \,(\bmod\ p).$$

We note that arithmetic in finite fields $GF(p)$ is done modulo p, and hence, the theorem holds for all integers a which are elements of a finite field $GF(p)$. The theorem can be stated in the form:

$$a^{p-1} \equiv 1 \,(\bmod\ p)$$

which is often useful in cryptography. One application is the computation of the inverse in a finite field. We can rewrite the equation as $a \cdot a^{p-2} \equiv 1 \,(\bmod\ p)$. This is exactly the definition of the multiplicative inverse. Thus, we immediately have a way for inverting an integer a modulo a prime:

$$a^{-1} \equiv a^{p-2} \,(\bmod\ p) \tag{6.7}$$

We note that this inversion method holds only if p is a prime. Let's look at an example:

Example 6.11. Let $p = 7$ and $a = 2$. We can compute the inverse of a as:

$$a^{p-2} = 2^5 = 32 \equiv 4 \bmod 7.$$

This is easy to verify: $2 \cdot 4 \equiv 1 \bmod 7$.

$\diamond$

Performing the exponentiation in Eq. (6.7) is usually slower than using the extended Euclidean algorithm. However, there are situations where it is advantageous to use Fermat's Little Theorem, e.g., on smart cards or other devices which have a hardware accelerator for fast exponentiation anyway. This is not uncommon because many public-key algorithms require exponentiation, as we will see in subsequent chapters.

A generalization of Fermat's Little Theorem to any integer moduli, i.e., moduli that are not necessarily primes, is *Euler's theorem*.

> **Theorem 6.3.3** Euler's Theorem
> *Let a and m be integers with $\gcd(a,m) = 1$, then:*
>
> $$a^{\Phi(m)} \equiv 1 \,(\bmod\ m).$$

Since it works modulo m, it is applicable to integer rings $\mathbb{Z}_m$. We show now an example for Euler's theorem with small values.

Example 6.12. Let $m = 12$ and $a = 5$. First, we compute Euler's phi function of m:

$$\Phi(12) = \Phi(2^2 \cdot 3) = (2^2 - 2^1)(3^1 - 3^0) = (4-2)(3-1) = 4.$$

Now we can verify Euler's theorem:

$$5^{\Phi(12)} = 5^4 = 25^2 = 625 \equiv 1 \bmod 12.$$

◇

It is easy to show that Fermat's Little Theorem is a special case of Euler's theorem. If p is a prime, it holds that $\Phi(p) = (p^1 - p^0) = p - 1$. If we use this value for Euler's theorem, we obtain: $a^{\Phi(p)} = a^{p-1} \equiv 1 \ (\bmod\ p)$, which is exactly Fermat's Little Theorem.

6.4 Discussion and Further Reading

Public-Key Cryptography in General Asymmetric cryptography was introduced in the landmark paper by Whitfield Diffie and Martin Hellman [58]. Ralph Merkle independently invented the concept of asymmetric cryptography but proposed an entirely different public-key algorithm [121]. There are a few good accounts of the history of public-key cryptography. The treatment in [57] by Diffie is recommended. Another good overview on public-key cryptography is [127]. A very instructive and detailed history of elliptic curve cryptography, including the relatively intense competition between RSA and ECC during the 1990s, is described in [100]. More recent development in asymmetric cryptography is tracked by the Workshop on Public-Key Cryptography (PKC) series.

Modular Arithmetic With respect to the mathematics introduced in this chapter, the introductory books on number theory recommended in Sect. 1.5 make good sources for further reading. In practical terms, the Extended Euclidean Algorithm (EEA) is the most crucial, since virtually all implementations of public-key schemes incorporate it, especially modular inversion. An important acceleration technique for the scheme is the binary EEA. Its advantage over the standard EEA is that it replaces divisions by bit shifts. This is in particular attractive for the very long numbers occurring in public-key schemes.

Alternative Public-Key Algorithms In addition to the three established families of asymmetric schemes, there exist several others. First, there are algorithms which have been broken or are believed to be insecure, e.g., knapsack schemes. Second, there are generalizations of the established algorithms, e.g., hyperelliptic curves, algebraic varieties or non-RSA factoring-based schemes. These schemes use the same one-way function, that is, integer factorization or the discrete logarithm in certain groups. Third, there are asymmetric algorithms which are based on different one-way functions. Four families of one-way function are of particular interest: hash-based, code-based, lattice-based and multivariate quadratic (MQ) public-key algorithms. There are, of course, reasons why they are not as widely used today.

In most cases, they have either practical drawbacks, such as very long keys (sometimes in the range of several megabytes), or the cryptographic strength is not well understood. Since about 2005, there has been growing interest in the cryptographic community in such asymmetric schemes. This is in part motivated by the fact that no quantum computing attacks are currently known against these four families of alternative asymmetric schemes. This is in contrast to RSA, discrete logarithm, and elliptic curve schemes and their variants, which are all vulnerable to attacks using quantum computers [153]. Even though it is not clear whether quantum computers will ever exist (the most optimistic estimates state that they are still several decades away), the alternative public-key algorithms are at times collectively referred to as *post-quantum cryptography*. A recent book [18] and a new workshop series [36, 35] provide more information about this area of active research.

6.5 Lessons Learned

- Public-key algorithms have capabilities that symmetric ciphers don't have, in particular digital signature and key establishment functions.
- Public-key algorithms are computationally intensive (a nice way of saying that they are *slow*), and hence are poorly suited for bulk data encryption.
- Only three families of public-key schemes are widely used. This is considerably fewer than in the case of symmetric algorithms.
- The extended Euclidean algorithm allows us to compute modular inverses quickly, which is important for almost all public-key schemes.
- Euler's phi function gives us the number of elements smaller than an integer n that are relatively prime to n. This is an important function for the RSA crypto scheme.

Problems

6.1. As we have seen in this chapter, public-key cryptography can be used for encryption and key exchange. Furthermore, it has some properties (such as nonrepudiation) which are not offered by secret key cryptography.

So why do we still use symmetric cryptography in current applications?

6.2. In this problem, we want to compare the computational performance of symmetric and asymmetric algorithms. Assume a fast public-key library such as OpenSSL [132] that can decrypt data at a rate of 100 Kbit/sec using the RSA algorithm on a modern PC. On the same machine, AES can decrypt at a rate of 17 Mbit/sec. Assume we want to decrypt a movie stored on a DVD. The movie requires 1 GByte of storage. How long does decryption take with either algorithm?

6.3. Assume a (small) company with 120 employees. A new security policy demands encrypted message exchange with a symmetric cipher. How many keys are required, if you are to ensure a secret communication for every possible pair of communicating parties?

6.4. The level of security in terms of the corresponding bit length directly influences the performance of the respective algorithm. We now analyze the influence of increasing the security level on the runtime.

Assume that a commercial Web server for an online shop can use either RSA or ECC for signature generation. Furthermore, assume that signature generation for RSA-1024 and ECC-160 takes 15.7 ms and 1.3 ms, respectively.

1. Determine the increase in runtime for signature generation if the security level from RSA is increased from 1024 bit to 3072 bit.
2. How does the runtime increase from 1024 bit to 15,360 bit?
3. Determine these numbers for the respective security levels of ECC.
4. Describe the difference between RSA and ECC when increasing the security level.

Hint: Recall that the computational complexity of both RSA and ECC grows with the cube of bit length. You may want to use Table 6.1 to determine the adequate bit length for ECC, given the security level of RSA.

6.5. Using the basic form of Euclid's algorithm, compute the greatest common divisor of

1. 7469 and 2464
2. 2689 and 4001

For this problem use only a pocket calculator. Show every iteration step of Euclid's algorithm, i.e., don't write just the answer, which is only a number. Also, for every gcd, provide the chain of gcd computations, i.e.,

$$\gcd(r_0, r_1) = \gcd(r_1, r_2) = \cdots.$$

6.6. Using the extended Euclidean algorithm, compute the greatest common divisor and the parameters s,t of

1. 198 and 243
2. 1819 and 3587

For every problem check if $s\,r_0 + t\,r_1 = \gcd(r_0, r_1)$ is actually fulfilled. The rules are the same as above: use a pocket calculator and show what happens in every iteration step.

6.7. With the Euclidean algorithm we finally have an efficient algorithm for finding the multiplicative inverse in Z_m that is much better than exhaustive search. Find the inverses in Z_m of the following elements a modulo m:

1. $a = 7$, $m = 26$ (affine cipher)
2. $a = 19$, $m = 999$

Note that the inverses must again be elements in Z_m and that you can easily verify your answers.

6.8. Determine $\phi(m)$, for $m = 12, 15, 26$, according to the definition: Check for each positive integer n smaller m whether $\gcd(n, m) = 1$. (You do *not* have to apply Euclid's algorithm.)

6.9. Develop formulae for $\phi(m)$ for the special cases when

1. m is a prime
2. $m = p \cdot q$, where p and q are primes. This case is of great importance for the RSA cryptosystem. Verify your formula for $m = 15, 26$ with the results from the previous problem.

6.10. Compute the inverse a^{-1} mod n with Fermat's Theorem (if applicable) or Euler's Theorem:

- $a = 4$, $n = 7$
- $a = 5$, $n = 12$
- $a = 6$, $n = 13$

6.11. Verify that Euler's Theorem holds in Z_m, $m = 6, 9$, for all elements a for which $\gcd(a, m) = 1$. Also verify that the theorem does not hold for elements a for which $\gcd(a, m) \neq 1$.

6.12. For the affine cipher in Chapter 1 the multiplicative inverse of an element modulo 26 can be found as
$$a^{-1} \equiv a^{11} \bmod 26.$$
Derive this relationship by using Euler's Theorem.

6.13. The extended Euclidean algorithm has the initial conditions $s_0 = 1, s_1 = 0, t_0 = 0, t_1 = 1$. *Derive* these conditions. It is helpful to look at how the general iteration formula for the Euclidean algorithm was derived in this chapter.

Chapter 7
The RSA Cryptosystem

After Whitfield Diffie and Martin Hellman introduced public-key cryptography in their landmark 1976 paper [58], a new branch of cryptography suddenly opened up. As a consequence, cryptologists started looking for methods with which public-key encryption could be realized. In 1977, Ronald Rivest, Adi Shamir and Leonard Adleman (cf. Fig. 7.1) proposed a scheme which became the most widely used asymmetric cryptographic scheme, RSA.

Fig. 7.1 An early picture of Adi Shamir, Ron Rivest, and Leonard Adleman (reproduced with permission from Ron Rivest)

In this chapter you will learn:

- How RSA works
- Practical aspects of RSA, such as computation of the parameters, and fast encryption and decryption
- Security estimations
- Implementational aspects

173

7.1 Introduction

The RSA crypto scheme, sometimes referred to as the Rivest–Shamir–Adleman algorithm, is currently the most widely used asymmetric cryptographic scheme, even though elliptic curves and discrete logarithm schemes are gaining ground. RSA was patented in the USA (but not in the rest of the world) until 2000.

There are many applications for RSA, but in practice it is most often used for:

- encryption of small pieces of data, especially for key transport
- digital signatures, which is discussed in Chap. 10, e.g., for digital certificates on the Internet

However, it should be noted that RSA encryption is not meant to replace symmetric ciphers because it is several times slower than ciphers such as AES. This is because of the many computations involved in performing RSA (or any other public-key algorithm) as we learn later in this chapter. Thus, the main use of the encryption feature is to securely exchange a key for a symmetric cipher (key transport). In practice, RSA is often used together with a symmetric cipher such as AES, where the symmetric cipher does the actual bulk data encryption.

The underlying one-way function of RSA is the integer factorization problem: Multiplying two large primes is computationally easy (in fact, you can do it with paper and pencil), but factoring the resulting product is very hard. Euler's theorem (Theorem 6.3.3) and Euler's phi function play important roles in RSA. In the following, we first describe how encryption, decryption and key generation work, then we talk about practical aspects of RSA.

7.2 Encryption and Decryption

RSA encryption and decryption is done in the integer ring $\mathbb{Z}_n$ and modular computations play a central role. Recall that rings and modular arithmetic in rings were introduced in Sect. 1.4.2. RSA encrypts plaintexts x, where we consider the bit string representing x to be an element in $\mathbb{Z}_n = \{0, 1, \ldots, n-1\}$. As a consequence the binary value of the plaintext x must be less than n. The same holds for the ciphertext. Encryption with the public key and decryption with the private key are as shown below:

RSA Encryption Given the public key $(n, e) = k_{pub}$ and the plaintext x, the encryption function is:

$$y = e_{k_{pub}}(x) \equiv x^e \bmod n \qquad (7.1)$$

where $x, y \in \mathbb{Z}_n$.

RSA Decryption Given the private key $d = k_{pr}$ and the ciphertext y, the decryption function is:

$$x = d_{k_{pr}}(y) \equiv y^d \bmod n \qquad (7.2)$$

where $x, y \in \mathbb{Z}_n$.

In practice, x, y, n and d are very long numbers, usually 1024 bit long or more. The value e is sometimes referred to as *encryption exponent* or *public exponent*, and the private key d is sometimes called *decryption exponent* or *private exponent*. If Alice wants to send an encrypted message to Bob, Alice needs to have his public key (n, e), and Bob decrypts with his private key d. We discuss in Sect. 7.3 how these three crucial parameters d, e, and n are generated.

Even without knowing more details, we can already state a few requirements for the RSA cryptosystem:

1. Since an attacker has access to the public key, it must be computationally infeasible to determine the private-key d given the public-key values e and n.
2. Since x is only unique up to the size of the modulus n, we cannot encrypt more than l bits with one RSA encryption, where l is the bit length of n.
3. It should be relatively easy to calculate $x^e \bmod n$, i.e., to encrypt, and $y^d \bmod n$, i.e., to decrypt. This means we need a method for fast exponentiation with very long numbers.
4. For a given n, there should be many private-key/public-key pairs, otherwise an attacker might be able to perform a brute-force attack. (It turns out that this requirement is easy to satisfy.)

7.3 Key Generation and Proof of Correctness

A distinctive feature of all asymmetric schemes is that there is a set-up phase during which the public and private key are computed. Depending on the public-key scheme, key generation can be quite complex. As a remark, we note that key generation is usually not an issue for block or stream ciphers.

Here are the steps involved in computing the public and private-key for an RSA cryptosystem.

RSA Key Generation
Output: public key: $k_{pub} = (n, e)$ and private key: $k_{pr} = (d)$
1. Choose two large primes p and q.
2. Compute $n = p \cdot q$.
3. Compute $\Phi(n) = (p - 1)(q - 1)$.
4. Select the public exponent $e \in \{1, 2, \ldots, \Phi(n) - 1\}$ such that

$$\gcd(e, \Phi(n)) = 1.$$

5. Compute the private key d such that

$$d \cdot e \equiv 1 \bmod \Phi(n)$$

The condition that $\gcd(e, \Phi(n)) = 1$ ensures that the inverse of e exists modulo $\Phi(n)$, so that there is always a private key d.

Two parts of the key generation are nontrivial: Step 1, in which the two large primes are chosen, as well as Steps 4 and 5 in which the public and private key are computed. The prime generation of Step 1 is quite involved and is addressed in Sect. 7.6. The computation of the keys d and e can be done at once using the extended Euclidean algorithm (EEA). In practice, one often starts by first selecting a public parameter e in the range $0 < e < \Phi(n)$. The value e must satisfy the condition $\gcd(e, \Phi(n)) = 1$. We apply the EEA with the input parameters n and e and obtain the relationship:

$$\gcd(\Phi(n), e) = s \cdot \Phi(n) + t \cdot e$$

If $\gcd(e, \Phi(n)) = 1$, we know that e is a valid public key. Moreover, we also know that the parameter t computed by the extended Euclidean algorithm is the inverse of e, and thus:

$$d = t \bmod \Phi(n)$$

In case that e and $\Phi(n)$ are not relatively prime, we simply select a new value for e and repeat the process. Note that the coefficient s of the EEA is not required for RSA and does not need to be computed.

We now see how RSA works by presenting a simple example.

Example 7.1. Alice wants to send an encrypted message to Bob. Bob first computes his RSA parameters in Steps 1–5. He then sends Alice his public key. Alice encrypts the message ($x = 4$) and sends the ciphertext y to Bob. Bob decrypts y using his private key.

Alice

message $x = 4$

$$\xleftarrow{\quad k_{pub}=(33,3)\quad}$$

$y = x^e \equiv 4^3 \equiv 31 \bmod 33$

$$\xrightarrow{\quad y=31 \quad}$$

Bob

1. choose $p = 3$ and $q = 11$
2. $n = p \cdot q = 33$
3. $\Phi(n) = (3-1)(11-1) = 20$
4. choose $e = 3$
5. $d \equiv e^{-1} \equiv 7 \bmod 20$

$y^d = 31^7 \equiv 4 = x \bmod 33$

Note that the private and public exponents fulfill the condition $e \cdot d = 3 \cdot 7 \equiv 1 \bmod \Phi(n)$.

◇

Practical RSA parameters are much, much larger. As can be seen from Table 6.1, the RSA modulus n should be at least 1024 bit long, which results in a bit length for p and q of 512. Here is an example of RSA parameters for this bit length:

$p = E0DFD2C2A288ACEBC705EFAB30E4447541A8C5A47A37185C5A9$
$\quad CB98389CE4DE19199AA3069B404FD98C801568CB9170EB712BF$
$\quad 10B4955CE9C9DC8CE6855C6123_h$

$q = EBE0FCF21866FD9A9F0D72F7994875A8D92E67AEE4B515136B2$
$\quad A778A8048B149828AEA30BD0BA34B977982A3D42168F594CA99$
$\quad F3981DDABFAB2369F229640115_h$

$n = CF33188211FDF6052BDBB1A37235E0ABB5978A45C71FD381A91$
$\quad AD12FC76DA0544C47568AC83D855D47CA8D8A779579AB72E635$
$\quad D0B0AAAC22D28341E998E90F82122A2C06090F43A37E0203C2B$
$\quad 72E401FD06890EC8EAD4F07E686E906F01B2468AE7B30CBD670$
$\quad 255C1FEDE1A2762CF4392C0759499CC0ABECFF008728D9A11ADF_h$

$e = 40B028E1E4CCF07537643101FF72444A0BE1D7682F1EDB553E3$
$\quad AB4F6DD8293CA1945DB12D796AE9244D60565C2EB692A89B888$
$\quad 1D58D278562ED60066DD8211E67315CF89857167206120405B0$
$\quad 8B54D10D4EC4ED4253C75FA74098FE3F7FB751FF5121353C554$
$\quad 391E114C85B56A9725E9BD5685D6C9C7EED8EE442366353DC39_h$

$d = C21A93EE751A8D4FBFD77285D79D6768C58EBF283743D2889A3$
$\quad 95F266C78F4A28E86F545960C2CE01EB8AD5246905163B28D0B$
$\quad 8BAABB959CC03F4EC499186168AE9ED6D88058898907E61C7CC$
$\quad CC584D65D801CFE32DFC983707F87F5AA6AE4B9E77B9CE630E2$
$\quad C0DF05841B5E4984D059A35D7270D500514891F7B77B804BED81_h$

What is interesting is that the message x is first raised to the eth power during encryption and the result y is raised to the dth power in the decryption, and the result of this is again equal to the message x. Expressed as an equation, this process is:

$$d_{k_{pr}}(y) = d_{k_{pr}}(e_{k_{pub}}(x)) \equiv (x^e)^d \equiv x^{de} \equiv x \bmod n. \tag{7.3}$$

This is the essence of RSA. We will now prove why the RSA scheme works.

Proof. We need to show that decryption is the inverse function of encryption, $d_{k_{pr}}(e_{k_{pub}}(x)) = x$. We start with the construction rule for the public and private key: $d \cdot e \equiv 1 \bmod \Phi(n)$. By definition of the modulo operator, this is equivalent to:

$$d \cdot e = 1 + t \cdot \Phi(n),$$

where t is some integer. Inserting this expression in Eq. (7.3):

$$d_{k_{pr}}(y) \equiv x^{de} \equiv x^{1+t \cdot \Phi(n)} \equiv x^{t \cdot \Phi(n)} \cdot x^1 \equiv (x^{\Phi(n)})^t \cdot x \bmod n. \tag{7.4}$$

This means we have to prove that $x \equiv (x^{\Phi(n)})^t \cdot x \bmod n$. We use now Euler's Theorem from Sect. 6.3.3, which states that if $\gcd(x,n) = 1$ then $1 \equiv x^{\Phi(n)} \bmod n$. A minor generalization immediately follows:

$$1 \equiv 1^t \equiv (x^{\Phi(n)})^t \bmod n, \tag{7.5}$$

where t is any integer. For the proof we distinguish two cases:

First case: $\gcd(x,n) = 1$
 Euler's Theorem holds here and we can insert Eq. (7.5) into (7.4):

$$d_{k_{pr}}(y) \equiv (x^{\Phi(n)})^t \cdot x \equiv 1 \cdot x \equiv x \bmod n. \qquad q.e.d.$$

This part of the proof establishes that decryption is actually the inverse function of encryption for plaintext values x which are relatively prime to the RSA modulus n. We provide now the proof for the other case.
Second case: $\gcd(x,n) = \gcd(x, p \cdot q) \neq 1$
 Since p and q are primes, x must have one of them as a factor:

$$x = r \cdot p \quad \text{or} \quad x = s \cdot q,$$

where r, s are integers such that $r < q$ and $s < p$. Without loss of generality we assume $x = r \cdot p$, from which follows that $\gcd(x, q) = 1$. Euler's Theorem holds in the following form:

$$1 \equiv 1^t \equiv (x^{\Phi(q)})^t \bmod q,$$

where t is any positive integer. We now look at the term $(x^{\Phi(n)})^t$ again:

$$(x^{\Phi(n)})^t \equiv (x^{(q-1)(p-1)})^t \equiv ((x^{\Phi(q)})^t)^{p-1} \equiv 1^{(p-1)} = 1 \bmod q.$$

Using the definition of the modulo operator, this is equivalent to:

$$(x^{\Phi(n)})^t = 1 + u \cdot q,$$

where u is some integer. We multiply this equation by x:

$$
\begin{aligned}
x \cdot (x^{\Phi(n)})^t &= x + x \cdot u \cdot q \\
&= x + (r \cdot p) \cdot u \cdot q \\
&= x + r \cdot u \cdot (p \cdot q) \\
&= x + r \cdot u \cdot n \\
x \cdot (x^{\Phi(n)})^t &\equiv x \ \mathrm{mod}\ n.
\end{aligned}
\tag{7.6}
$$

Inserting Eq. (7.6) into Eq. (7.4) yields the desired result:

$$d_{k_{pr}} = (x^{\Phi(n)})^t \cdot x \equiv x \ \mathrm{mod}\ n.$$

□

If this proof seems somewhat lengthy, please remember that the correctness of RSA is simply assured by Step 5 of the RSA key generation phase. The proof becomes simpler by using the Chinese Remainder Theorem which we have not introduced.

7.4 Encryption and Decryption: Fast Exponentiation

Unlike symmetric algorithms such as AES, DES or stream ciphers, public-key algorithms are based on arithmetic with very long numbers. Unless we pay close attention to how to realize the necessary computations, we can easily end up with schemes that are too slow for practical use. If we look at RSA encryption and decryption in Eqs. (7.1) and (7.2), we see that both are based on modular exponentiation. We restate both operations here for convenience:

$$
\begin{aligned}
y &= e_{k_{pub}}(x) \equiv x^e \ \mathrm{mod}\ n \quad \text{(encryption)} \\
x &= d_{k_{pr}}(y) \equiv y^d \ \mathrm{mod}\ n \quad \text{(decryption)}
\end{aligned}
$$

A straightforward way of exponentiation looks like this:

$$x \xrightarrow{SQ} x^2 \xrightarrow{MUL} x^3 \xrightarrow{MUL} x^4 \xrightarrow{MUL} x^5 \cdots$$

where SQ denotes squaring and MUL multiplication. Unfortunately, the exponents e and d are in general very large numbers. The exponents are typically chosen in the range of 1024–3072 bit or even larger. (The public exponent e is sometimes chosen to be a small value, but d is always very long.) Straightforward exponentiation as shown above would thus require around 2^{1024} or more multiplications. Since the number of atoms in the visible universe is estimated to be around 2^{300}, computing 2^{1024} multiplications to set up one secure session for our Web browser is not too tempting. The central question is whether there are considerably faster methods for exponentiation available. The answer is, luckily, yes. Otherwise we could forget about RSA and pretty much all other public-key cryptosystems in use today, since they all rely on exponentiation. One such method is the *square-and-multiply algorithm*. We first show a few illustrative examples with small numbers before presenting the actual algorithm.

Example 7.2. Let's look at how many multiplications are required to compute the simple exponentiation x^8. With the straightforward method:

$$x \xrightarrow{SQ} x^2 \xrightarrow{MUL} x^3 \xrightarrow{MUL} x^4 \xrightarrow{MUL} x^5 \xrightarrow{MUL} x^6 \xrightarrow{MUL} x^7 \xrightarrow{MUL} x^8$$

we need seven multiplications and squarings. Alternatively, we can do something faster:

$$x \xrightarrow{SQ} x^2 \xrightarrow{SQ} x^4 \xrightarrow{SQ} x^8$$

which requires only three squarings that are roughly as complex as a multiplication.
◇

This fast method works fine but is restricted to exponents that are powers of 2, i.e., values e and d of the form 2^i. Now the question is, whether we can extend the method to arbitrary exponents? Let us look at another example:

Example 7.3. This time we have the more general exponent 26, i.e., we want to compute x^{26}. Again, the naïve method would require 25 multiplications. A faster way is as follows:

$$x \xrightarrow{SQ} x^2 \xrightarrow{MUL} x^3 \xrightarrow{SQ} x^6 \xrightarrow{SQ} x^{12} \xrightarrow{MUL} x^{13} \xrightarrow{SQ} x^{26}.$$

This approach takes a total of six operations, two multiplications and four squarings.
◇

Looking at the last example, we see that we can achieve the desired result by performing two basic operations:

1. *squaring* the current result,
2. *multiplying* the current result by the base element x.

In the example above we computed the sequence SQ, MUL, SQ, SQ, MUL, SQ. However, we do not know the sequence in which the squarings and multiplications have to be performed for other exponents. One solution is the *square-and-multiply* algorithm. It provides a systematic way for finding the sequence in which we have to perform squarings and multiplications by x for computing x^H. Roughly speaking, the algorithm works as follows:

The algorithm is based on scanning the bit of the exponent from the left (the most significant bit) to the right (the least significant bit). In every iteration, i.e., for every exponent bit, the current result is squared. If and only if the currently

scanned exponent bit has the value 1, a multiplication of the current result by
x is executed following the squaring.

This seems like a simple if somewhat odd rule. For better understanding, let's revisit the example from above. This time, let's pay close attention to the exponent bits.

Example 7.4. We again consider the exponentiation x^{26}. For the square-and-multiply algorithm, the binary representation of the exponent is crucial:

$$x^{26} = x^{11010_2} = x^{(h_4 h_3 h_2 h_1 h_0)_2}.$$

The algorithm scans the exponent bits, starting on the left with h_4 and ending with the rightmost bit h_0.

Step
#0 $x = x^{1_2}$ initial setting, bit processed: $h_4 = 1$

#1a $(x^1)^2 = x^2 = x^{10_2}$ SQ, bit processed: h_3
#1b $x^2 \cdot x = x^3 = x^{10_2} x^{1_2} = x^{11_2}$ MUL, since $h_3 = 1$

#2a $(x^3)^2 = x^6 = (x^{11_2})^2 = x^{110_2}$ SQ, bit processed: h_2
#2b   no MUL, since $h_2 = 0$

#3a $(x^6)^2 = x^{12} = (x^{110_2})^2 = x^{1100_2}$ SQ, bit processed: h_1
#3b $x^{12} \cdot x = x^{13} = x^{1100_2} x^{1_2} = x^{1101_2}$ MUL, since $h_1 = 1$

#4a $(x^{13})^2 = x^{26} = (x^{1101_2})^2 = x^{11010_2}$ SQ, bit processed: h_0
#4b   no MUL, since $h_0 = 0$

To understand the algorithm it is helpful to closely observe how the binary representation of the exponent evolves. We see that the first basic operation, squaring, results in a left shift of the exponent, with a 0 put in the rightmost position. The other basic operation, multiplication by x, results in filling a 1 into the rightmost position of the exponent. Compare how the highlighted exponents change from iteration to iteration.

◇

Here is the pseudo code for the square-and-multiply algorithm:

Square-and-Multiply for Modular Exponentiation
Input:
base element x
exponent $H = \sum_{i=0}^{t} h_i 2^i$ with $h_i \in 0, 1$ and $h_t = 1$
and modulus n
Output: $x^H \bmod n$
Initialization: $r = x$
Algorithm:

1   FOR $i = t - 1$ DOWNTO 0
1.1 $r = r^2 \bmod n$
         IF $h_i = 1$
1.2 $r = r \cdot x \bmod n$
2   RETURN (r)

The modulo reduction is applied after each multiplication and squaring operation in order to keep the intermediate results small. It is helpful to compare this pseudo code with the verbal description of the algorithm above.

We determine now the complexity of the square-and-multiply algorithm for an exponent H with a bit length of $t + 1$, i.e., $\lceil \log_2 H \rceil = t + 1$. The number of squarings is independent of the actual value of H, but the number of multiplications is equal to the Hamming weight, i.e., the number of ones in its binary representation. Thus, we provide here the average number of multiplication, denoted by $\overline{MUL}$:

$$\#SQ = t$$
$$\#\overline{MUL} = 0.5t$$

Because the exponents used in cryptography have often good random properties, assuming that half of their bits have the value one is often a valid approximation.

Example 7.5. How many operations are required on average for an exponentiation with a 1024-bit exponent?

Straightforward exponentiation takes $2^{1024} \approx 10^{300}$ multiplications. That is completely impossible, no matter what computer resources we might have at hand. However, the square-and-multiply algorithm requires only

$$1.5 \cdot 1024 = 1536$$

squarings and multiplications on average. This is an impressive example for the difference of an algorithm with linear complexity (straightforward exponentiation) and logarithmic complexity (square-and-multiply algorithm). Remember, though, that each of the 1536 individual squarings and multiplications involves 1024-bit numbers. That means the number of integer operations on a CPU is much higher than 1536, but certainly doable on modern computers.

◇

7.5 Speed-up Techniques for RSA

As we learned in Sect. 7.4, RSA involves exponentiation with very long numbers. Even if the low-level arithmetic involving modular multiplication and squaring as well as the square-and-multiply algorithm are implemented carefully, performing a full RSA exponentiation with operands of length 1024 bit or beyond is computationally intensive. Thus, people have studied speed-up techniques for RSA since its invention. We introduce two of the most popular general acceleration techniques in the following.

7.5.1 Fast Encryption with Short Public Exponents

A surprisingly simple and very powerful trick can be used when RSA operations with the public key e are concerned. This is in practice encryption and, as we'll learn later, verification of an RSA digital signature. In this situation, the public key e can be chosen to be a very small value. In practice, the three values $e = 3$, $e = 17$ and $e = 2^{16} + 1$ are of particular importance. The resulting complexities when using these public keys are given in Table 7.1.

Table 7.1 Complexity of RSA exponentiation with short public exponents

Public key e	e as binary string	#MUL + #SQ
3	11_2	2
17	10001_2	5
$2^{16} + 1$	$1\,0000\,0000\,0000\,0001_2$	17

These complexities should be compared to the $1.5t$ multiplications and squarings that are required for exponents of full length. Here $t + 1$ is the bit length of the RSA modulus n, i.e., $\lceil \log_2 n \rceil = t + 1$. We note that all three exponents listed above have a low Hamming weight, i.e., number of ones in the binary representation. This results in a particularly low number of operations for performing an exponentiation. Interestingly, RSA is still secure if such short exponents are being used. Note that the private key d still has in general the full bit length $t + 1$ even though e is short.

An important consequence of the use of short public exponents is that encryption of a message and verification of an RSA signature is a very fast operation. In fact, for these two operations, RSA is in almost all practical cases the fastest public-key scheme available. Unfortunately, there is no such easy way to accelerate RSA when the private key d is involved, i.e., for decryption and signature generation. Hence, these two operations tend to be slow. Other public-key algorithms, in particular elliptic curves, are often much faster for these two operations. The following section shows how we can achieve a more moderate speed-up when using the private exponent d.

7.5.2 Fast Decryption with the Chinese Remainder Theorem

We cannot choose a short private key without compromising the security for RSA. If we were to select keys d as short as we did in the case of encryption in the section above, an attacker could simply brute-force all possible numbers up to a given bit length, i.e., 50 bit. But even if the numbers are larger, say 128 bit, there are key recovery attacks. In fact, it can be shown that the private key must have a length of at least $0.3t$ bit, where t is the bit length of the modulus n. In practice, e is often chosen short and d has full bit length. What one does instead is to apply a method which is based on the Chinese Remainder Theorem (CRT). We do not introduce the CRT itself here but merely how it applies to accelerate RSA decryption and signature generation.

Our goal is to perform the exponentiation $x^d \bmod n$ efficiently. First we note that the party who possesses the private key also knows the primes p and q. The basic idea of the CRT is that rather than doing arithmetic with one "long" modulus n, we do two individual exponentiations modulo the two "short" primes p and q. This is a type of transformation arithmetic. Like any transform, there are three steps: transforming into the CRT domain, computation in the CRT domain, and inverse transformation of the result. Those three steps are explained below.

Transformation of the Input into the CRT Domain

We simply reduce the base element x modulo the two factors p and q of the modulus n, and obtain what is called the modular representation of x.

$$x_p \equiv x \bmod p$$
$$x_q \equiv x \bmod q$$

Exponentiation in the CRT Domain

With the reduced versions of x we perform the following two exponentiations:

$$y_p = x_p^{d_p} \bmod p$$
$$y_q = x_q^{d_q} \bmod q$$

where the two new exponents are given by:

$$d_p \equiv d \bmod (p-1)$$
$$d_q \equiv d \bmod (q-1)$$

Note that both exponents in the transform domain, d_p and d_q, are bounded by p and q, respectively. The same holds for the transformed results y_p and y_q. Since the two primes are in practice chosen to have roughly the same bit length, the two exponents as well as y_p and y_q have about half the bit length of n.

Inverse Transformation into the Problem Domain

The remaining step is now to assemble the final result y from its modular representation (y_p, y_q). This follows from the CRT and can be done as:

$$y \equiv [q\,c_p]\,y_p + [p\,c_q]\,y_q \bmod n \qquad (7.7)$$

where the coefficients c_p and c_q are computed as:

$$c_p \equiv q^{-1} \bmod p, \qquad c_q \equiv p^{-1} \bmod q$$

Since the primes change very infrequently for a given RSA implementation, the two expressions in brackets in Eq. (7.7) can be precomputed. After the precomputations, the entire reverse transformation is achieved with merely two modular multiplications and one modular addition.

Before we consider the complexity of RSA with CRT, let's have a look at an example.

Example 7.6. Let the RSA parameters be given by:

$$
\begin{aligned}
p &= 11 & e &= 7 \\
q &= 13 & d &\equiv e^{-1} \equiv 103 \bmod 120 \\
n &= p \cdot q = 143
\end{aligned}
$$

We now compute an RSA decryption for the ciphertext $y = 15$ using the CRT, i.e., the value $y^d = 15^{103} \bmod 143$. In the first step, we compute the modular representation of y:

$$
\begin{aligned}
y_p &\equiv 15 \equiv 4 \ \bmod 11 \\
y_p &\equiv 15 \equiv 2 \ \bmod 13
\end{aligned}
$$

In the second step, we perform the exponentiation in the transform domain with the short exponents. These are:

$$
\begin{aligned}
d_p &\equiv 103 \equiv 3 \ \bmod 10 \\
d_q &\equiv 103 \equiv 7 \ \bmod 12
\end{aligned}
$$

Here are the exponentiations:

$$
\begin{aligned}
x_p &\equiv y_p^{d_p} = 4^3 = 64 \equiv 9 \bmod 11 \\
x_q &\equiv y_q^{d_q} = 2^7 = 128 \equiv 11 \bmod 13
\end{aligned}
$$

In the last step, we have to compute x from its modular representation (x_p, x_q). For this, we need the coefficients:

$$c_p = 13^{-1} \equiv 2^{-1} \equiv 6 \bmod 11 \qquad c_q = 11^{-1} \equiv 6 \bmod 13$$

The plaintext x follows now as:

$$x \equiv [qc_p]x_p + [pc_q]x_q \bmod n$$
$$x \equiv [13 \cdot 6]9 + [11 \cdot 6]11 \bmod 143$$
$$x \equiv 702 + 726 = 1428 \equiv 141 \bmod 143$$

◇

If you want to verify the result, you can compute $y^d \bmod 143$ using the square-and-multiply algorithm.

We will now establish the computational complexity of the CRT method. If we look at the three steps involved in the CRT-based exponentiation, we conclude that for a practical complexity analysis the transformation and inverse transformation can be ignored since the operations involved are negligible compared to the actual exponentiations in the transform domain. For convenience, we restate these CRT exponentiations here:

$$y_p = x_p^{d_p} \bmod p$$
$$y_q = x_q^{d_q} \bmod q$$

If we assume that n has $t + 1$ bit, both p and q are about $t/2$ bit long. All numbers involved in the CRT exponentiations, i.e., x_p, x_q, d_p and d_q, are bound in size by p and q, respectively, and thus also have a length of about $t/2$ bit. If we use the square-and-multiply algorithm for the two exponentiations, each requires on average approximately $1.5t/2$ modular multiplications and squarings. Together, the number of multiplications and squarings is thus:

$$\#SQ + \#MUL = 2 \cdot 1.5t/2 = 1.5t$$

This appears to be exactly the same computational complexity as regular exponentiation without the CRT. However, each multiplication and squaring involves numbers which have a length of only $t/2$ bit. This is in contrast to the operations without CRT, where each multiplication was performed with t-bit variables. Since the complexity of multiplication decreases quadratically with the bit length, each $t/2$-bit multiplication is four times faster than a t-bit multiplication.[1] Thus, *the total speed-up obtained through the CRT is a factor of 4*. This speed-up by four can be very valuable in practice. Since there are hardly any drawbacks involved, CRT-based exponentiations are used in many cryptographic products, e.g., for Web browser encryption. The method is also particularly valuable for implementations on smart

---

[1] The reason for the quadratic complexity is easy to see with the following example. If we multiply a 4-digit decimal number *abcd* by another number *wxyz*, we multiply each digit from the first operand with each digit of the second operand, for a total of $4^2 = 16$ digit multiplications. On the other hand, if we multiply two numbers with two digits, i.e., *ab* times *wx*, only $2^2 = 4$ elementary multiplications are needed.

cards, e.g., for banking applications, which are only equipped with a small micro-processor. Here, digital signing is often needed, which involves the secret key d. By applying the CRT for signature computation, the smart card is four times as fast. For example, if a regular 1024-bit RSA exponentiation takes 3 sec, using the CRT reduces that time to 0.75 sec. This acceleration might make the difference between a product with high customer acceptance (0.75 sec) and a product with a delay that is not acceptable for many applications (3 sec). This example is a good demonstration how basic number theory can have direct impact in the real world.

7.6 Finding Large Primes

There is one important practical aspect of RSA which we have not discussed yet: generating the primes p and q in Step 1 of the key generation. Since their product is the RSA modulus $n = p \cdot q$, the two primes should have about half the bit length of n. For instance, if we want to set up RSA with a modulus of length $\lceil \log_2 n \rceil =$ 1024, p and q should have a bit length of about 512 bit. The general approach is to generate integers at random which are then checked for primality, as depicted in Fig. 7.2, where RNG stands for random number generator. The RNG should be non predictable because if an attacker can compute or guess one of the two primes, RSA can be broken easily as we will see later in this chapter.

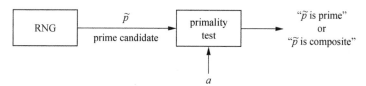

Fig. 7.2 Principal approach to generating primes for RSA

In order to make this approach work, we have to answer two questions:

1. How many random integers do we have to test before we have a prime? (If the likelihood of a prime is too small, it might take too long.)
2. How fast can we check whether a random integer is prime? (Again, if the test is too slow, the approach is impractical.)

It turns out that both steps are reasonably fast, as is discussed in the following.

7.6.1 How Common Are Primes?

Now we'll answer the question whether the likelihood that a randomly picked inte-ger p is a prime is sufficiently high. We know from looking at the first few positive integers that primes become less dense as the value increases:

$$2,3,5,7,11,13,17,19,23,29,31,37,\ldots$$

The question is whether there is still a reasonable chance that a random number with, say, 512 bit, is a prime. Luckily, this is the case. The chance that a randomly picked integer $\tilde{p}$ is a prime follows from the famous prime number theorem and is approximately $1/\ln(\tilde{p})$. In practice, we only test odd numbers so that the likelihood doubles. Thus, the probability for a random odd number $\tilde{p}$ to be prime is

$$P(\tilde{p} \text{ is prime}) \approx \frac{2}{\ln(\tilde{p})}.$$

In order to get a better feeling for what this probability means for RSA primes, let's look at an example:

Example 7.7. For RSA with a 1024-bit modulus n, the primes p and q each should have a length of about 512 bits, i.e., $p,q \approx 2^{512}$. The probability that a random odd number $\tilde{p}$ is a prime is

$$P(\tilde{p} \text{ is prime}) \approx \frac{2}{\ln(2^{512})} = \frac{2}{512 \ln(2)} \approx \frac{1}{177}.$$

This means that we expect to test 177 random numbers before we find one that is a prime.

◇

The likelihood of integers being primes decreases slowly, proportional to the bit length of the integer. This means that even for very long RSA parameters, say with 4096 bit, the density of primes is still sufficiently high.

7.6.2 Primality Tests

The other step we have to do is to decide whether the randomly generated integers $\tilde{p}$ are primes. A first idea could be to factor the number in question. However, for the numbers used in RSA, factorization is not possible since p and q are too large. (In fact, we especially choose numbers that cannot be factored because factoring n is the best known attack against RSA.) The situation is not hopeless, though. Remember that we are *not* interested in the factorization of $\tilde{p}$. Instead we merely need the statement whether the number being tested is a prime or not. It turns out that such primality tests are computationally much easier than factorization. Examples for primality tests are the Fermat test, the Miller–Rabin test or variants of them. We introduce primality test algorithms in this section.

Practical primality tests behave somewhat unusually: if the integer $\tilde{p}$ in question is being fed into a primality test algorithm, the answer is either

1. "$\tilde{p}$ is composite" (i.e., not a prime), which is always a true statement, or
2. "$\tilde{p}$ is prime", which is only true with a high probability.

If the algorithm output is "composite", the situation is clear: The integer in question is not a prime and can be discarded. If the output statement is "prime", $\tilde{p}$ is probably a prime. In rare cases, however, an integers prompts a "prime" statement but it *lies*, i.e., it yields an incorrect positive answer. There is way to deal with this behavior. Practical primality tests are *probabilistic algorithms*. That means they have a second parameter a as input which can be chosen at random. If a composite number $\tilde{p}$ together with a parameter a yields the incorrect statement "$\tilde{p}$ is prime", we repeat the test a second time with a different value for a. The general strategy is to test a prime candidate $\tilde{p}$ so often with several different random values a that the likelihood that the pair $(\tilde{p}, a)$ lies every single time is sufficiently small, say, less than 2^{-80}. Remember that as soon as the statement "$\tilde{p}$ is composite" occurs, we know for certain that $\tilde{p}$ is not a prime and we can discard it.

Fermat Primality Test

One primality test is based on Fermat's Little Theorem, Theorem (6.3.2).

Fermat Primality Test
Input: prime candidate $\tilde{p}$ and security parameter s
Output: statement "$\tilde{p}$ is composite" or "$\tilde{p}$ is likely prime"
Algorithm:

1 FOR $i = 1$ TO s
1.1   choose random $a \in \{2, 3, \ldots, \tilde{p} - 2\}$
1.2   IF $a^{\tilde{p}-1} \not\equiv 1$
1.3   RETURN ("$\tilde{p}$ is composite")
2   RETURN ("$\tilde{p}$ is likely prime")

The idea behind the test is that Fermat's theorem holds for all primes. Hence, if a number is found for which $a^{\tilde{p}-1} \not\equiv 1$ in Step 1.2, it is certainly not a prime. However, the reverse is not true. There could be composite numbers which in fact fulfill the condition $a^{\tilde{p}-1} \equiv 1$. In order to detect them, the algorithm is run s times with different values of a.

Unfortunately, there are certain composite integers which behave like primes in the Fermat test for many values of a. These are the *Carmichael numbers*. Given a Carmichael number C, the following expression holds for all integers a for which $\gcd(a, C) = 1$:

$$a^{C-1} \equiv 1 \bmod C$$

Such special composites are very rare. For instance, there exist approximately only $100,000$ Carmichael numbers below 10^{15}.

Example 7.8. Carmichael Number
$n = 561 = 3 \cdot 11 \cdot 17$ is a Carmichael number since

$$a^{560} \equiv 1 \bmod 561$$

for all $\gcd(a, 561) = 1$.

$\diamond$

If the prime factors of a Carmichael numbers are all large, there are only few bases a for which Fermat's test detects that the number is actually composite. For this reason, in practice the more powerful Miller–Rabin test is often used to generate RSA primes.

Miller–Rabin Primality Test

In contrast to Fermat's test, the Miller–Rabin test does not have any composite numbers for which a large number of base elements a yield the statement "prime". The test is based on the following theorem:

Theorem 7.6.1 *Given the decomposition of an odd prime candidate $\tilde{p}$*

$$\tilde{p} - 1 = 2^u r$$

where r is odd. If we can find an integer a such that

$$a^r \not\equiv 1 \bmod \tilde{p} \qquad and \qquad a^{r 2^j} \not\equiv \tilde{p} - 1 \bmod \tilde{p}$$

for all $j = \{0, 1, \ldots, u - 1\}$, then $\tilde{p}$ is composite. Otherwise, it is probably a prime.

We can turn this into an efficient primality test.

Miller–Rabin Primality Test
Input: prime candidate $\tilde{p}$ with $\tilde{p} - 1 = 2^u r$ and security parameter s
Output: statement "$\tilde{p}$ is composite" or "$\tilde{p}$ is likely prime"
Algorithm:

1   FOR $i = 1$ TO s
        choose random $a \in \{2, 3, \ldots, \tilde{p} - 2\}$
1.2 $z \equiv a^r \bmod \tilde{p}$
1.3   IF $z \not\equiv 1$ and $z \not\equiv \tilde{p} - 1$
1.4   FOR $j = 1$ TO $u - 1$
                $z \equiv z^2 \bmod \tilde{p}$
                IF $z = 1$
                    RETURN ("$\tilde{p}$ is composite")
1.5   IF $z \neq \tilde{p} - 1$
                RETURN ("$\tilde{p}$ is composite")
2   RETURN ("$\tilde{p}$ is likely prime")

Step 1.2 is computed by using the square-and-multiply algorithm. The IF statement in Step 1.3 tests the theorem for the case $j = 0$. The FOR loop 1.4 and the IF statement 1.5 test the right-hand side of the theorem for the values $j = 1, \ldots, u - 1$.

It can still happen that a composite number $\tilde{p}$ gives the incorrect statement "prime". However, the likelihood of this rapidly decreases as we run the test with several different random base elements a. The number of runs is given by the security parameter s in the Miller–Rabin test. Table 7.2 shows how many different values a must be chosen in order to have a probability of less than 2^{-80} that a composite is incorrectly detected as a prime.

Table 7.2 Number of runs within the Miller–Rabin primality test for an error probability of less than 2^{-80}

Bit lengths of $\tilde{p}$	Security parameter s
250	11
300	9
400	6
500	5
600	3

Example 7.9. Miller–Rabin Test
Let $\tilde{p} = 91$. Write $\tilde{p}$ as $\tilde{p} - 1 = 2^1 \cdot 45$. We select a security parameter of $s = 4$. Now, choose s times a random value a:

1. Let $a = 12$: $z = 12^{45} \equiv 90 \bmod 91$, hence, $\tilde{p}$ is likely prime.
2. Let $a = 17$: $z = 17^{45} \equiv 90 \bmod 91$, hence, $\tilde{p}$ is likely prime.
3. Let $a = 38$: $z = 38^{45} \equiv 90 \bmod 91$, hence, $\tilde{p}$ is likely prime.

4. Let $a = 39$: $z = 39^{45} \equiv 78 \bmod 91$, hence, $\tilde{p}$ is composite.

Since the numbers 12, 17 and 38 give incorrect statements for the prime candidate $\tilde{p} = 91$, they are called "liars for 91".

◇

7.7 RSA in Practice: Padding

What we described so far is the so-called "schoolbook RSA" system which has several weaknesses. In practice RSA has to be used with a padding scheme. Padding schemes are extremely important, and if not implemented properly, an RSA implementation may be insecure. The following properties of schoolbook RSA encryption are problematic:

■ RSA encryption is deterministic, i.e., for a specific key, a particular plaintext is always mapped to a particular ciphertext. An attacker can derive statistical properties of the plaintext from the ciphertext. Furthermore, given some pairs of plaintext–ciphertext, partial information can be derived from new ciphertexts which are encrypted with the same key.
■ Plaintext values $x = 0$, $x = 1$, or $x = -1$ produce ciphertexts equal to 0, 1, or -1.
■ Small public exponents e and small plaintexts x might be subject to attacks if no padding or weak padding is used. However, there is no known attack against small public exponents such as $e = 3$.

RSA has another undesirable property, namely that it is *malleable*. A crypto scheme is said to be malleable if the attacker Oscar is capable of transforming the ciphertext into another ciphertext which leads to a known transformation of the plaintext. Note that the attacker does not decrypt the ciphertext but is merely capable of manipulating the plaintext in a predictable manner. This is easily achieved in the case of RSA if the attacker replaces the ciphertext y by $s^e y$, where s is some integer. If the receiver decrypts the manipulated ciphertext, he computes:

$$(s^e y)^d \equiv s^{ed} x^{ed} \equiv s x \bmod n.$$

Even though Oscar is not able to decrypt the ciphertext, such targeted manipulations can still do harm. For instance, if x were an amount of money which is to be transferred or the value of a contract, by choosing $s = 2$ Oscar could exactly double the amount in a way that goes undetected by the receiver.

A possible solution to all these problems is the use of padding, which embeds a random structure into the plaintext before encryption and avoids the above mentioned problems. Modern techniques such as *Optimal Asymmetric Encryption Padding (OAEP)* for padding RSA messages are specified and standardized in Public Key Cryptography Standard #1 (PKCS #1).

Let M be the message to be padded, let k be the length of the modulus n in bytes, let $|H|$ be the length of the hash function output in bytes and let $|M|$ be the length of the message in bytes. A hash function computes a message digest of fixed length (e.g., 160 or 256 bit) for every input. More about hash functions is found in Chap. 11. Furthermore, let L be an optional label associated with the message (otherwise, L is an empty string as default). According to the most recent version PKCS#1 (v2.1), padding a message within the RSA encryption scheme is done in the following way:

1. Generate a string PS of length $k - |M| - 2|H| - 2$ of zeroed bytes. The length of PS may be zero.
2. Concatenate $Hash(L)$, PS, a single byte with hexadecimal value 0x01, and the message M to form a data block DB of length $k - |H| - 1$ bytes as

$$DB = Hash(L)||PS||0x01||M.$$

3. Generate a random byte string $seed$ of length $|H|$.
4. Let $dbMask = MGF(seed, k - |H| - 1)$, where MGF is the mask generation function. In practice, a hash function such as SHA-1 is often used as MFG.
5. Let $maskedDB = DB \oplus dbMask$.
6. Let $seedMask = MGF(maskedDB, |H|)$.
7. Let $maskedSeed = seed \oplus seedMask$.
8. Concatenate a single byte with hexadecimal value 0x00, $maskedSeed$ and $maskedDB$ to form an encoded message EM of length k bytes as

$$EM = 0x00||maskedSeed||maskedDB.$$

Figure 7.3 shows the structure of a padded message M.

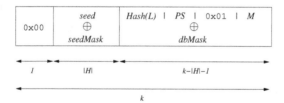

Fig. 7.3 RSA encryption of a message M with Optimal Asymmetric Encryption Padding (OAEP)

On the decryption side, the structure of the decrypted message has to be verified. For instance, if there is no byte with hexadecimal value 0x01 to separate PS from M, a decryption error occurred. In any case, returning a decryption error to the user (or a potential attacker!) should not reveal any information about the plaintext.

7.8 Attacks

There have been numerous attacks proposed against RSA since it was invented in 1977. None of the attacks are serious, and moreover, they typically exploit weaknesses in the way RSA is implemented or used rather than the RSA algorithm itself. There are three general attack families against RSA:

1. Protocol attacks
2. Mathematical attacks
3. Side-channel attacks

We comment on each of them in the following.

Protocol Attacks

Protocol attacks exploit weaknesses in the way RSA is being used. There have been several protocol attacks over the years. Among the better known ones are the attacks that exploit the malleability of RSA, which was introduced in the previous section. Many of them can be avoided by using padding. Modern security standards describe exactly how RSA should be used, and if one follows those guidelines, protocol attacks should not be possible.

Mathematical Attacks

The best mathematical cryptanalytical method we know is factoring the modulus. An attacker, Oscar, knows the modulus n, the public key e and the ciphertext y. His goal is to compute the private key d which has the property that $e \cdot d \equiv \mod \Phi(n)$. It seems that he could simply apply the extended Euclidean algorithm and compute d. However, he does not know the value of $\Phi(n)$. At this point factoring comes in: the best way to obtain this value is to decompose n into its primes p and q. If Oscar can do this, the attack succeeds in three steps:

$$\Phi(n) = (p-1)(q-1)$$
$$d^{-1} \equiv e \mod \Phi(n)$$
$$x \equiv y^d \mod n.$$

In order to prevent this attack, the modulus must be sufficiently large. This is the sole reason why moduli of 1024 or more bit are needed for a RSA. The proposal of the RSA scheme in 1977 sparked much interest in the old problem of integer factorization. In fact, the major progress that has been made in factorization in the last three decades would most likely not have happened if it weren't for RSA. Table 7.3 shows a summary of the RSA factoring records that have occurred since the beginning of the 1990s. These advances have been possible mainly due to improvements in factoring algorithms, and to a lesser extent due to improved computer technology.

Even though factoring has become easier than the RSA designers had assumed 30 years ago, factoring RSA moduli beyond a certain size still is out of reach.

Table 7.3 Summary of RSA factoring records since 1991

Decimal digits	Bit length	Date
100	330	April 1991
110	364	April 1992
120	397	June 1993
129	426	April 1994
140	463	February 1999
155	512	August 1999
200	664	May 2005

Of historical interest is the 129-digit modulus which was published in a column by Martin Gardner in *Scientific American* in 1997. It was estimated that the best factoring algorithms of that time would take 40 trillion $(4 \cdot 10^{13})$ years. However, factoring methods improved considerably, particularly during the 1980s and 1990s, and it took in fact less than 30 years.

Which exact length the RSA modulus should have is the topic of much discussion. Until recently, many RSA applications used a bit length of 1024 bits as default. Today it is believed that it might be possible to factor 1024-bit numbers within a period of about 10–15 years, and intelligence organizations might be capable of doing it possibly even earlier. Hence, it is recommended to choose RSA parameters in the range of 2048–4096 bits for long-term security.

Side-Channel Attacks

A third and entirely different family of attacks are side-channel attacks. They exploit information about the private key which is leaked through physical channels such as the power consumption or the timing behavior. In order to observe such channels, an attacker must typically have direct access to the RSA implementation, e.g., in a cell phone or a smart card. Even though side-channel attacks are a large and active field of research in modern cryptography and beyond the scope of this book, we show one particularly impressive such attack against RSA in the following.

Figure 7.4 shows the power trace of an RSA implementation on a microprocessor. More precisely, it shows the electric current drawn by the processor over time. Our goal is to extract the private key d which is used during the RSA decryption. We clearly see intervals of high activity between short periods of less activity. Since the main computational load of RSA is the squarings and multiplication during the exponentiation, we conclude that the high-activity intervals correspond to those two operations. If we look more closely at the power trace, we see that there are high activity intervals which are short and others which are longer. In fact, the longer ones appear to be about twice as long. This behavior is explained by the square-and-multiply algorithm. If an exponent bit has the value 0, only a squaring is performed. If an exponent bit has the value 1, a squaring together with a multiplication is computed. But this timing behavior reveals immediately the key: A long period of activity corresponds to the bit value 1 of the secret key, and a short period to a key bit with value 0. As shown in the figure, by simply looking at the power trace we can identify the secret exponent. Thus we can learn the following 12 bits of the private key by looking at the trace:

$$\text{operations:} \quad S \; SM \; SM \; S \; SM \; S \; S \; SM \; SM \; SM \; S \; SM$$
$$\text{private key:} \quad 0 \quad 1 \quad 1 \quad 0 \quad 1 \quad 0 \; 0 \quad 1 \quad 1 \quad 1 \; 0 \quad 1$$

Obviously, in real-life we can also find all 1024 or 2048 bits of a full private key. During the short periods with low activity, the square-and-multiply algorithm scans and processes the exponent bits before it triggers the next squaring or squaring-and-multiplication sequence.

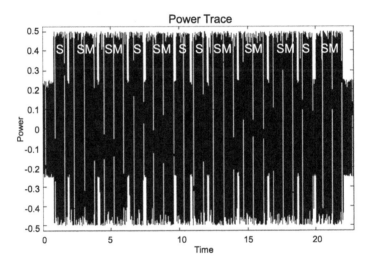

Fig. 7.4 The power trace of an RSA implementation

This specific attack is classified as simple power analysis or SPA. There are several countermeasures available to prevent the attack. A simple one is to execute a multiplication with dummy variables after a squaring that corresponds to an exponent bit 0. This results in a power profile (and a run time) which is independent of the exponent. However, countermeasures against more advanced side-channel attacks are not as straightforward.

7.9 Implementation in Software and Hardware

RSA is the prime example (almost literally) for a public-key algorithm that is very computationally intensive. Hence, the implementation of public-key algorithms is much more crucial than that of symmetric ciphers like 3DES and AES, which are significantly faster. In order to get an appreciation for the computational load, we develop a rough estimate for the number of integer multiplications needed for an RSA operation.

We assume a 2048-bit RSA modulus. For decryption we need on average 3072 squaring and multiplications, each of which involves 2048-bit operands. Let's assume a 32-bit CPU so that each operand is represented by $2048/32 = 64$ registers. A single long-number multiplication requires now $64^2 = 4096$ integer multiplications since we have to multiply every register of the first operand with every register of the second operand. In addition, we have to modulo reduce each of these multiplications. The best algorithms for doing this also require roughly $64^2 = 4096$ integer multiplications. Thus, in total, the CPU has to perform about $4096 + 4096 = 8192$ integer multiplications for a single long-number multiplication. Since we have 3072 of these, the number of integer multiplications for one decryption is:

$$\#(\text{32-bit mult}) = 3072 \times 8192 = 25,165,824$$

Of course, using a smaller modulus results in fewer operations, but given that integer multiplications are among the most costly operations on current CPUs, it is probably clear that the computational demand is quite impressive. Note that most other public key schemes have a similar complexity.

The extremely high computational demand of RSA was, in fact, a serious hindrance to its adoption in practice after it had been invented. Doing hundreds of thousands of integer multiplications was out of question with 1970s-style computers. The only option for RSA implementations with an acceptable run time was to realize RSA on special hardware chips until the mid- to late 1980s. Even the RSA inventors investigated hardware architecture in the early days of the algorithm. Since then much research has focused on ways to quickly perform modular integer arithmetic. Given the enormous capabilities of state-of-the-art VLSI chips, an RSA operation can today be done in the range of 100 μs on high-speed hardware.

Similarly, due to Moore's Law, RSA implementations in software have become possible since the late 1980s. Today, a typical decryption operation on a 2 GHz CPU takes around 10 ms for 2048-bit RSA. Even though this is sufficient for many PC applications, the throughput is about $100 \times 2048 = 204,800$ bit/s if one uses RSA for encryption of large amounts of data. This is quite slow compared to the speed of many of today's networks. For this reason RSA and other public-key algorithms are not used for bulk data encryption. Rather, symmetric algorithms are used that are often faster by a factor of 1000 or so.

7.10 Discussion and Further Reading

RSA and Variants The RSA cryptosystem is widely used in practice and is well standardized in bodies such as PKCS#1 [149]. Over the years several variants have been proposed. One generalization is to use a modulus which is composed of more than two primes. Also proposed have been multipower moduli of the form $n = p^2 q$ [162] as well as multifactor ones where $n = p q r$ [45]. In both cases speed-ups by a factor of approximately 2–3 are possible.

There are also several other crypto schemes which are based on the integer factorization problem. A prominent one is the Rabin scheme [140]. In contrast to RSA, it can be shown that the Rabin scheme is equivalent to factoring. Thus, it is said that the cryptosystem is *provable secure*. Other schemes which rely on the hardness of integer factorization include the probabilistic encryption scheme by Blum–Goldwasser [28] and the Blum Blum Shub pseudo-random number generator [27]. The *Handbook of Applied Cryptography* [120] describes all the schemes mentioned in a coherent form.

Implementation The actual performance of an RSA implementation heavily depends on the efficiency of the arithmetic used. Generally speaking, speed-ups are possible at two levels. On the higher level, improvements of the square-and-multiply algorithm are an option. One of the fastest methods is the sliding window exponentiation which gives an improvement of about 25% over the square-and-multiply algorithm. A good compilation of exponentiation methods is given in [120, Chap. 14]. On the lower layer, modular multiplication and squaring with long numbers can be improved. One set of techniques deals with efficient algorithms for modular reduction. In practice, Montgomery reduction is the most popular choice; see [41] for a good treatment of software techniques and [72] for hardware. Several alternatives to the Montgomery method have also been proposed over the years [123]; [120, Chap. 14]. Another angle to accelerate long number arithmetic is to apply fast multiplication methods. Spectral techniques such as the fast Fourier transform (FFT) are usually not applicable because the operands are still too short, but methods such as the Karatsuba algorithm [99] are very useful. Reference [17] gives a comprehensive but fairly mathematical treatment of the area of multiplication algorithms, and [172] describes the Karatsuba method from a practical viewpoint.

Attacks Breaking RSA analytically has been a subject of intense investigation for the last 30 years. Especially during the 1980s, major progress in factorization algorithms was made, which was not in small part motivated by RSA. There have been numerous other attempts to mathematically break RSA, including attacks against short private exponents. A good survey is given in [32]. More recently, proposals have been made to build special computers whose sole purpose is to break RSA. Proposals include an optoelectronic factoring machine [151] and several other architectures based on conventional semiconductor technology [152, 79].

Side channel attacks have been systematically studied in academia and industry since the mid- to late 1990s. RSA, as well as most other symmetric and asymmetric schemes, are vulnerable against differential power analysis (DPA), which is more powerful than the simple power analysis (SPA) shown in this section. On the other hand, numerous countermeasures against DPA are known. Good references are *The Side Channel Cryptanalysis Lounge* [70] and the excellent book on DPA [113]. Related implementation-based attacks are *fault injection attacks* and *timing attacks*. It is important to stress that a cryptosystem can be mathematically very strong but still be vulnerable to side-channel attacks.

7.11 Lessons Learned

- RSA is the most widely used public-key cryptosystem. In the future, elliptic curve cryptosystems will probably catch up in popularity.
- RSA is mainly used for key transport (i.e., encryption of keys) and digital signatures.
- The public key e can be a short integer. The private key d needs to have the full length of the modulus. Hence, encryption can be significantly faster than decryption.
- RSA relies on the integer factorization problem. Currently, 1024-bit (about 310 decimal digits) numbers cannot be factored. Progress in factorization algorithms and factorization hardware is hard to predict. It is advisable to use RSA with a 2048-bit modulus if one needs reasonable long-term security, especially with respect to extremely well funded attackers.
- "Schoolbook RSA" allows several attacks, and in practice RSA should be used together with padding.

Problems

7.1. Let the two primes $p = 41$ and $q = 17$ be given as set-up parameters for RSA.

1. Which of the parameters $e_1 = 32, e_2 = 49$ is a valid RSA exponent? Justify your choice.
2. Compute the corresponding private key $K_{pr} = (p, q, d)$. Use the extended Euclidean algorithm for the inversion and point out every calculation step.

7.2. Computing modular exponentiation efficiently is inevitable for the practicability of RSA. Compute the following exponentiations $x^e \bmod m$ applying the square-and-multiply algorithm:

1. $x = 2$, $e = 79$, $m = 101$
2. $x = 3$, $e = 197$, $m = 101$

After every iteration step, show the exponent of the intermediate result in binary notation.

7.3. Encrypt and decrypt by means of the RSA algorithm with the following system parameters:

1. $p = 3$, $q = 11$, $d = 7$, $x = 5$
2. $p = 5$, $q = 11$, $e = 3$, $x = 9$

Only use a pocket calculator at this stage.

7.4. One major drawback of public-key algorithms is that they are relatively slow. In Sect. 7.5.1 we learned that an acceleration technique is to use short exponents e. Now we study short exponents in this problem in more detail.

1. Assume that in an implementation of the RSA cryptosystem one modular squaring takes 75% of the time of a modular multiplication. How much quicker is one encryption on average if instead of a 2048-bit public key the short exponent $e = 2^{16} + 1$ is used? Assume that the square-and-multiply algorithm is being used in both cases.
2. Most short exponents are of the form $e = 2^n + 1$. Would it be advantageous to use exponents of the form $2^n - 1$? Justify your answer.
3. Compute the exponentiation $x^e \bmod 29$ of $x = 5$ with both variants of e from above for $n = 4$. Use the square-and-multiply algorithm and show each step of your computation.

7.5. In practice the short exponents $e = 3$, 17 and $2^{16} + 1$ are widely used.

1. Why can't we use these three short exponents as values for the exponent d in applications where we want to accelerate decryption?
2. Suggest a minimum bit length for the exponent d and explain your answer.

7.6. Verify the RSA with CRT example in the chapter by computing $y^d = 15^{103} \bmod 143$ using the square-and-multiply algorithm.

7.7. An RSA encryption scheme has the set-up parameters $p = 31$ and $q = 37$. The public key is $e = 17$.

1. Decrypt the ciphertext $y = 2$ using the CRT.
2. Verify your result by encrypting the plaintext without using the CRT.

7.8. Popular RSA modulus sizes are 1024, 2048, 3072 and 4092 bit.

1. How many random odd integers do we have to test on average until we expect to find one that is a prime?
2. Derive a simple formula for any arbitrary RSA modulus size.

7.9. One of the most attractive applications of public-key algorithms is the establishment of a secure session key for a private-key algorithm such as AES over an insecure channel.

Assume Bob has a pair of public/private keys for the RSA cryptosystem. Develop a simple protocol using RSA which allows the two parties Alice and Bob to agree on a shared secret key. Who determines the key in this protocol, Alice, Bob, or both?

7.10. In practice, it is sometimes desirable that both communication parties influence the selection of the session key. For instance, this prevents the other party from choosing a key which is a *weak key* for a symmetric algorithm. Many block ciphers such as DES and IDEA have weak keys. Messages encrypted with weak keys can be recovered relatively easily from the ciphertext.

Develop a protocol similar to the one above in which both parties influence the key. Assume that both Alice and Bob have a pair of public/private keys for the RSA cryptosystem. Please note that there are several valid approaches to this problem. Show just one.

7.11. In this exercise, you are asked to attack an RSA encrypted message. Imagine being the attacker: You obtain the ciphertext $y = 1141$ by eavesdropping on a certain connection. The public key is $k_{pub} = (n, e) = (2623, 2111)$.

1. Consider the encryption formula. All variables except the plaintext x are known. Why can't you simply solve the equation for x?
2. In order to determine the private key d, you have to calculate $d \equiv e^{-1} \mod \Phi(n)$. There is an efficient expression for calculating $\Phi(n)$. Can we use this formula here?
3. Calculate the plaintext x by computing the private key d through factoring $n = p \cdot q$. Does this approach remain suitable for numbers with a length of 1024 bit or more?

7.12. We now show how an attack with chosen ciphertext can be used to break an RSA encryption.

1. Show that the *multiplicative property* holds for RSA, i.e., show that the product of two ciphertexts is equal to the encryption of the product of the two respective plaintexts.

2. This property can under certain circumstances lead to an attack. Assume that Bob first receives an encrypted message y_1 from Alice which Oscar obtains by eavesdropping. At a later point in time, we assume that Oscar can send an innocent looking ciphertext y_2 to Bob, and that Oscar can obtain the decryption of y_2. In practice this could, for instance, happen if Oscar manages to hack into Bob's system such that he can get access to decrypted plaintext for a limited period of time.

7.13. In this exercise, we illustrate the problem of using nonprobabilistic cryptosystems, such as schoolbook RSA, imprudently. Nonprobabilistic means that the same sequence of plaintext letters maps to the same ciphertext. This allows traffic analysis (i.e., to draw some conclusion about the cleartext by merely observing the ciphertext) and in some cases even to the total break of the cryptoystem. The latter holds especially if the number of possible plaintexts is small. Suppose the following situation:

Alice wants to send a message to Bob encrypted with his public key pair (n, e). Therefore, she decides to use the ASCII table to assign a number to each character ($Space \rightarrow 32$, $! \rightarrow 33$, ..., $A \rightarrow 65$, $B \rightarrow 66$, ..., $\sim \rightarrow 126$) and to encrypt them separately.

1. Oscar eavesdrops on the transferred ciphertext. Describe how he can successfully decrypt the message by exploiting the nonprobabilistic property of RSA.
2. Bob's RSA public key is $(n, e) = (3763, 11)$. Decrypt the ciphertext

$$y = 2514, 1125, 333, 3696, 2514, 2929, 3368, 2514$$

with the attack proposed in 1. For simplification, assume that Alice only chose capital letters A–Z during the encryption.
3. Is the attack still possible if we use the OAEP padding? Exactly explain your answer.

7.14. The modulus of RSA has been enlarged over the years in order to thwart improved attacks. As one would assume, public-key algorithms become slower as the modulus length increases. We study the relation between modulus length and performance in this problem. The performance of RSA, and of almost any other public-key algorithm, is dependent on how fast modulo exponentiation with large numbers can be performed.

1. Assume that one modulo multiplication or squaring with k-bit numbers takes $c \cdot k^2$ clock cycles, where c is a constant. How much slower is RSA encryption/decryption with 1024 bits compared to RSA with 512 bits on average? Only consider the encryption/decryption itself with an exponent of full length and the square-and-multiply algorithm.
2. In practice, the Karatsuba algorithm, which has an asymptotical complexity that is proportional to $k^{\log_2 3}$, is often used for long number multiplication in cryptography. Assume that this more advanced technique requires $c' \cdot k^{\log_2 3} = c' \cdot k^{1.585}$ clock cycles for multiplication or squaring where c' is a constant. What is the

ratio between RSA encryption with 1024 bit and RSA with 512 bit if the Karat-suba algorithm is used in both cases? Again, assume that full-length exponents are being used.

7.15. (Advanced problem!) There are ways to improve the square-and-multiply algorithm, that is, to reduce the number of operations required. Although the number of squarings is fixed, the number of multiplications can be reduced. Your task is to come up with a modified version of the square-and-multiply algorithm which requires fewer multiplications. Give a detailed description of how the new algorithm works and what the complexity is (number of operations).

Hint: Try to develop a generalization of the square-and-multiply algorithm which processes more than one bit at a time. The basic idea is to handle k (e.g., $k = 3$) exponent bit per iteration rather than one bit in the original square-and-multiply algorithm.

7.16. Let us now investigate side-channel attacks against RSA. In a simple implementation of RSA without any countermeasures against side-channel leakage, the analysis of the current consumption of the microcontroller in the decryption part directly yields the private exponent. Figure 7.5 shows the power consumption of an implementation of the square-and-multiply algorithm. If the microcontroller computes a squaring or a multiplication, the power consumption increases. Due to the small intervals in between the loops, every iteration can be identified. Furthermore, for each round we can identify whether a single squaring (short duration) or a squaring followed by a multiplication (long duration) is being computed.

1. Identify the respective rounds in the figure and mark these with S for squaring or SM for squaring and multiplication.
2. Assume the square-and-multiply algorithm has been implemented such that the exponent is being scanned from left to right. Furthermore, assume that the starting values have been initialized. What is the private exponent d?
3. This key belongs to the RSA setup with the primes $p = 67$ and $q = 103$ and $e = 257$. Verify your result. (Note that in practice an attacker wouldn't know the values of p and q.)

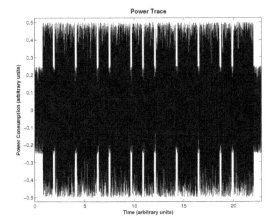

Fig. 7.5 Power consumption of an RSA decryption

Chapter 8
Public-Key Cryptosystems Based on the Discrete Logarithm Problem

In the previous chapter we learned about the RSA public-key scheme. As we have seen, RSA is based on the hardness of factoring large integers. The integer factorization problem is said to be the *one-way function* of RSA. As we saw earlier, roughly speaking a function is *one-way* if it is computationally easy to compute the function $f(x) = y$, but computationally infeasible to invert the function: $f^{-1}(y) = x$. The question is whether we can find other one-way functions for building asymmetric crypto schemes. It turns out that most non-RSA public-key algorithms with practical relevance are based on another one-way function, the discrete logarithm problem.

In this chapter you will learn:

■ The Diffie–Hellman key exchange
■ Cyclic groups which are important for a deeper understanding of Diffie–Hellman key exchange
■ The discrete logarithm problem, which is of fundamental importance for many practical public-key algorithms
■ Encryption using the Elgamal scheme

The security of many cryptographic schemes relies on the computational intractability of finding solutions to the *Discrete Logarithm Problem (DLP)*. Well-known examples of such schemes are the Diffie–Hellman key exchange and the Elgamal encryption scheme, both of which will be introduced in this chapter. Also, the Elgamal digital signature scheme (cf. Section 8.5.1) and the digital signature algorithm (cf. Section 10.2) are based on the DLP, as are cryptosystems based on elliptic curves (Section 9.3).

We start with the basic Diffie–Hellman protocol, which is surprisingly simple and powerful. The discrete logarithm problem is defined in what are called *cyclic groups*. The concept of this algebraic structure is introduced in Section 8.2. A formal definition of the DLP as well as some illustrating examples are provided, followed by a brief description of attack algorithms for the DLP. With this knowledge we will revisit the Diffie–Hellman protocol and more formally talk about its security. We will then develop a method for encrypting data using the DLP that is known as the Elgamal cryptosystem.

8.1 Diffie–Hellman Key Exchange

The *Diffie–Hellman key exchange (DHKE)*, proposed by Whitfield Diffie and Martin Hellman in 1976 [58], was the first asymmetric scheme published in the open literature. The two inventors were also influenced by the work of Ralph Merkle. It provides a practical solution to the key distribution problem, i.e., it enables two parties to derive a common secret key by communicating over an insecure channel[1]. The DHKE is a very impressive application of the discrete logarithm problem that we'll study in the subsequent sections. This fundamental key agreement technique is implemented in many open and commercial cryptographic protocols like Secure Shell (SSH), Transport Layer Security (TLS), and Internet Protocol Security (IPSec).The basic idea behind the DHKE is that exponentiation in $\mathbb{Z}_p^*$, p prime, is a one-way function and that exponentiation is commutative, i.e.,

$$k = (\alpha^x)^y \equiv (\alpha^y)^x \bmod p$$

The value $k \equiv (\alpha^x)^y \equiv (\alpha^y)^x \bmod p$ is the joint secret which can be used as the session key between the two parties.

Let us now consider how the Diffie–Hellman key exchange protocol over $\mathbb{Z}_p^*$ works. In this protocol we have two parties, Alice and Bob, who would like to establish a shared secret key. There is possibly a trusted third party that properly chooses the public parameters which are needed for the key exchange. However, it is also possible that Alice or Bob generate the public parameters. Strictly speaking, the DHKE consists of two protocols, the set-up protocol and the main protocol, which performs the actual key exchange. The set-up protocol consists of the following steps:

Diffie–Hellman Set-up

1. Choose a large prime p.
2. Choose an integer $\alpha \in \{2, 3, \ldots, p-2\}$.
3. Publish p and α.

These two values are sometimes referred to as *domain parameters*. If Alice and Bob both know the public parameters p and α computed in the set-up phase, they can generate a joint secret key k with the following key-exchange protocol:

---

[1] The channel needs to be authenticated, but that will be discussed later in this book.

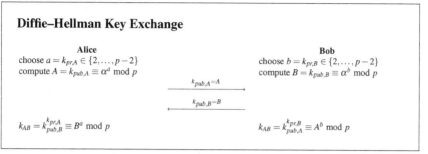

Here is the proof that this surprisingly simple protocol is correct, i.e., that Alice and Bob in fact compute the same session key k_{AB}.

Proof. Alice computes

$$B^a \equiv (\alpha^b)^a \equiv \alpha^{ab} \bmod p$$

while Bob computes

$$A^b \equiv (\alpha^a)^b \equiv \alpha^{ab} \bmod p$$

and thus Alice and Bob both share the session key $k_{AB} \equiv \alpha^{ab} \bmod p$. The key can now be used to establish a secure communication between Alice and Bob, e.g., by using k_{AB} as key for a symmetric algorithm like AES or 3DES. □

We'll look now at a simple example with small numbers.

Example 8.1. The Diffie–Hellman domain parameters are $p = 29$ and $\alpha = 2$. The protocol proceeds as follows:

Alice		**Bob**
choose $a = k_{pr,A} = 5$		choose $b = k_{pr,B} = 12$
$A = k_{pub,A} = 2^5 \equiv 3 \bmod 29$		$B = k_{pub,B} = 2^{12} \equiv 7 \bmod 29$
	$\xrightarrow{\quad A=3 \quad}$	
	$\xleftarrow{\quad B=7 \quad}$	
$k_{AB} = B^a \equiv 7^5 = 16 \bmod 29$		$k_{AB} = A^b = 3^{12} \equiv 16 \bmod 29$

As one can see, both parties compute the value $k_{AB} = 16$, which can be used as a joint secret, e.g., as a session key for symmetric encryption.
◇

The computational aspects of the DHKE are quite similar to those of RSA. During the set-up phase, we generate p using the probabilistic prime-finding algorithms discussed in Section 7.6. As shown in Table 6.1, p should have a similar length as the RSA modulus n, i.e., 1024 or beyond, in order to provide strong security. The integer α needs to have a special property: It should be a primitive element, a topic which we discuss in the following sections. The session key k_{AB} that is being computed in the protocol has the same bit length as p. If we want to use it as a symmetric key for algorithms such as AES, we can simply take the 128 most significant bits. Alternatively, a hash function is sometimes applied to k_{AB} and the output is then used as a symmetric key.

During the actual protocol, we first have to choose the private keys a and b. They should stem from a true random generator in order to prevent an attacker from guessing them. For computing the public keys A and B as well as for computing the session key, both parties can make use of the square-and-multiply algorithm. The public keys are typically precomputed. The main computation that needs to be done for a key exchange is thus the exponentiation for the session key. In general, since the bit lengths and the computations of RSA and the DHKE are very similar, they require a similar effort. However, the trick of using short public exponents that was shown in Section 7.5 is not applicable to the DHKE.

What we showed so far is the classic Diffie–Hellman key exchange protocol in the group $\mathbb{Z}_p^*$, where p is a prime. The protocol can be generalized, in particular to groups of elliptic curves. This gives rise to elliptic curve cryptography, which has become a very popular asymmetric scheme in practice. In order to better understand elliptic curves and schemes such as Elgamal encryption, which are also closely related to the DHKE, we introduce the discrete logarithm problem in the following sections. This problem is the mathematical basis for the DHKE. After we have introduced the discrete logarithm problem, we will revisit the DHKE and discuss its security.

8.2 Some Algebra

This section introduces some fundamentals of abstract algebra, in particular the notion of groups, subgroups, finite groups and cyclic groups, which are essential for understanding discrete logarithm public-key algorithms.

8.2.1 Groups

For convenience, we restate here the definition of groups which was introduced in the Chapter 4:

> **Definition 8.2.1** Group
>
> A group *is a set of elements G together with an operation ∘ which combines two elements of G. A group has the following properties.*
>
> 1. *The group operation ∘ is* closed. *That is, for all $a, b, \in G$, it holds that $a \circ b = c \in G$.*
> 2. *The group operation is* associative. *That is, $a \circ (b \circ c) = (a \circ b) \circ c$ for all $a, b, c \in G$.*
> 3. *There is an element $1 \in G$, called the* neutral element *(or* identity element*), such that $a \circ 1 = 1 \circ a = a$ for all $a \in G$.*
> 4. *For each $a \in G$ there exists an element $a^{-1} \in G$, called the* inverse *of a, such that $a \circ a^{-1} = a^{-1} \circ a = 1$.*
> 5. *A group G is* abelian (or commutative) *if, furthermore, $a \circ b = b \circ a$ for all $a, b \in G$.*

Note that in cryptography we use both multiplicative groups, i.e., the operation "∘" denotes multiplication, and additive groups where "∘" denotes addition. The latter notation is used for elliptic curves as we'll see later.

Example 8.2. To illustrate the definition of groups we consider the following examples.

- $(\mathbb{Z}, +)$ is a group, i.e., the set of integers $\mathbb{Z} = \{\ldots, -2, -1, 0, 1, 2, \ldots\}$ together with the usual addition forms an abelian group, where $e = 0$ is the identity element and $-a$ is the inverse of an element $a \in \mathbb{Z}$.
- $(\mathbb{Z} \text{ without } 0, \cdot)$ is **not** a group, i.e., the set of integers $\mathbb{Z}$ (without the element 0) and the usual multiplication does not form a group since there exists no inverse a^{-1} for an element $a \in \mathbb{Z}$ with the exception of the elements -1 and 1.
- $(\mathbb{C}, \cdot)$ is a group, i.e., the set of complex numbers $u + iv$ with $u, v \in \mathbb{R}$ and $i^2 = -1$ together with the complex multiplication defined by

$$(u_1 + iv_1) \cdot (u_2 + iv_2) = (u_1 u_2 - v_1 v_2) + i(u_1 v_2 + v_1 u_2)$$

forms an abelian group. The identity element of this group is $e = 1$, and the inverse a^{-1} of an element $a = u + iv \in \mathbb{C}$ is given by $a^{-1} = (u - i)/(u^2 + v^2)$.

◇

However, all of these groups do not play a significant role in cryptography because we need groups with a finite number of elements. Let us now consider the group $\mathbb{Z}_n^*$ which is very important for many cryptographic schemes such as DHKE, Elgamal encryption, digital signature algorithm and many others.

> **Theorem 8.2.1**
> The set $\mathbb{Z}_n^*$ which consists of all integers $i = 0, 1, \ldots, n-1$ for which
> $\gcd(i, n) = 1$ forms an abelian group under multiplication modulo
> n. The identity element is $e = 1$.

Let us verify the validity of the theorem by considering the following example:

Example 8.3. If we choose $n = 9$, $\mathbb{Z}_n^*$ consists of the elements $\{1, 2, 4, 5, 7, 8\}$.

Table 8.1 Multiplication table for $\mathbb{Z}_9^*$

$\times \bmod 9$	1	2	4	5	7	8
1	1	2	4	5	7	8
2	2	4	8	1	5	7
4	4	8	7	2	1	5
5	5	1	2	7	8	4
7	7	5	1	8	4	2
8	8	7	5	4	2	1

By computing the *multiplication table* for $\mathbb{Z}_9^*$, depicted in Table 8.1, we can easily check most conditions from Definition 8.2.1. Condition 1 (closure) is satisfied since the table only consists of integers which are elements of $\mathbb{Z}_9^*$. For this group Conditions 3 (identity) and 4 (inverse) also hold since each row and each column of the table is a permutation of the elements of $\mathbb{Z}_9^*$. From the symmetry along the main diagonal, i.e., the element at row i and column j equals the element at row j and column i, we can see that Condition 5 (commutativity) is satisfied. Condition 2 (associativity) cannot be directly derived from the shape of the table but follows immediately from the associativity of the usual multiplication in $\mathbb{Z}_n$.

◇

Finally, the reader should remember from Section 6.3.1 that the inverse a^{-1} of each element $a \in \mathbb{Z}_n^*$ can be computed by using the extended Euclidean algorithm.

8.2.2 Cyclic Groups

In cryptography we are almost always concerned with *finite* structures. For instance, for AES we needed a finite field. We provide now the straightforward definition of a finite group:

Definition 8.2.2 Finite Group

A group $(G, \circ)$ is finite if it has a finite number of elements. We denote the cardinality or order of the group G by $|G|$.

Example 8.4. Examples of finite groups are:

- $(\mathbb{Z}_n, +)$: the cardinality of $\mathbb{Z}_n$ is $|\mathbb{Z}_n| = n$ since $\mathbb{Z}_n = \{0, 1, 2, \ldots, n-1\}$.
- $(\mathbb{Z}_n^*, \cdot)$: remember that $\mathbb{Z}_n^*$ is defined as the set of positive integers smaller than n which are relatively prime to n. Thus, the cardinality of $\mathbb{Z}_n^*$ equals Euler's phi function evaluated for n, i.e., $|\mathbb{Z}_n^*| = \Phi(n)$. For instance, the group $\mathbb{Z}_9^*$ has a cardinality of $\Phi(9) = 3^2 - 3^1 = 6$. This can be verified by the earlier example where we saw that the group consist of the six elements $\{1, 2, 4, 5, 7, 8\}$.

◇

The remainder of this section deals with a special type of groups, namely cyclic groups, which are the basis for discrete logarithm-based cryptosystems. We start with the following definition:

Definition 8.2.3 Order of an element

The order $ord(a)$ of an element a of a group $(G, \circ)$ is the smallest positive integer k such that

$$a^k = \underbrace{a \circ a \circ \ldots \circ a}_{k \ times} = 1,$$

where 1 is the identity element of G.

We'll examine this definition by looking at an example.

Example 8.5. We try to determine the order of $a = 3$ in the group $\mathbb{Z}_{11}^*$. For this, we keep computing powers of a until we obtain the identity element 1.

$$
\begin{aligned}
a^1 &= 3 \\
a^2 &= a \cdot a = 3 \cdot 3 = 9 \\
a^3 &= a^2 \cdot a = 9 \cdot 3 = 27 \equiv 5 \bmod 11 \\
a^4 &= a^3 \cdot a = 5 \cdot 3 = 15 \equiv 4 \bmod 11 \\
a^5 &= a^4 \cdot a = 4 \cdot 3 = 12 \equiv 1 \bmod 11
\end{aligned}
$$

From the last line it follows that $ord(3) = 5$.

◇

It is very interesting to look at what happens if we keep multiplying the result by a:

$$
\begin{aligned}
a^6 &= a^5 \cdot a \equiv 1 \cdot a \equiv 3 \bmod 11 \\
a^7 &= a^5 \cdot a^2 \equiv 1 \cdot a^2 \equiv 9 \bmod 11 \\
a^8 &= a^5 \cdot a^3 \equiv 1 \cdot a^3 \equiv 5 \bmod 11 \\
a^9 &= a^5 \cdot a^4 \equiv 1 \cdot a^4 \equiv 4 \bmod 11 \\
a^{10} &= a^5 \cdot a^5 \equiv 1 \cdot 1 \equiv 1 \bmod 11 \\
a^{11} &= a^{10} \cdot a \equiv 1 \cdot a \equiv 3 \bmod 11 \\
&\vdots
\end{aligned}
$$

We see that from this point on, the powers of a run through the sequence $\{3,9,5,4,1\}$ indefinitely. This cyclic behavior gives rise to following definition:

Definition 8.2.4 Cyclic Group

A group G which contains an element α with maximum order $ord(\alpha) = |G|$ is said to be cyclic. *Elements with maximum order are called* primitive elements *or* generators.

An element α of a group G with maximum order is called a generator since every element a of G can be written as a power $\alpha^i = a$ of this element for some i, i.e., α *generates* the entire group. Let us verify these properties by considering the following example.

Example 8.6. We want to check whether $a = 2$ happens to be a primitive element of $\mathbb{Z}_{11}^* = \{1,2,3,4,5,6,7,8,9,10\}$. Note that the cardinality of the group is $|\mathbb{Z}_{11}^*| = 10$. Let's look at all the elements that are generated by powers of the element $a = 2$:

$$
\begin{aligned}
a &= 2 & a^6 &\equiv 9 \bmod 11 \\
a^2 &= 4 & a^7 &\equiv 7 \bmod 11 \\
a^3 &= 8 & a^8 &\equiv 3 \bmod 11 \\
a^4 &\equiv 5 \bmod 11 & a^9 &\equiv 6 \bmod 11 \\
a^5 &\equiv 10 \bmod 11 & a^{10} &\equiv 1 \bmod 11
\end{aligned}
$$

From the last result it follows that

$$
ord(a) = 10 = |\mathbb{Z}_{11}^*|.
$$

This implies that (i) $a = 2$ is a primitive element and (ii) $|\mathbb{Z}_{11}^*|$ is cyclic.

We now want to verify whether the powers of $a = 2$ actually generate all elements of the group $\mathbb{Z}_{11}^*$. Let's look again at all the elements that are generated by powers of two.

i	1	2	3	4	5	6	7	8	9	10
a^i	2	4	8	5	10	9	7	3	6	1

By looking at the bottom row, we see that that the powers 2^i in fact generate all elements of the group $\mathbb{Z}_{11}^*$. We note that the order in which they are generated looks quite arbitrary. This seemingly random relationship between the exponent i and the group elements is the basis for cryptosystems such as the Diffie–Hellman key exchange.

◇

From this example we see that the group $\mathbb{Z}_{11}^*$ has the element 2 as a generator. It is important to stress that the number 2 is not necessarily a generator in other cyclic groups $\mathbb{Z}_n^*$. For instance, in $\mathbb{Z}_7^*$, $\text{ord}(2) = 3$, and the element 2 is thus not a generator in that group.

Cyclic groups have interesting properties. The most important ones for cryptographic applications are given in the following theorems.

Theorem 8.2.2 *For every prime p, $(\mathbb{Z}_p^*, \cdot)$ is an abelian finite cyclic group.*

This theorem states that the multiplicative group of every prime field is cyclic. This has far reaching consequences in cryptography, where these groups are the most popular ones for building discrete logarithm cryptosystems. In order to underline the practical relevance of these somewhat esoteric looking theorem, consider that almost every Web browser has a cryptosystem over $\mathbb{Z}_p^*$ built in.

Theorem 8.2.3

Let G be a finite group. Then for every $a \in G$ it holds that:

1. $a^{|G|} = 1$
2. $\text{ord}(a)$ divides $|G|$

The first property is a generalization of Fermat's Little Theorem for all cyclic groups. The second property is very useful in practice. It says that in a cyclic group only element orders which divide the group cardinality exist.

Example 8.7. We consider again the group $\mathbb{Z}_{11}^*$ which has a cardinality of $|\mathbb{Z}_{11}^*| = 10$. The only element orders in this group are 1, 2, 5, and 10, since these are the only integers that divide 10. We verify this property by looking at the order of all elements in the group:

$$\begin{array}{ll}
\text{ord}(1) = 1 & \text{ord}(6) \ = 10 \\
\text{ord}(2) = 10 & \text{ord}(7) \ = 10 \\
\text{ord}(3) = 5 & \text{ord}(8) \ = 10 \\
\text{ord}(4) = 5 & \text{ord}(9) \ = 5 \\
\text{ord}(5) = 5 & \text{ord}(10) = 2
\end{array}$$

Indeed, only orders that divide 10 occur.

◇

> **Theorem 8.2.4** *Let G be a finite cyclic group. Then it holds that*
> 1. *The number of primitive elements of G is $\Phi(|G|)$.*
> 2. *If $|G|$ is prime, then all elements $a \neq 1 \in G$ are primitive.*

The first property can be verified by the example above. Since $\Phi(10) = (5 - 1)(2 - 1) = 4$, the number of primitive elements is four, which are the elements 2, 6, 7 and 8. The second property follows from the previous theorem. If the group cardinality is prime, the only possible element orders are 1 and the cardinality itself. Since only the element 1 can have an order of one, all other elements have order p.

8.2.3 Subgroups

In this section we consider subsets of (cyclic) groups which are groups themselves. Such sets are referred to as *subgroups*. In order to check whether a subset H of a group G is a subgroup, one can verify if all the properties of our group definition in Section 8.2.1 also hold for H. In the case of cyclic groups, there is an easy way to generate subgroups which follows from this theorem:

> **Theorem 8.2.5 *Cyclic Subgroup Theorem***
> *Let $(G, \circ)$ be a cyclic group. Then every element $a \in G$ with $ord(a) = s$ is the primitive element of a cyclic subgroup with s elements.*

This theorem tells us that any element of a cyclic group is the generator of a subgroup which in turn is also cyclic.

Example 8.8. Let us verify the above theorem by considering a subgroup of $G = \mathbb{Z}_{11}^*$. In an earlier example we saw that $ord(3) = 5$, and the powers of 3 generate the subset $H = \{1, 3, 4, 5, 9\}$ according to Theorem 8.2.5. We verify now whether this set is actually a group by having a look at its multiplication table:

Table 8.2 Multiplication table for the subgroup $H = \{1, 3, 4, 5, 9\}$

$\times$ mod 11	1 3 4 5 9
1	1 3 4 5 9
3	3 9 1 4 5
4	4 1 5 9 3
5	5 4 9 3 1
9	9 5 3 1 4

H is closed under multiplication modulo 11 (Condition 1) since the table only consists of integers which are elements of H. The group operation is obviously associative and commutative since it follows regular multiplication rules (Conditions 2 and 5, respectively). The neutral element is 1 (Condition 3), and for every element $a \in H$ there exists an inverse $a^{-1} \in H$ which is also an element of H (Condition 4). This can be seen from the fact that every row and every column of the table contains the identity element. Thus, H is a subgroup of $\mathbb{Z}_{11}^*$ (depicted in Figure 8.1).

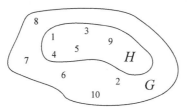

Fig. 8.1 Subgroup H of the cyclic group $G = \mathbb{Z}_{11}^*$

More precisely, it is a subgroup of prime order 5. It should also be noted that 3 is not the only generator of H but also 4, 5 and 9, which follows from Theorem 8.2.4.
◇

An important special case are subgroups of prime order. If this group cardinality is denoted by q, all non-one elements have order q according to Theorem 8.2.4.

From the Cyclic Subgroup Theorem we know that each element $a \in G$ of a group G generates some subgroup H. By using Theorem 8.2.3, the following theorem follows.

Theorem 8.2.6 Lagrange's theorem
Let H be a subgroup of G. Then $|H|$ divides $|G|$.

Let us now consider an application of Lagrange's theorem:

Example 8.9. The cyclic group $\mathbb{Z}_{11}^*$ has cardinality $|\mathbb{Z}_{11}^*| = 10 = 1 \cdot 2 \cdot 5$. Thus, it follows that the subgroups of $\mathbb{Z}_{11}^*$ have cardinalities 1, 2, 5 and 10 since these are all possible divisors of 10. All subgroups H of $\mathbb{Z}_{11}^*$ and their generators α are given below:

subgroup	elements	primitive elements
H_1	$\{1\}$	$\alpha = 1$
H_2	$\{1, 10\}$	$\alpha = 10$
H_3	$\{1, 3, 4, 5, 9\}$	$\alpha = 3, 4, 5, 9$

◇

The following final theorem of this section fully characterizes the subgroups of a finite cyclic group:

> **Theorem 8.2.7**
> *Let G be a finite cyclic group of order n and let α be a generator of G. Then for every integer k that divides n there exists exactly one cyclic subgroup H of G of order k. This subgroup is generated by $\alpha^{n/k}$. H consists exactly of the elements $a \in G$ which satisfy the condition $a^k = 1$. There are no other subgroups.*

This theorem gives us immediately a construction method for a subgroup from a given cyclic group. The only thing we need is a primitive element and the group cardinality *n*. One can now simple compute $\alpha^{n/k}$ and obtains a generator of the subgroup with *k* elements.

Example 8.10. We again consider the cyclic group $\mathbb{Z}_{11}^*$. We saw earlier that $\alpha = 8$ is a primitive element in the group. If we want to have a generator β for the subgroup of order 2, we compute:

$$\beta = \alpha^{n/k} = 8^{10/2} = 8^5 = 32768 \equiv 10 \bmod 11.$$

We can now verify that the element 10 in fact generates the subgroup with two elements: $\beta^1 = 10$, $\beta^2 = 100 \equiv 1 \bmod 11$, $\beta^3 \equiv 10 \bmod 11$, etc.

   Remark: Of course, there are smarter ways of computing $8^5 \bmod 11$, e.g., through $8^5 = 8^2 \, 8^2 \, 8 \equiv (-2)(-2)8 \equiv 32 \equiv 10 \bmod 11$.

   $\diamond$

8.3 The Discrete Logarithm Problem

After the somewhat lengthy introduction to cyclic groups one might wonder how they are related to the rather straightforward DHKE protocol. It turns out that the underlying one-way function of the DHKE, the discrete logarithm problem (DLP), can directly be explained using cyclic groups.

8.3.1 The Discrete Logarithm Problem in Prime Fields

We start with the DLP over $\mathbb{Z}_p^*$, where *p* is a prime.

Definition 8.3.1 Discrete Logarithm Problem (DLP) in $\mathbb{Z}_p^*$

Given is the finite cyclic group $\mathbb{Z}_p^*$ of order $p-1$ and a primitive element $\alpha \in \mathbb{Z}_p^*$ and another element $\beta \in \mathbb{Z}_p^*$. The DLP is the problem of determining the integer $1 \leq x \leq p-1$ such that:

$$\alpha^x \equiv \beta \mod p$$

Remember from Section 8.2.2 that such an integer x must exist since α is a primitive element and each group element can be expressed as a power of any primitive element. This integer x is called the *discrete logarithm of β to the base α*, and we can formally write:

$$x = \log_\alpha \beta \mod p.$$

Computing discrete logarithms modulo a prime is a very hard problem if the parameters are sufficiently large. Since exponentiation $\alpha^x \equiv \beta \mod p$ is computationally easy, this forms a one-way function.

Example 8.11. We consider a discrete logarithm in the group $\mathbb{Z}_{47}^*$, in which $\alpha = 5$ is a primitive element. For $\beta = 41$ the discrete logarithm problem is: Find the positive integer x such that

$$5^x \equiv 41 \mod 47$$

Even for such small numbers, determining x is not entirely straightforward. By using a brute-force attack, i.e., systematically trying all possible values for x, we obtain the solution $x = 15$.

$\diamond$

In practice, it is often desirable to have a DLP in groups with prime cardinality in order to prevent the Pohlig–Hellman attack (cf. Section 8.3.3). Since groups $\mathbb{Z}_p^*$ have cardinality $p-1$, which is obviously not prime, one often uses DLPs in subgroups of $\mathbb{Z}_p^*$ with prime order, rather than using the group $\mathbb{Z}_p^*$ itself. We illustrate this with an example.

Example 8.12. We consider the group $\mathbb{Z}_{47}^*$ which has order 46. The subgroups in $\mathbb{Z}_{47}^*$ have thus a cardinality of 23, 2 and 1. $\alpha = 2$ is an element in the subgroup with 23 elements, and since 23 is a prime, α is a primitive element in the subgroup. A possible discrete logarithm problem is given for $\beta = 36$ (which is also in the subgroup): Find the positive integer x, $1 \leq x \leq 23$, such that

$$2^x \equiv 36 \mod 47$$

By using a brute-force attack, we obtain a solution for $x = 17$.

$\diamond$

8.3.2 The Generalized Discrete Logarithm Problem

The feature that makes the DLP particularly useful in cryptography is that it is not restricted to the multiplicative group $\mathbb{Z}_p^*$, p prime, but can be defined over any cyclic groups. This is called the *generalized discrete logarithm problem (GDLP)* and can be stated as follows.

Definition 8.3.2 Generalized Discrete Logarithm Problem
Given is a finite cyclic group G with the group operation $\circ$ and cardinality n. We consider a primitive element $\alpha \in G$ and another element $\beta \in G$. The discrete logarithm problem is finding the integer x, where $1 \leq x \leq n$, such that:

$$\beta = \underbrace{\alpha \circ \alpha \circ \ldots \circ \alpha}_{x \; times} = \alpha^x$$

As in the case of the DLP in $\mathbb{Z}_p^*$, such an integer x must exist since α is a primitive element, and thus each element of the group G can be generated by repeated application of the group operation on α.

It is important to realize that there are cyclic groups in which the DLP is *not* difficult. Such groups cannot be used for a public-key cryptosystem since the DLP is not a one-way function. Consider the following example.

Example 8.13. This time we consider the additive group of integers modulo a prime. For instance, if we choose the prime $p = 11$, $G = (\mathbb{Z}_{11}, +)$ is a finite cyclic group with the primitive element $\alpha = 2$. Here is how α generates the group:

i	1	2	3	4	5	6	7	8	9	10	11
$i\alpha$	2	4	6	8	10	1	3	5	7	9	0

We try now to solve the DLP for the element $\beta = 3$, i.e., we have to compute the integer $1 \leq x \leq 11$ such that

$$x \cdot 2 = \underbrace{2 + 2 + \ldots + 2}_{x \; times} \equiv 3 \bmod 11$$

Here is how an "attack" against this DLP works. Even though the group operation is addition, we can express the relationship between α, β and the discrete logarithm x in terms of *multiplication*:

$$x \cdot 2 \equiv 3 \bmod 11.$$

In order to solve for x, we simply have to invert the primitive element α:

$$x \equiv 2^{-1} 3 \bmod 11$$

Using, e.g., the extended Euclidean algorithm, we can compute $2^{-1} \equiv 6 \bmod 11$ from which the discrete logarithm follows as:

$$x \equiv 2^{-1} 3 \equiv 7 \bmod 11$$

The discrete logarithm can be verified by looking at the small table provided above.

We can generalize the above trick to any group $(\mathbb{Z}_n, +)$ for arbitrary n and elements $\alpha, \beta \in \mathbb{Z}_n$. Hence, we conclude that the generalized DLP is computationally easy over $\mathbb{Z}_n$. The reason why the DLP can be solved here easily is that we have mathematical operations which are not in the additive group, namely multiplication and inversion.

◇

After this counterexample we now list discrete logarithm problems that have been proposed for use in cryptography:

1. The multiplicative group of the prime field $\mathbb{Z}_p$ or a subgroup of it. For instance, the classical DHKE uses this group, but also Elgamal encryption or the Digital Signature Algorithm (DSA). These are the oldest and most widely used types of discrete logarithm systems.
2. The cyclic group formed by an elliptic curve. Elliptic curve cryptosystems are introduced in Chapter 9. They have become popular in practice over the last decade.
3. The multiplicative group of a Galois field $GF(2^m)$ or a subgroup of it. These groups can be used completely analogous to multiplicative groups of prime fields, and schemes such as the DHKE can be realized with them. They are not as popular in practice because the attacks against them are somewhat more powerful than those against the DLP in $\mathbb{Z}_p$. Hence DLPs over $GF(2^m)$ require somewhat higher bit lengths for providing the same level of security than those over $\mathbb{Z}_p$.
4. Hyperelliptic curves or algebraic varieties, which can be viewed as generalization as elliptic curves. They are currently rarely used in practice, but in particular hyperelliptic curves have some advantages such as short operand lengths.

There have been proposals for other DLP-based cryptosystems over the years, but none of them have really been of interest in practice. Often it was found that the underlying DL problem was not difficult enough.

8.3.3 Attacks Against the Discrete Logarithm Problem

This section introduce methods for solving discrete logarithm problems. Readers only interested in the constructive use of DL schemes can skip this section.

As we have seen, the security of many asymmetric primitives is based on the difficulty of computing the DLP in cyclic groups, i.e., to compute x for a given α and β in G such that

$$\beta = \underbrace{\alpha \circ \alpha \circ \ldots \circ \alpha}_{x \text{ times}} = \alpha^x$$

holds. We still do not know the exact difficulty of computing the discrete logarithm x in any given actual group. What we mean by this is that even though some attacks are known, one does not know whether there are any better, more powerful algorithms for solving the DLP. This situation is similar to the hardness of integer factorization, which is the one-way function underlying RSA. Nobody really knows what the *best possible* factorization method is. For the DLP some interesting general results exist regarding its computational hardness. This section gives a brief overview of algorithms for computing discrete logarithms which can be classified into *generic algorithms* and *nongeneric algorithms* and which will be discussed in a little more detail.

Generic Algorithms

Generic DL algorithms are methods which only use the group operation and no other algebraic structure of the group under consideration. Since they do not exploit special properties of the group, they work in any cyclic group. Generic algorithms for the discrete logarithm problem can be subdivided into two classes. The first class encompasses algorithms whose running time depends on the size of the cyclic group, like the *brute-force search*, the *baby-step giant-step* algorithm and *Pollard's rho* method. The second class are algorithms whose running time depends on the size of the prime factors of the group order, like the *Pohlig–Hellman* algorithm.

Brute-Force Search

A brute-force search is the most naïve and computationally costly way for computing the discrete logarithm $\log_\alpha \beta$. We simply compute powers of the generator α successively until the result equals β:

$$\alpha^1 \overset{?}{=} \beta$$
$$\alpha^2 \overset{?}{=} \beta$$
$$\vdots$$
$$\alpha^x \overset{?}{=} \beta$$

For a random logarithm x, we do expect to find the correct solution after checking half of all possible x. This gives us a complexity of $\mathcal{O}(|G|)$ steps[2], where $|G|$ is the cardinality of the group.

To avoid brute-force attacks on DL-based cryptosystems in practice, the cardinality $|G|$ of the underlying group must thus be sufficiently large. For instance, in the case of the group $\mathbb{Z}_p^*$, p prime, which is the basis for the DHKE, $(p-1)/2$ tests are required on average to compute a discrete logarithm. Thus, $|G| = p - 1$ should be at least in the order of 2^{80} to make a brute-force search infeasible using today's computer technology. Of course, this consideration only holds if a brute-force attack is the only feasible attack which is never the case. There exist much more powerful algorithms to solve discrete logarithms as we will see below.

Shanks' Baby-Step Giant-Step Method

Shanks' algorithm is a time-memory tradeoff method, which reduces the time of a brute-force search at the cost of extra storage. The idea is based on rewriting the discrete logarithm $x = \log_\alpha \beta$ in a two-digit representation:

$$x = x_g \, m + x_b \qquad \text{for } 0 \leq x_g, x_b < m. \tag{8.1}$$

The value m is chosen to be of the size of the square root of the group order, i.e., $m = \lceil \sqrt{|G|} \rceil$. We can now write the discrete logarithm as $\beta = \alpha^x = \alpha^{x_g \, m + x_b}$ which leads to

$$\beta \cdot (\alpha^{-m})^{x_g} = \alpha^{x_b}. \tag{8.2}$$

The idea of the algorithm is to find a solution (x_g, x_b) for Eq. (8.2), from which the discrete logarithm then follows directly according to Eq. (8.1). The core idea for the algorithm is that Eq. (8.2) can be solved by searching for x_g and x_b separatedly, i.e., using a divide-and-conquer approach. In the first phase of the algorithm we compute and store all values α^{x_b}, where $0 \leq x_b < m$. This is the *baby-step phase* that requires $m \approx \sqrt{|G|}$ steps (group operations) and needs to store $m \approx \sqrt{|G|}$ group elements.

In the *giant-step phase*, the algorithm checks for all x_g in the range $0 \leq x_g < m$ whether the following condition is fulfilled:

$$\beta \cdot (\alpha^{-m})^{x_g} \stackrel{?}{=} \alpha^{x_b}$$

for some stored entry α^{x_b} that was computed during the baby-step phase. In case of a match, i.e., $\beta \cdot (\alpha^{-m})^{x_{g,0}} = \alpha^{x_{b,0}}$ for some pair $(x_{g,0}, x_{b,0})$, the discrete logarithm is given by

$$x = x_{g,0} \, m + x_{b,0}.$$

The baby-step giant-step method requires $\mathcal{O}(\sqrt{|G|})$ computational steps and an equal amount of memory. In a group of order 2^{80}, an attacker would only need

---

[2] We use the popular "big-Oh" notation here. A complexity function $f(x)$ has big-Oh notation $\mathcal{O}(g(x))$ if $f(x) \leq c \cdot g(x)$ for some constant c and for input values x greater than some value x_0.

approximately $2^{40} = \sqrt{2^{80}}$ computations and memory locations, which is easily achievable with today's PCs and hard disks. Thus, in order to obtain an attack complexity of 2^{80}, a group must have a cardinality of at least $|G| \geq 2^{160}$. In the case of groups $G = \mathbb{Z}_p^*$, the prime p should thus have at least a length of 160 bit. However, as we see below, there are more powerful attacks against DLPs in $\mathbb{Z}_p^*$ which forces even larger bit lengths of p.

Pollard's Rho Method

Pollard's rho method has the same expected run time $\mathcal{O}(\sqrt{|G|})$ as the baby-step giant-step algorithm but only negligible space requirements. The method is a probabilistic algorithm which is based on the birthday paradox (cf. Section 11.2.3). We will only sketch the algorithm here. The basic idea is to pseudorandomly generate group elements of the form $\alpha^i \cdot \beta^j$. For every element we keep track of the values i and j. We continue until we obtain a collision of two elements, i.e., until we have:

$$\alpha^{i_1} \cdot \beta^{j_1} = \alpha^{i_2} \cdot \beta^{j_2}. \tag{8.3}$$

If we substitute $\beta = \alpha^x$ and compare the exponents on both sides of the equation, the collision leads to the relation $i_1 + xj_1 \equiv i_2 + xj_2 \bmod |G|$. (Note that we are in a cyclic group with $|G|$ elements and have to take the exponent modulo $|G|$.) From here the discrete logarithm can easily computed as:

$$x \equiv \frac{i_2 - i_1}{j_1 - j_2} \bmod |G|$$

An important detail, which we omit here, is the exact way to find the collision (8.3). In any case, the pseudorandom generation of the elements is a random walk through the group. This can be illustrated by the shape of the Greek letter rho, hence the name of this attack.

Pollard's rho method is of great practical importance because it is currently the best known algorithm for computing discrete logarithms in elliptic curve groups. Since the method has an attack complexity of $\mathcal{O}(\sqrt{|G|})$ computations, elliptic curve groups should have a size of at least 2^{160}. In fact, elliptic curve cryptosystems with 160-bit operands are very popular in practice.

There are still much more powerful attacks known for the DLP in $\mathbb{Z}_p^*$, as we will see below.

Pohlig–Hellman Algorithm

The Pohlig–Hellman method is an algorithm which is based on the Chinese Remainder Theorem (not introduced in this book); it exploits a possible factorization of the order of a group. It is typically not used by itself but in conjunction with any of the other DLP attack algorithms in this section. Let

$$|G| = p_1^{e_1} \cdot p_2^{e_2} \cdot \ldots \cdot p_l^{e_l}$$

be the prime factorization of the group order $|G|$. Again, we attempt to compute a discrete logarithm $x = \log_\alpha \beta$ in G. This is also a divide-and-conquer algorithm. The basic idea is that rather than dealing with the large group G, one computes smaller discrete logarithms $x_i \equiv x \bmod p_i^{e_i}$ in the subgroups of order $p_i^{e_i}$. The desired discrete logarithm x can then be computed from all x_i, $i = 1, \ldots, l$, by using the Chinese Remainder Theorem. Each individual small DLP x_i can be computed using Pollard's rho method or the baby-step giant-step algorithm.

The run time of the algorithm clearly depends on the prime factors of the group order. To prevent the attack, the group order must have its largest prime factor in the range of 2^{160}. An important practical consequence of the Pohlig–Hellman algorithm is that one needs to know the prime factorization of the group order. Especially in the case of elliptic curve cryptosystems, computing the order of the cyclic group is not always easy.

Nongeneric Algorithms: The Index-Calculus Method

All algorithms introduced so far are completely independent of the group being attacked, i.e., they work for discrete logarithms defined over any cyclic group. Nongeneric algorithms efficiently exploit special properties, i.e., the inherent structure, of certain groups. This can lead to much more powerful DL algorithms. The most important nongeneric algorithm is the index-calculus method.

Both the baby-step giant-step algorithm and Pollard's rho method have a run time which is exponential in the bit length of the group order, namely of about $2^{n/2}$ steps, where n is the bit length of $|G|$. This greatly favors the crypto designer over the cryptanalyst. For instance, increasing the group order by a mere 20 bit increases the attack effort by a factor of $1024 = 2^{10}$. This is a major reason why elliptic curves have better long-term security behavior than RSA or cryptosystems based on the DLP in $\mathbb{Z}_p^*$. The question is whether there are more powerful algorithms for DLPs in certain specific groups. The answer is yes.

The index-calculus method is a very efficient algorithm for computing discrete logarithms in the cyclic groups $\mathbb{Z}_p^*$ and $GF(2^m)^*$. It has a subexponential running time. We will not introduce the method here but just provide a very brief description. The index-calculus method depends on the property that a significant fraction of elements of G can be efficiently expressed as products of elements of a small subset of G. For the group $\mathbb{Z}_p^*$ this means that many elements should be expressable as a product of small primes. This property is satisfied by the groups $\mathbb{Z}_p^*$ and $GF(2^m)^*$. However, one has not found a way to do the same for elliptic curve groups. The index calculus is so powerful that in order to provide a security of 80 bit, i.e., an attacker has to perform 2^{80} steps, the prime p of a DLP in $\mathbb{Z}_p^*$ should be at least 1024 bit long. Table 8.3 gives an overview on the DLP records achieved since the early 1990s. The index-calculus method is somewhat more powerful for solving the DLP in $GF(2^m)^*$. Hence the bit lengths have to be chosen somewhat longer to achieve the same level of security. For that reason, DLP schems in $GF(2^m)^*$ are not as widely used in practice.

Table 8.3 Summary of records for computing discrete logarithms in $\mathbb{Z}_p^*$

Decimal digits	Bit length	Date
58	193	1991
65	216	1996
85	282	1998
100	332	1999
120	399	2001
135	448	2006
160	532	2007

8.4 Security of the Diffie–Hellman Key Exchange

After the introduction of the discrete logarithm problem, we are now well prepared to discuss the security of the DHKE from Section 8.1. First, it should be noted that a protocol that uses the basic version of the DHKE is not secure against active attacks. This means if an attacker Oscar can either modify messages or generate false messages, Oscar can defeat the protocol. This is called *man-in-the-middle attack* and is described in Section 13.3.

Let's now consider the possibilities of a passive adversary, i.e., Oscar can only listen but not alter messages. His goal is to compute the session key k_{AB} shared by Alice and Bob. Which information does Oscar get from observing the protocol? Certainly, Oscar knows α and p because these are public parameters chosen during the set-up protocol. In addition, Oscar can obtain the values $A = k_{pub,A}$ and $B = k_{pub,B}$ by eavesdropping on the channel during an execution of the key-exchange protocol. Thus, the question is whether he is capable of computing $k = \alpha^{ab}$ from $\alpha, p, A \equiv \alpha^a \bmod p$ and $B \equiv \alpha^b \bmod p$. This problem is called the *Diffie–Hellman problem* (DHP). Like the discrete logarithm problem it can be generalized to arbitrary finite cyclic groups. Here is a more formal statement of the DHP:

Definition 8.4.1 Generalized Diffie–Hellman Problem (DHP)
Given is a finite cyclic group G of order n, a primitive element $\alpha \in G$ and two elements $A = \alpha^a$ and $B = \alpha^b$ in G. The Diffie–Hellman problem is to find the group element α^{ab}.

One general approach to solving the Diffie–Hellman problem is as follows. For illustration purposes we consider the DHP in the multiplicative group $\mathbb{Z}_p^*$. Suppose—and that's a big "suppose"—Oscar knows an efficient method for computing discrete logarithms in $\mathbb{Z}_p^*$. Then he could also solve the Diffie–Hellman problem and obtain the key k_{AB} via the following two steps:

1. Compute Alice's private key $a = k_{pr,A}$ by solving the discrete logarithm problem: $a \equiv \log_\alpha A \bmod p$.
2. Compute the session key $k_{AB} \equiv B^a \bmod p$.

But as we know from Section 8.3.3, even though this looks easy, computing the discrete logarithm problem is infeasible if p is sufficiently large.

At this point it is important to note that it is not known whether solving the DLP is the only way to solve the DHP. In theory, it is possible that there exists another method for solving the DHP *without* computing the discrete logarithm. We note that the situation is analogous to RSA, where it is also not known whether factoring is the best way of breaking RSA. However, even though it is not proven in a mathematical sense, it is often assumed that solving the DLP efficiently is the only way for solving the DHP efficiently.

Hence, in order to assure the security of the DHKE in practice, we have to ascertain that the corresponding DLP cannot be solved. This is achieved by choosing p

large enough so that the index-calculus method cannot compute the DLP. By consulting Table 6.1 we see that a security level of 80 bit is achieved by primes of lengths 1024 bit, and for 128 bit security we need about 3072 bit. An additional requirement is that in order to prevent the Pohlig–Hellman attack, the order $p-1$ of the cyclic group must not factor in only small prime factors. Each of the subgroups formed by the factors of $p-1$ can be attacked using the baby-step giant-step method or Pollards's rho method, but not by the index-calculus method. Hence, the smallest prime factor of $p-1$ must be at least 160 bit long for an 80-bit security level, and at least 256 bit long for a security level of 128 bit.

8.5 The Elgamal Encryption Scheme

The *Elgamal encryption scheme* was proposed by Taher Elgamal in 1985 [73]. It is also often referred to as Elgamal encryption. It can be viewed as an extension of the DHKE protocol. Not surprisingly, its security is also based on the intractability of the discrete logarithm problem and the Diffie–Hellman problem. We consider the Elgamal encryption scheme over the group $\mathbb{Z}_p^*$, where p is a prime. However, it can be applied to other cyclic groups too in which the DL and DH problem is intractable, for instance, in the multiplicative group of a Galois field $GF(2^m)$.

8.5.1 From Diffie–Hellman Key Exchange to Elgamal Encryption

In order to understand the Elgamal scheme, it is very helpful to see how it follows almost immediately from the DHKE. We consider two parties, Alice and Bob. If Alice wants to send an encrypted message x to Bob, both parties first perform a Diffie–Hellman key exchange to derive a shared key k_M. For this we assume that a large prime p and a primitive element α have been generated. Now, the new idea is that Alice uses this key as a multiplicative mask to encrypt x as $y \equiv x \cdot k_M \bmod p$. This process is depicted below.

Principle of Elgamal Encryption

Alice		Bob

Bob

(a) choose $d = k_{pr,B} \in \{2,\ldots,p-2\}$
(b) compute $\beta = k_{pub,B} \equiv \alpha^d \bmod p$

$$\xleftarrow{\hspace{2cm} \beta \hspace{2cm}}$$

Alice

(c) choose $i = k_{pr,A} \in \{2,\ldots,p-2\}$
(d) compute $k_E = k_{pub,A} \equiv \alpha^i \bmod p$

$$\xrightarrow{\hspace{2cm} k_E \hspace{2cm}}$$

(e) compute $k_M \equiv \beta^i \bmod p$ (f) compute $k_M \equiv k_E^d \bmod p$
(g) encrypt message $x \in \mathbb{Z}_p^*$
    $y \equiv x \cdot k_M \bmod p$

$$\xrightarrow{\hspace{2cm} y \hspace{2cm}}$$

(h) decrypt $x \equiv y \cdot k_M^{-1} \bmod p$

The protocol consists of two phases, the classical DHKE (Steps a–f) which is followed by the message encryption and decryption (Steps g and h, respectively). Bob computes his private key d and public key β. This key pair does not change, i.e., it can be used for encrypting many messages. Alice, however, has to generate a new public–private key pair for the encryption of every message. Her private key is denoted by i and her public key by k_E. The latter is an ephemeral (existing only temporarily) key, hence the index "E". The joint key is denoted by k_M because it is used for masking the plaintext.

For the actual encryption, Alice simply multiplies the plaintext message x by the masking key k_M in $\mathbb{Z}_p^*$. On the receiving side, Bob reverses the encryption by multipliying with the inverse mask. Note that one property of cyclic groups is that, given any key $k_M \in \mathbb{Z}_p^*$, every messages x maps to another ciphertext if the two values are multiplied. Moreover, if the key k_M is randomly drawn from $\mathbb{Z}_p^*$, every ciphertext $y \in \{1,2,\ldots,p-1\}$ is equally likely.

8.5.2 The Elgamal Protocol

We provide now a somewhat more formal description of the scheme. We distinguish three phases. The set-up phase is executed once by the party who issues the public key and who will receive the message. The encryption phase and the decryption phase are executed every time a message is being sent. In contrast to the DHKE, no trusted third party is needed to choose a prime and primitive element. Bob generates them and makes them public, by placing them in a database or on his website.

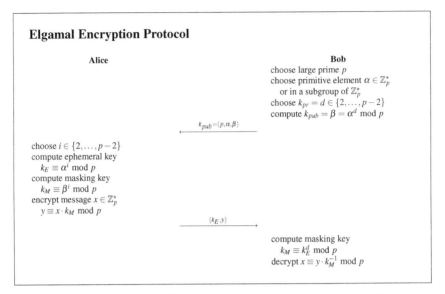

Elgamal Encryption Protocol

Alice	Bob
	choose large prime p
	choose primitive element $\alpha \in \mathbb{Z}_p^*$
	or in a subgroup of $\mathbb{Z}_p^*$
	choose $k_{pr} = d \in \{2,\dots,p-2\}$
	compute $k_{pub} = \beta = \alpha^d \bmod p$

$$\xleftarrow{\quad k_{pub}=(p,\alpha,\beta) \quad}$$

choose $i \in \{2,\dots,p-2\}$
compute ephemeral key
  $k_E \equiv \alpha^i \bmod p$
compute masking key
  $k_M \equiv \beta^i \bmod p$
encrypt message $x \in \mathbb{Z}_p^*$
  $y \equiv x \cdot k_M \bmod p$

$$\xrightarrow{\quad (k_E,y) \quad}$$

compute masking key
  $k_M \equiv k_E^d \bmod p$
decrypt $x \equiv y \cdot k_M^{-1} \bmod p$

The actual Elgamal encryption protocol rearranges the sequence of operations from the naïve Diffie–Hellman inspired approach we saw before. The reason for this is that Alice has to send only one message to Bob, as opposed to two messages in the earlier protocol.

The ciphertext consists of two parts, the ephemeral key, k_E, and the masked plaintext, y. Since in general all parameters have a bit length of $\lceil \log_2 p \rceil$, the ciphertext (k_E, y) is twice as long as the message. Thus, the *message expansion factor* of Elgamal encryption is two.

We prove now the correctness of the Elgamal protocol.

Proof. We have to show that $d_{k_{pr}}(k_E, y)$ actually yields the original message x.

$$
\begin{aligned}
d_{k_{pr}}(k_E, y) &\equiv y \cdot (k_M)^{-1} \bmod p \\
&\equiv [x \cdot k_M] \cdot (k_E^d)^{-1} \bmod p \\
&\equiv [x \cdot (\alpha^d)^i][(\alpha^i)^d]^{-1} \bmod p \\
&\equiv x \cdot \alpha^{d \cdot i - d \cdot i} \equiv x \bmod p
\end{aligned}
$$

$\square$

Let's look at an example with small numbers.

Example 8.14. In this example, Bob generates the Elgamal keys and Alice encrypts the message $x = 26$.

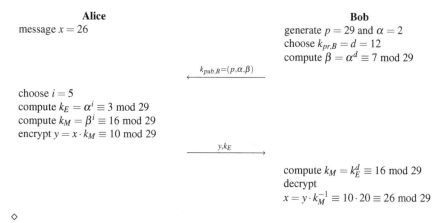

Alice
message $x = 26$

Bob
generate $p = 29$ and $\alpha = 2$
choose $k_{pr,B} = d = 12$
compute $\beta = \alpha^d \equiv 7 \bmod 29$

$$\xleftarrow{\quad k_{pub,B}=(p,\alpha,\beta) \quad}$$

choose $i = 5$
compute $k_E = \alpha^i \equiv 3 \bmod 29$
compute $k_M = \beta^i \equiv 16 \bmod 29$
encrypt $y = x \cdot k_M \equiv 10 \bmod 29$

$$\xrightarrow{\quad y,k_E \quad}$$

compute $k_M = k_E^d \equiv 16 \bmod 29$
decrypt
$x = y \cdot k_M^{-1} \equiv 10 \cdot 20 \equiv 26 \bmod 29$

◇

It is important to note that, unlike the schoolbook version of the RSA scheme, Elgamal is a *probabilistic encryption scheme*, i.e., encrypting two identical messages x_1 and x_2, where $x_1, x_2 \in \mathbb{Z}_p^*$ using the same public key results (with extremely high likelihood) in two different ciphertexts $y_1 \neq y_2$. This is because i is chosen at random from $\{2, 3, \cdots, p-2\}$ for each encryption, and thus also the session key $k_M = \beta^i$ used for encryption is chosen at random for each encryption. In this way a brute-force search for x is avoided a priori.

8.5.3 Computational Aspects

Key Generation During the key generation by the receiver (Bob in our example), a prime p must be generated, and the public and private have to be computed. Since the security of Elgamal also depends on the discrete logarithm problem, p needs to have the properties discussed in Section 8.3.3. In particular, it should have a length of at least 1024 bits. To generate such a prime, the prime-finding algorithms discussed in Section 7.6 can be used. The private key should be generated by a true random number generater. The public key requires one exponentiation for which the square-and-multiply algorithm is used.

Encryption Within the encryption procedure, two modular exponentiations and one modular multiplication are required for computing the ephemeral and the masking key, as well as for the message encryption. All operands involved have a bit length of $\lceil \log_2 p \rceil$. For efficient exponentiation, one should apply the square-and-multiply algorithm that was introduced in Section 7.4. It is important to note that the two exponentiations, which constitute almost all computations necessary, are independent of the plaintext. Hence, in some applications they can be precomputed at times of low computational load, stored and used when the actual encryption is needed. This can be a major advantage in practice.

Decryption The main steps of the decryption are first an exponentiation $k_M = k^d \bmod p$, using the square-and-multiply algorithm, followed by an inversion of k_M

that is performed with the extended Euclidean algorithm. However, there is a short-cut based on Fermat's Little Theorem that combines these two steps in a single one. From the theorem, which was introduced in Section 6.3.4, follows that

$$k_E^{p-1} \equiv 1 \bmod p$$

for all $k_E \in \mathbb{Z}_p^*$. We can now merge Step 1 and 2 of the decryption as follows:

$$
\begin{aligned}
k_M^{-1} &\equiv (k_E^d)^{-1} \bmod p \\
&\equiv (k_E^d)^{-1} k_E^{p-1} \bmod p \\
&\equiv k_E^{p-d-1} \bmod p \qquad\qquad (8.4)
\end{aligned}
$$

The equivalence relation (8.4) allows us to compute the inverse of the masking key using a single exponentiation with the exponent $(p - d - 1)$. After that, one modular multiplication is required to recover $x \equiv y \cdot k_M^{-1} \bmod p$. As a consequence, decryption essentially requires one execution of the square-and-multiply algorithm followed by a single modular multiplication for recovering the plaintext.

8.5.4 Security

If we want to assess the security of the Elgamal encryption scheme it is important to distinguish between passive, i.e., listen-only, and active attacks, which allow Oscar to generate and alter messages.

Passive Attacks

The security of the Elgamal encryption scheme against passive attacks, i.e., recovering x from the information p, α, $\beta = \alpha^d$, $k_E = \alpha^i$ and $y = x \cdot \beta^i$ obtained by eavesdropping, relies on the hardness of the Diffie–Hellman problem (cf. Section 8.4). Currently there is no other method known for solving the DHP than computing discrete logarithms. If we assume Oscar has supernatural powers and can in fact compute DLPs, he would have two ways of attacking the Elgamal scheme.

■ Recover x by finding Bob's secret key d:

$$d = \log_\alpha \beta \bmod p.$$

This step solves the DLP, which is computationally infeasible if the parameters are chosen correctly. However, if Oscar succeeds with it, he can decrypt the plaintext by performing the same steps as the receiver, Bob:

$$x \equiv y \cdot (k_E^d)^{-1} \bmod p.$$

- Alternatively, instead of computing Bob's secret exponent d, Oscar could attempt to recover Alice's random exponent i:

$$i = \log_\alpha k \bmod p.$$

Again, this step is solving the discrete logarithm problem. Should Oscar succeed with it, he can compute the plaintext as:

$$x \equiv y \cdot (\beta^i)^{-1} \bmod p.$$

In both cases Oscar has to solve the DL problem in the finite cyclic group $\mathbb{Z}_p^*$. In contrast to elliptic curves, the more powerful index-calculus method (Section 8.3.3) can be applied here. Thus, in order to guarantee the security of the Elgamal scheme over $\mathbb{Z}_p^*$ today, p should at least have a length of 1024 bits.

Just as in the DHKE protocol, we have to be careful that we do not fall vicitim to what is a called a *small subgroup attack*. In order to counter this attack, in practice primitive elements α are used which generate a subgroup of prime order. In such groups, all elements are primitive and small subgroups do not exist. One of the problems illustrates the pitfalls of a small subgroup attack with an example.

Active Attacks

Like in every other asymmetric scheme, it must be assured that the public keys are authentic. This means that the encrypting party, Alice in our example, in fact has the public key that belongs to Bob. If Oscar manages to convince Alice that his key is Bob's, he can easily attack the scheme. In order to prevent the attack, certificates can be used, a topic that is discussed in Chapter 13.

Another weakness, if not necessarily an attack that requires any direct action by Oscar, of the Elgamal encryption is that the secret exponent i should not be reused. Assume Alice used the value i for encrypting two subsequent messages, x_1 and x_2. In this case, the two masking keys would be the same, namely $k_M = \beta^i$. Also, the two ephemeral keys would be identical. She would send the two ciphertexts (y_1, k_E) and (y_1, k_E) over the channel. If Oscar knows or can guess the first message, he can compute the masking key as $k_M \equiv y_1 x_1^{-1} \bmod p$. With this, he can decrypt x_2 :

$$x_2 \equiv y_2 k_M^{-1} \bmod p$$

Any other message encrypted with the same i value can also be recovered this way. As a consequence of this attack, one has to take care that the secret exponent i does not repeat. For instance, if one would use a cryptographically secure PRNG, as introduced in Section 2.2.1, but with the same seed value every time a session is initiated, the same sequence of i values would be used for every encryption, a fact that could be exploited by Oscar. Note that Oscar can detect the re-use of secret exponents because they lead to identical ephemeral keys.

Another active attack against Elgamal exploits its malleability. If Oscar observes the ciphertext (k_E, y), he can replace it by

$$(k_E, sy),$$

where s is some integer. The receiver would compute

$$d_{k_{pr}}(k_E, sy) \equiv sy \cdot k_M^{-1} \bmod p$$
$$\equiv s(x \cdot k_M) \cdot k_M^{-1} \bmod p$$
$$\equiv sx \bmod p$$

Thus, the decrypted text is also a multiple of s. The situation is exactly the same as for the attack that exploited the malleability of RSA, which was introduced in Section 7.7. Oscar is not able to decrypt the ciphertext, but he can manipulated it in a specific way. For instance, he could double or triple the integer value of the decryption result by choosing s equal to 2 or 3, respectively. As it was the case for RSA, schoolbook Elgamal encryption is often not used in practice, but some padding is introduced to prevent these types of attacks.

8.6 Discussion and Further Reading

Diffie–Hellman Key Exchange and Elgamal Encryption The DHKE was introduced in the landmark paper [58], which also described the concept of public-key cryptography. Due to the independent discovery of asymmetric cryptography by Ralph Merkle, Hellman suggested in 2003 that the algorithm should be named "Diffie–Hellman–Merkle key exchange". In Chapter 13 of this book, a more detailed treatment of key exchanges based on the DHKE is provided. The scheme is standardized in ANSI X9.42 [5] and is used in numerous security protocols such as TLS. One of the attractive features of DHKE is that it can be generalized to any cyclic group, not only to the multiplicative group of a prime field that was shown in this chapter. In practice, the most popular group in addition to $\mathbb{Z}_p^*$ is the DHKE over an elliptic curve that is presented in Section 9.3.

The DHKE is a two-party protocol. It can be extended to a group key agreement in which more than two parties establish a joint Diffie–Hellman key, see, e.g., [38].

The Elgamal encryption as proposed in 1985 by Tahar Elgamal [73] is widely used. For example, Elgamal is part of the free GNU Privacy Guard (GnuPG), OpenSSL, Pretty Good Privacy (PGP) and other crypto software. Active attacks against the Elgamal encryption scheme such as those discussed in Section 8.5.4 have quite strong requirements that have to be fulfilled, which is quite difficult in reality. There exist schemes which are related to Elgamal but have stronger security properties. These include, e.g., the *Cramer–Shoup System* [49] and the *DHAES* [1] scheme proposed by Abdalla, Bellare and Rogaway; these are secure against chosen ciphertext attacks under certain assumptions.

Discrete Logarithm Problem This chapter sketched the most important attack algorithms for solving discrete logarithm problems. A good overview on these, including further references, are given in [168, p. 164 ff.]. We also discussed the relationship between the Diffie–Hellman problem (DHP) and the discrete logarithm problem (DLP). This relationship is a matter of great importance for the foundations of cryptography. Key contributions which study this are [31, 118].

The idea of using the DLP in groups other than $\mathbb{Z}_p^\star$ is exploited in elliptic curve cryptography, a topic that is treated in Chapter 9. Other cryptoystems based on the generalized DLP include hyperelliptic curves, a comprehensive treatment of which can be found in [44]. Rather than using the prime field $\mathbb{Z}_p^\star$ it is also possible to use certain extension fields which offer computational advantages. Two of the better-studied DL systems over extension fields are Lucas-Based Cryptosystems [26] and Efficient and Compact Subgroup Trace Representation (XTR) [109].

8.7 Lessons Learned

- The Diffie–Hellman protocol is a widely used method for key exchange. It is based on cyclic groups.
- The discrete logarithm problem is one of the most important one-way functions in modern asymmetric cryptography. Many public-key algorithms are based on it.
- In practice, the multiplicative group of the prime field $\mathbb{Z}_p$ or the group of an elliptic curve are used most often.
- For the Diffie–Hellman protocol in $\mathbb{Z}_p^*$, the prime p should be at least 1024 bits long. This provides a security roughly equivalent to an 80-bit symmetric cipher. For a better long-term security, a prime of length 2048 bits should be chosen.
- The Elgamal scheme is an extension of the DHKE where the derived session key is used as a multiplicative masked to encrypt a message.
- Elgamal is a probabilistic encryption scheme, i.e., encrypting two identical messages does not yield two identical ciphertexts.
- For the Elgamal encryption scheme over $\mathbb{Z}_p^*$, the prime p should be at least 1024 bits long, i.e., $p > 2^{1000}$.

Problems

8.1. Understanding the functionality of groups, cyclic groups and subgroups is important for the use of public-key cryptosystems based on the discrete logarithm problem. That's why we are going to practice some arithmetic in such structures in this set of problems.

Let's start with an easy one. Determine the order of all elements of the multiplicative groups of:

1. $\mathbb{Z}_5^*$
2. $\mathbb{Z}_7^*$
3. $\mathbb{Z}_{13}^*$

Create a list with two columns for every group, where each row contains an element a and the order $\text{ord}(a)$.

(Hint: In order to get familiar with cyclic groups and their properties, it is a good idea to compute all orders "by hand", i.e., use only a pocket calculator. If you want to refresh your mental arithmetic skills, try not to use a calculator whenever possible, in particular for the first two groups.)

8.2. We consider the group $\mathbb{Z}_{53}^*$. What are the possible element orders? How many elements exist for each order?

8.3. We now study the groups from Problem 8.2.

1. How many elements does each of the multiplicative groups have?
2. Do all orders from above divide the number of elements in the corresponding multiplicative group?
3. Which of the elements from Problem 8.1 are primitive elements?
4. Verify for the groups that the number of primitive elements is given by $\phi(|\mathbb{Z}_p^*|)$.

8.4. In this exercise we want to identify primitive elements (generators) of a multiplicative group since they play a big role in the DHKE and and many other public-key schemes based on the DL problem. You are given a prime $p = 4969$ and the corresponding multiplicative group $\mathbb{Z}_{4969}^*$.

1. Determine how many generators exist in $\mathbb{Z}_{4969}^*$.
2. What is the probability of a randomly chosen element $a \in \mathbb{Z}_{4969}^*$ being a generator?
3. Determine the smallest generator $a \in \mathbb{Z}_{4969}^*$ with $a > 1000$.
   Hint: The identification can be done naïvely through testing *all* possible factors of the group cardinality $p - 1$, or more efficiently by checking the premise that $a^{(p-1)/q_i} \neq 1 \bmod p$ for all prime factors q_i with $p - 1 = \prod q_i^{e_i}$. You can simply start with $a = 1001$ and repeat these steps until you find a respective generator of $\mathbb{Z}_{4969}^*$.
4. What measures can be taken in order to simplify the search for generators for arbitrary groups $\mathbb{Z}_p^*$?

8.5. Compute the two public keys and the common key for the DHKE scheme with the parameters $p = 467$, $\alpha = 2$, and

1. $a = 3, b = 5$
2. $a = 400, b = 134$
3. $a = 228, b = 57$

In all cases, perform the computation of the common key for Alice *and* Bob. This is also a perfect check of your results.

8.6. We now design another DHKE scheme with the same prime $p = 467$ as in Problem 8.5. This time, however, we use the element $\alpha = 4$. The element 4 has order 233 and generates thus a subgroup with 233 elements. Compute k_{AB} for

1. $a = 400, b = 134$
2. $a = 167, b = 134$

Why are the session keys identical?

8.7. In the DHKE protocol, the private keys are chosen from the set

$$\{2, \ldots, p-2\}.$$

Why are the values 1 and $p - 1$ excluded? Describe the weakness of these two values.

8.8. Given is a DHKE algorithm. The modulus p has 1024 bit and α is a generator of a subgroup where $\text{ord}(\alpha) \approx 2^{160}$.

1. What is the maximum value that the private keys should have?
2. How long does the computation of the session key take on average if one modular multiplication takes 700 μs, and one modular squaring 400 μs? Assume that the public keys have already been computed.
3. One well-known acceleration technique for discrete logarithm systems uses short primitive elements. We assume now that α is such a short element (e.g., a 16-bit integer). Assume that modular multiplication with α takes now only 30 μs. How long does the computation of the public key take now? Why is the time for one modular squaring still the same as above if we apply the square-and-multiply algorithm?

8.9. We now want to consider the importance of the proper choice of generators in multiplicative groups.

1. Show that the order of an element $a \in \mathbb{Z}_p$ with $a = p - 1$ is always 2.
2. What subgroup is generated by a?
3. Briefly describe a simple attack on the DHKE which exploits this property.

8.10. We consider a DHKE protocol over a Galois fields $GF(2^m)$. All arithmetic is done in $GF(2^5)$ with $P(x) = x^5 + x^2 + 1$ as an irreducible field polynomial. The primitive element for the Diffie–Hellman scheme is $\alpha = x^2$. The private keys are $a = 3$ and $b = 12$. What is the session key k_{AB}?

8.11. In this chapter, we saw that the Diffie–Hellman protocol is as secure as the Diffie–Hellman problem which is probably as hard as the DL problem in the group $\mathbb{Z}_p^*$. However, this only holds for passive attacks, i.e., if Oscar is only capable of eavesdropping. If Oscar can manipulate messages between Alice and Bob, the key agreement protocol can easily be broken! Develop an active attack against the Diffie–Hellman key agreement protocol with Oscar being the man in the middle.

8.12. Write a program which computes the discrete logarithm in $\mathbb{Z}_p^*$ by exhaustive search. The input parameters for your program are p, α, β. The program computes x where $\beta = \alpha^x \bmod p$.

Compute the solution to $\log_{106} 12375$ in $\mathbb{Z}_{24691}$.

8.13. Encrypt the following messages with the Elgamal scheme ($p = 467$ and $\alpha = 2$):

1. $k_{pr} = d = 105, i = 213, x = 33$
2. $k_{pr} = d = 105, i = 123, x = 33$
3. $k_{pr} = d = 300, i = 45, x = 248$
4. $k_{pr} = d = 300, i = 47, x = 248$

Now decrypt every ciphertext and show all steps.

8.14. Assume Bob sends an Elgamal encrypted message to Alice. Wrongly, Bob uses the same parameter i for all messages. Moreover, we know that each of Bob's cleartexts start with the number $x_1 = 21$ (Bob's ID). We now obtain the following ciphertexts

$$(k_{E,1} = 6, y_1 = 17),$$
$$(k_{E,2} = 6, y_2 = 25).$$

The Elgamal parameters are $p = 31, \alpha = 3, \beta = 18$. Determine the second plaintext x_2.

8.15. Given is an Elgamal crypto system. Bob tries to be especially smart and chooses the following pseudorandom generator to compute new i values:

$$i_j = i_{j-1} + f(j) \ , \ 1 \le j \tag{8.5}$$

where $f(j)$ is a "complicated" but known pseudorandom function (for instance, $f(j)$ could be a cryptographic hash function such as SHA or RIPE-MD160). i_0 is a true random number that is not known to Oscar.

Bob encrypts n messages x_j as follows:

$$k_{E_j} = \alpha^{i_j} \bmod p,$$
$$y_j = x_j \cdot \beta^{i_j} \bmod p,$$

where $1 \leq j \leq n$. Assume that the last cleartext x_n is known to Oscar and all ciphertext.

Provide a formula with which Oscar can compute any of the messages x_j, $1 \leq j \leq n-1$. Of course, following Kerckhoffs' principle, Oscar knows the construction method shown above, including the function $f()$.

8.16. Given an Elgamal encryption scheme with public parameters $k_{pub} = (p, \alpha, \beta)$ and an unknown private key $k_{pr} = d$. Due to an erroneous implementation of the random number generator of the encrypting party, the following relation holds for two temporary keys:

$$k_{M,j+1} = k_{M,j}^2 \quad \bmod p.$$

Given n consecutive ciphertexts

$$(k_{E_1}, y_1), (k_{E_2}, y_2), ..., (k_{E_n}, y_n)$$

to the plaintexts

$$x_1, x_2, ..., x_n.$$

Furthermore, the first plaintext x_1 is known (e.g., header information).

1. Describe how an attacker can compute the plaintexts $x_1, x_2, ..., x_n$ from the given quantities.
2. Can an attacker compute the private key d from the given information? Give reasons for your answer.

8.17. Considering the four examples from Problem 8.13, we see that the Elgamal scheme is nondeterministic: A given plaintext x has many valid ciphertexts, e.g., both $x = 33$ and $x = 248$ have the same ciphertext in the problem above.

1. Why is the Elgamal signature scheme nondeterministic?
2. How many valid ciphertexts exist for each message x (general expression)?
   How many are there for the system in Problem 8.13 (numerical answer)?
3. Is the RSA crypto system nondeterministic once the public key has been chosen?

8.18. We investigate the weaknesses that arise in Elgamal encryption if a public key of small order is used. We look at the following example. Assume Bob uses the group $\mathbb{Z}_{29}^*$ with the primitive element $\alpha = 2$. His public key is $\beta = 28$.

1. What is the order of the public key?
2. Which masking keys k_M are possible?
3. Alice encrypts a text message. Every character is encoded according to the simple rule a $\rightarrow$ 0,..., z $\rightarrow$ 25. There are three additional ciphertext symbols: ä $\rightarrow$ 26, ö $\rightarrow$ 27, ü $\rightarrow$ 28. She transmits the following 11 ciphertexts (k_E, y):

$$(3,15), (19,14), (6,15), (1,24), (22,13), (4,7),$$
$$(13,24), (3,21), (18,12), (26,5), (7,12)$$

Decrypt the message without computing Bob's private key. Just look at the ciphertext and use the fact that there are only very few masking keys and a bit of guesswork.

Chapter 9
Elliptic Curve Cryptosystems

Elliptic Curve Cryptography (ECC) is the newest member of the three families of established public-key algorithms of practical relevance introduced in Sect. 6.2.3. However, ECC has been around since the mid-1980s.

ECC provides the same level of security as RSA or discrete logarithm systems with considerably shorter operands (approximately 160–256 bit vs. 1024–3072 bit). ECC is based on the generalized discrete logarithm problem, and thus DL-protocols such as the Diffie–Hellman key exchange can also be realized using elliptic curves. In many cases, ECC has performance advantages (fewer computations) and bandwidth advantages (shorter signatures and keys) over RSA and Discrete Logarithm (DL) schemes. However, RSA operations which involve short public keys as introduced in Sect. 7.5.1 are still much faster than ECC operations.

The mathematics of elliptic curves are considerably more involved than those of RSA and DL schemes. Some topics, e.g., counting points on elliptic curves, go far beyond the scope of this book. Thus, the focus of this chapter is to explain the basics of ECC in a clear fashion without too much mathematical overhead, so that the reader gains an understanding of the most important functions of cryptosystems based on elliptic curves.

In this chapter, you will learn:

- The basic pros and cons of ECC vs. RSA and DL schemes.
- What an elliptic curve is and how to compute with it.
- How to build a DL problem with an elliptic curve.
- Protocols that can be realized with elliptic curves.
- Current security estimations of cryptosystems based on elliptic curves.

9.1 How to Compute with Elliptic Curves

We start by giving a short introduction to the mathematical concept of elliptic curves, independent of their cryptographic applications. ECC is based on the generalized discrete logarithm problem. Hence, what we try to do first is to find a cyclic group on which we can build our cryptosystem. Of course, the mere existence of a cyclic group is not sufficient. The DL problem in this group must also be computationally hard, which means that it must have good one-way properties.

We start by considering certain polynomials (e.g., functions with sums of exponents of x and y), and we plot them over the real numbers.

Example 9.1. Let's look at the polynomial equation $x^2 + y^2 = r^2$ over the real numbers $\mathbb{R}$. If we plot all the pairs (x, y) which fulfill this equation in a coordinate sys-

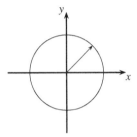

Fig. 9.1 Plot of all points (x, y) which fulfill the equation $x^2 + y^2 = r^2$ over $\mathbb{R}$

tem, we obtain a circle as shown in Fig. 9.1.

◇

We now look at other polynomial equations over the real numbers.

Example 9.2. A slight generalization of the circle equation is to introduce coefficients to the two terms x^2 and y^2, i.e., we look at the set of solutions to the equation $a \cdot x^2 + b \cdot y^2 = c$ over the real numbers. It turns out that we obtain an ellipse, as

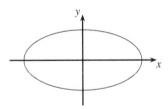

Fig. 9.2 Plot of all points (x, y) which fulfill the equation $a \cdot x^2 + b \cdot y^2 = c$ over $\mathbb{R}$

shown in Figure 9.2.

◇

9.1.1 Definition of Elliptic Curves

From the two examples above, we conclude that we can form certain types of curves from polynomial equations. By "curves", we mean the set of points (x,y) which are solutions of the equations. For example, the point $(x = r, y = 0)$ fulfills the equation of a circle and is, thus, in the set. The point $(x = r/2, y = r/2)$ is not a solution to the polynomial $x^2 + y^2 = r^2$ and is, thus, not a set member. An *elliptic curve* is a special type of polynomial equation. For cryptographic use, we need to consider the curve not over the real numbers but over a finite field. The most popular choice is prime fields $GF(p)$ (cf. Sect. 4.2), where all arithmetic is performed modulo a prime p.

Definition 9.1.1 Elliptic Curve
The elliptic curve over $\mathbb{Z}_p$, $p > 3$, is the set of all pairs $(x,y) \in \mathbb{Z}_p$ which fulfill

$$y^2 \equiv x^3 + a \cdot x + b \bmod p \qquad (9.1)$$

together with an imaginary point of infinity $\mathcal{O}$, where

$$a, b \in \mathbb{Z}_p$$

and the condition $4 \cdot a^3 + 27 \cdot b^2 \neq 0 \bmod p$.

The definition of elliptic curve requires that the curve is nonsingular. Geometrically speaking, this means that the plot has no self-intersections or vertices, which is achieved if the discriminant of the curve $-16(4a^3 + 27b^2)$ is nonzero.

For cryptographic use we are interested in studying the curve over a prime field as in the definition. However, if we plot such an elliptic curve over $\mathbb{Z}_p$, we do not get anything remotely resembling a curve. However, nothing prevents us from taking an elliptic curve equation and plotting it over the set of real numbers.

Example 9.3. In Figure 9.3 the elliptic curve $y^2 = x^3 - 3x + 3$ is shown over the real numbers.

◇

We notice several things from this elliptic curve plot.[1] First, the elliptic curve is symmetric with respect to the x-axis. This follows directly from the fact that for all values x_i which are on the elliptic curve, both $y_i = \sqrt{x_i^3 + a \cdot x_i + b}$ and $y_i' = -\sqrt{x_i^3 + a \cdot x_i + b}$ are solutions. Second, there is one intersection with the x-axis. This follows from the fact that it is a cubic equation if we solve for $y = 0$ which has one real solution (the intersection with the x-axis) and two complex solutions (which do not show up in the plot). There are also elliptic curves with three intersections with the x-axis.

---

[1] Note that elliptic curves are not ellipses. They play a role in determining the circumference of ellipses, hence the name.

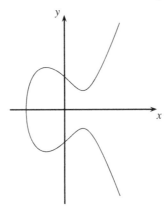

Fig. 9.3 $y^2 = x^3 - 3x + 3$ over $\mathbb{R}$

We now return to our original goal of finding a curve with a large cyclic group, which is needed for constructing a discrete logarithm problem. The first task for finding a group is done, namely identifying a set of elements. In the elliptic curve case, the group elements are the points that fulfill Eq. (9.1). The next question at hand is: How do we define a group operation with those points? Of course, we have to make sure that the group laws from Definition 4.3.1 in Sect. 4.2 hold for the operation.

9.1.2 Group Operations on Elliptic Curves

Let's denote the group operation with the addition symbol[2] "+". "Addition" means that given two points and their coordinates, say $P = (x_1, y_1)$ and $Q = (x_2, y_2)$, we have to compute the coordinates of a third point R such that:

$$P + Q = R$$
$$(x_1, y_1) + (x_2, y_2) = (x_3, y_3)$$

As we will see below, it turns out that this addition operation looks quite arbitrary. Luckily, there is a nice geometric interpretation of the addition operation if we consider a curve defined over the real numbers. For this geometric interpretation, we have to distinguish two cases: the addition of two distinct points (named point addition) and the addition of one point to itself (named point doubling).

Point Addition $P + Q$ This is the case where we compute $R = P + Q$ and $P \neq Q$. The construction works as follows: Draw a line through P and Q and obtain a third point of intersection between the elliptic curve and the line. Mirror this third

---

[2] Note that the choice of naming the operation "addition" is completely arbitrary; we could have also called it multiplication.

intersection point along the x-axis. This mirrored point is, by definition, the point R. Figure 9.4 shows the point addition on an elliptic curve over the real numbers.

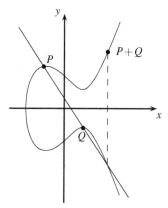

Fig. 9.4 Point addition on an elliptic curve over the real numbers

Point Doubling $P + P$ This is the case where we compute $P + Q$ but $P = Q$. Hence, we can write $R = P + P = 2P$. We need a slightly different construction here. We draw the tangent line through P and obtain a second point of intersection between this line and the elliptic curve. We mirror the point of the second intersection along the x-axis. This mirrored point is the result R of the doubling. Figure 9.5 shows the

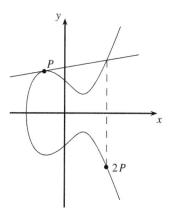

Fig. 9.5 Point doubling on an elliptic curve over the real numbers

doubling of a point on an elliptic curve over the real numbers.

You might wonder why the group operations have such an arbitrary looking form. Historically, this *tangent-and-chord* method was used to construct a third point if two points were already known, while only using the four standard algebraic operations add, subtract, multiply and divide. It turns out that if points on the elliptic curve are *added* in this very way, the set of points also fulfill most conditions necessary for a group, that is, closure, associativity, existence of an identity element and existence of an inverse.

Of course, in a cryptosystem we cannot perform geometric constructions. However, by applying simple coordinate geometry, we can express both of the geometric constructions from above through analytic expressions, i.e., formulae. As stated above, these formulae only involve the four basic algebraic operations. These operations can be performed in any field, not only over the field of the real numbers (cf. Sect. 4.2). In particular, we can take the curve equation from above, but we now consider it over prime fields $GF(p)$ rather than over the real numbers. This yields the following analytical expressions for the group operation.

Elliptic Curve Point Addition and Point Doubling

$$x_3 = s^2 - x_1 - x_2 \bmod p$$
$$y_3 = s(x_1 - x_3) - y_1 \bmod p$$

where

$$s = \begin{cases} \frac{y_2 - y_1}{x_2 - x_1} \bmod p \; ; \text{if } P \neq Q \text{ (point addition)} \\ \frac{3x_1^2 + a}{2y_1} \bmod p \; ; \text{if } P = Q \text{ (point doubling)} \end{cases}$$

Note that the parameter s is the slope of the line through P and Q in the case of point addition, or the slope of the tangent through P in the case of point doubling.

Even though we made major headway towards the establishment of a finite group, we are not there yet. One thing that is still missing is an identity (or neutral) element $\mathcal{O}$ such that:

$$P + \mathcal{O} = P$$

for all points P on the elliptic curve. It turns out that there isn't any point (x, y) that fulfills the condition. Instead we **define** an abstract *point at infinity* as the neutral element $\mathcal{O}$. This point at infinity can be visualized as a point that is located towards "plus" infinity along the y-axis or towards "minus" infinity along the y-axis.

According the group definition, we can now also define the inverse $-P$ of any group element P as:

$$P + (-P) = \mathcal{O}.$$

The question is how do we find $-P$? If we apply the tangent-and-chord method from above, it turns out that the inverse of the point $P = (x_p, y_p)$ is the point $-P = (x_p, -y_p)$, i.e., the point that is reflected along the x-axis. Figure 9.6 shows the point P together with its inverse. Note that finding the inverse of a point $P = (x_p, y_p)$ is now trivial. We simply take the negative of its y coordinate. In the case of elliptic curves over a prime field $GF(p)$ (the most interesting case in cryptography), this is easily achieved since $-y_p \equiv p - y_p \bmod p$, hence

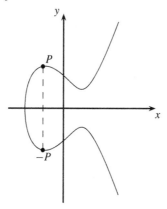

Fig. 9.6 The inverse of a point P on an elliptic curve

$$-P = (x_p, p - y_p).$$

Now that we have defined all group properties for elliptic curves, we now look at an example for the group operation.

Example 9.4. We consider a curve over the small field $\mathbb{Z}_{17}$:

$$E : y^2 \equiv x^3 + 2x + 2 \bmod 17.$$

We want to double the point $P = (5, 1)$.

$$2P = P + P = (5, 1) + (5, 1) = (x_3, y_3)$$
$$s = \frac{3x_1^2 + a}{2y_1} = (2 \cdot 1)^{-1}(3 \cdot 5^2 + 2) = 2^{-1} \cdot 9 \equiv 9 \cdot 9 \equiv 13 \bmod 17$$
$$x_3 = s^2 - x_1 - x_2 = 13^2 - 5 - 5 = 159 \equiv 6 \bmod 17$$
$$y_3 = s(x_1 - x_3) - y_1 = 13(5 - 6) - 1 = -14 \equiv 3 \bmod 17$$
$$2P = (5, 1) + (5, 1) = (6, 3)$$

For illustrative purposes we check whether the result $2P = (6, 3)$ is actually a point on the curve by inserting the coordinates into the curve equation:

$$y^2 \equiv x^3 + 2 \cdot x + 2 \bmod 17$$
$$3^2 \equiv 6^3 + 2 \cdot 6 + 2 \bmod 17$$
$$9 = 230 \equiv 9 \bmod 17$$

◇

9.2 Building a Discrete Logarithm Problem with Elliptic Curves

What we have done so far is to establish the group operations (point addition and doubling), we have provided an identity element, and we have shown a way of finding the inverse for any point on the curve. Thus, we now have all necessary requirements in place to motivate the following theorem:

> **Theorem 9.2.1** *The points on an elliptic curve together with $\mathcal{O}$ have cyclic subgroups. Under certain conditions all points on an elliptic curve form a cyclic group.*

Please note that we have not proved the theorem. This theorem is extremely useful because we have a good understanding of the properties of cyclic groups. In particular, we know that by definition a primitive element must exist such that its powers generate the entire group. Moreover, we know quite well how to build cryptosystems from cyclic groups. Here is an example for the cyclic group of an elliptic curve.

Example 9.5. We want to find all points on the curve:

$$E : y^2 \equiv x^3 + 2 \cdot x + 2 \bmod 17.$$

It happens that all points on the curve form a cyclic group and that the order is $\#E = 19$. For this specific curve the group order is a prime and, according to Theorem 8.2.4, every element is primitive.

As in the previous example we start with the primitive element $P = (5,1)$. We compute now all "powers" of P. More precisely, since the group operation is addition, we compute $P, 2P, \ldots, (\#E) P$. Here is a list of the elements that we obtain:

$$
\begin{aligned}
2P &= (5,1) + (5,1) = (6,3) & 11P &= (13,10) \\
3P &= 2P + P = (10,6) & 12P &= (0,11) \\
4P &= (3,1) & 13P &= (16,4) \\
5P &= (9,16) & 14P &= (9,1) \\
6P &= (16,13) & 15P &= (3,16) \\
7P &= (0,6) & 16P &= (10,11) \\
8P &= (13,7) & 17P &= (6,14) \\
9P &= (7,6) & 18P &= (5,16) \\
10P &= (7,11) & 19P &= \mathcal{O}
\end{aligned}
$$

From now on, the cyclic structure becomes visible since:

$$
\begin{aligned}
20P &= 19P + P = \mathcal{O} + P = P \\
21P &= 2P
\end{aligned}
$$

$$\vdots$$

It is also instructive to look at the last computation above, which yielded:

$$18P + P = \mathcal{O}.$$

This means that $P = (5,1)$ is the inverse of $18P = (5,16)$, and vice versa. This is easy to verify. We have to check whether the two x coordinates are identical and that the two y coordinates are each other's additive inverse modulo 17. The first condition obviously hold and the second one too, since

$$-1 \equiv 16 \bmod 17.$$

◇

To set up DL cryptosystems it is important to know the order of the group. Even though knowing the exact number of points on a curve is an elaborate task, we know the approximate number due to *Hasse's theorem*.

Theorem 9.2.2 Hasse's theorem
Given an elliptic curve E modulo p, the number of points on the curve is denoted by #E and is bounded by:

$$p + 1 - 2\sqrt{p} \leq \#E \leq p + 1 + 2\sqrt{p}.$$

Hasse's theorem, which is also known as *Hasse's bound*, states that the number of points is roughly in the range of the prime p. This has major practical implications. For instance, if we need an elliptic curve with 2^{160} elements, we have to use a prime of length of about 160 bit.

Let's now turn our attention to the details of setting up the discrete logarithm problem. For this, we can strictly proceed as described in Chapter 8.

Definition 9.2.1 Elliptic Curved Discrete Logarithm Problem (ECDLP)
Given is an elliptic curve E. We consider a primitive element P and another element T. The DL problem is finding the integer d, where $1 \leq d \leq \#E$, such that:

$$\underbrace{P + P + \cdots + P}_{d \ times} = dP = T. \tag{9.2}$$

In cryptosystems, d is the private key which is an integer, while the public key T is a point on the curve with coordinates $T = (x_T, y_T)$. In contrast, in the case of the DL problem in $\mathbb{Z}_p^*$, both keys were integers. The operation in Eq. (9.2) is called *point multiplication*, since we can formally write $T = dP$. This terminology can be misleading, however, since we cannot directly multiply the integer d with a curve point P. Instead, dP is merely a convenient notation for the repeated application of the group operation in Equation (9.2).[3] Let's now look at an example for an ECDLP.

Example 9.6. We perform a point multiplication on the curve $y^2 \equiv x^3 + 2x + 2$ mod 17 that was also used in the previous example. We want to compute

$$13P = P + P + \ldots + P$$

where $P = (5, 1)$. In this case, we can simply use the table that was compiled earlier:

$$13P = (16, 4).$$

◇

Point multiplication is analog to exponentiation in multiplicative groups. In order to do it efficiently, we can directly adopt the square-and-multiply algorithm. The only difference is that squaring becomes doubling and multiplication becomes addition of P. Here is the algorithm:

Double-and-Add Algorithm for Point Multiplication
Input: elliptic curve E together with an elliptic curve point P
a scalar $d = \sum_{i=0}^{t} d_i 2^i$ with $d_i \in 0, 1$ and $d_t = 1$
Output: $T = dP$
Initialization:
$T = P$
Algorithm:
1   FOR $i = t - 1$ DOWNTO 0
1.1 $T = T + T$ mod n
            IF $d_i = 1$
1.2 $T = T + P$ mod n
2   RETURN (T)

For a random scalar of length of $t + 1$ bit, the algorithm requires on average $1.5t$ point doubles and additions. Verbally expressed, the algorithm scans the bit representation of the scalar d from left to right. It performs a doubling in every iteration, and only if the current bit has the value 1 does it perform an addition of P. Let's look at an example.

Example 9.7. We consider the scalar multiplication $26P$, which has the following binary representation:

$$26P = (11010_2)P = (d_4 d_3 d_2 d_1 d_0)_2 P.$$

---

[3] Note that the symbol "+" was chosen arbitrarily to denote the group operation. If we had chosen a multiplicative notation instead, the ECDLP would have had the form $P^d = T$, which would have been more consistent with the conventional DL problem in $\mathbb{Z}_p^*$.

The algorithm scans the scalar bits starting on the left with d_4 and ending with the rightmost bit d_0.

Step		
#0	$P = 1_2 P$	inital setting, bit processed: $d_4 = 1$
#1a	$P + P = 2P = 10_2 P$	DOUBLE, bit processed: d_3
#1b	$2P + P = 3P = 10_2 P + 1_2 P = 11_2 P$	ADD, since $d_3 = 1$
#2a	$3P + 3P = 6P = 2(11_2 P) = 110_2 P$	DOUBLE, bit processed: d_2
#2b		no ADD, since $d_2 = 0$
#3a	$6P + 6P = 12P = 2(110_2 P) = 1100_2 P$	DOUBLE, bit processed: d_1
#3b	$12P + P = 13P = 1100_2 P + 1_2 P = 1101_2 P$	ADD, since $d_1 = 1$
#4a	$13P + 13P = 26P = 2(1101_2 P) = 11010_2 P$	DOUBLE, bit processed: d_0
#4b		no ADD, since $d_0 = 0$

It is instructive to observe how the binary representation of the exponent evolves. We see that doubling results in a left shift of the scalar, with a 0 put in the rightmost position. By performing addition with P, a 1 is inserted into the rightmost position of the scalar. Compare how the highlighted exponents change from iteration to iteration.

◇

If we go back to elliptic curves over the real numbers, there is a nice geometric interpretation for the ECDLP: given a starting point P, we compute $2P$, $3P$, ..., $dP = T$, effectively hopping back and forth on the elliptic curve. We then publish the starting point P (a public parameter) and the final point T (the public key). In order to break the cryptosystem, an attacker has to figure out how often we "jumped" on the elliptic curve. The number of hops is the secret d, the private key.

9.3 Diffie–Hellman Key Exchange with Elliptic Curves

In complete analogy to the conventional Diffie–Hellman key exchange (DHKE) introduced in Sect. 8.1, we can now realize a key exchange using elliptic curves. This is referred to as elliptic curve Diffie–Hellman key exchange, or ECDH. First we have to agree on domain parameters, that is, a suitable elliptic curve over which we can work and a primitive element on this curve.

ECDH Domain Parameters

1. Choose a prime p and the elliptic curve

$$E : y^2 \equiv x^3 + a \cdot x + b \bmod p$$

2. Choose a primitive element $P = (x_P, y_P)$
The prime p, the curve given by its coefficients a, b, and the primitive element P are the domain parameters.

Note that in practice finding a suitable elliptic curve is a relatively difficult task. The curves have to show certain properties in order to be secure. More about this is said below. The actual key exchange is done the same way it was done for the conventional Diffie–Hellman protocol.

Elliptic Curve Diffie–Hellman Key Exchange (ECDH)

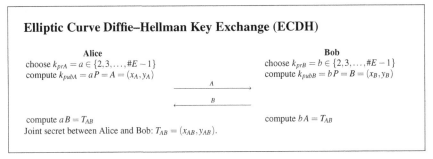

The correctness of the protocol is easy to prove.

Proof. Alice computes

$$a B = a (b P)$$

while Bob computes

$$b A = b (a P).$$

Since point addition is associative (remember that associativity is one of the group properties), both parties compute the same result, namely the point $T_{AB} = abP$. □

As can be seen in the protocol, Alice and Bob choose the private keys a and b, respectively, which are two large integers. With the private keys both generate their respective public keys A and B, which are points on the curve. The public keys are computed by point multiplication. The two parties exchange these public parameters with each other. The joint secret T_{AB} is then computed by both Alice and Bob by performing a second point multiplication involving the public key they received and their own secret parameter. The joint secret T_{AB} can be used to derive a session key, e.g., as input for the AES algorithm. Note that the two coordinates (x_{AB}, y_{AB}) are not independent of each other: Given x_{AB}, the other coordinate can be computed by simply inserting the x value in the elliptic curve equation. Thus, only one of the two coordinates should be used for the derivation of a session key. Let's look at an example with small numbers:

Example 9.8. We consider the ECDH with the following domain parameters. The elliptic curve is $y^2 \equiv x^3 + 2x + 2 \bmod 17$, which forms a cyclic group of order $\#E = 19$. The base point is $P = (5,1)$. The protocol proceeds as follows:

Alice	**Bob**
choose $a = k_{pr,A} = 3$	choose $b = k_{pr,B} = 10$
$A = k_{pub,A} = 3P = (10,6)$	$B = k_{pub,B} = 10P = (7,11)$

$$\xrightarrow{\hspace{3cm} A \hspace{3cm}}$$

$$\xleftarrow{\hspace{3cm} B \hspace{3cm}}$$

$$T_{AB} = aB = 3(7,11) = (13,10) \qquad\qquad T_{AB} = bA = 10(10,6) = (13,10)$$

The two scalar multiplications that each Alice and Bob perform require the Double-and-Add algorithm.

$\diamond$

One of the coordinates of the joint secret T_{AB} can now be used as session key. In practice, often the x-coordinate is hashed and then used as a symmetric key. Typically, not all bits are needed. For instance, in a 160-bit ECC scheme, hashing the x-coordinate with SHA-1 results in a 160-bit output of which only 128 would be used as an AES key.

Please note that elliptic curves are not restricted to the DHKE. In fact, almost all other discrete logarithm protocols, in particular digital signatures and encryption, e.g., variants of Elgamal, can also be realized. The widely used elliptic curve digital signature algorithms (ECDSA) will be introduced in Sect. 10.5.1.

9.4 Security

The reason we use elliptic curves is that the ECDLP has very good one-way characteristics. If an attacker Oscar wants to break the ECDH, he has the following information: E, p, P, A, and B. He wants to compute the joint secret between Alice and Bob $T_{AB} = a \cdot b \cdot P$. This is called the elliptic curve Diffie–Hellman problem (ECDHP). There appears to be only one way to compute the ECDHP, namely to solve either of the discrete logarithm problems:

$$a = \log_P A$$

or

$$b = \log_P B$$

If the elliptic curve is chosen with care, the best known attacks against the ECDLP are considerably weaker than the best algorithms for solving the DL problem modulo p, and the best factoring algorithms which are used for RSA attacks. In particular, the index-calculus algorithms, which are powerful attacks against the DLP modulo p, are not applicable against elliptic curves. For carefully selected elliptic curves, the only remaining attacks are generic DL algorithms, that is Shanks' baby-step giant-step method and Pollard's rho method, which were described in Sect. 8.3.3. Since the number of steps required for such an attack is roughly equal to the square root of the group cardinality, a group order of at least 2^{160} should be used. According to Hasse's theorem, this requires that the prime p used for the elliptic curve must be roughly 160-bit long. If we attack such a group with generic algorithms, we need around $\sqrt{2^{160}} = 2^{80}$ steps. A security level of 80 bit provides medium-term security. In practice, elliptic curve bit lengths up to 256 bit are commonly used, which provide security levels of up to 128 bit.

It should be stressed that this security is only achieved if cryptographically strong elliptic curves are used. There are several families of curves that possess cryptographic weaknesses, e.g., supersingular curves. They are relatively easy to spot, however. In practice, often standardized curves such as ones proposed by the National Institute of Standards and Technology (NIST) are being used.

9.5 Implementation in Software and Hardware

Before using ECC, a curve with good cryptographic properties needs to be identified. In practice, a core requirement is that the cyclic group (or subgroup) formed by the curve points has prime order. Moreover, certain mathematical properties that lead to cryptographic weaknesses must be ruled out. Since assuring all these properties is a nontrivial and computationally demanding task, often standardized curves are used in practice.

When implementing elliptic curves it is useful to view an ECC scheme as a structure with four layers. On the bottom layer modular arithmetic, i.e., arithmetic in the prime field $GF(p)$, is performed. We need all four field operations: addition, subtraction, multiplication and inversion. On the next layer, the two group operations, point doubling and point addition, are realized. They make use of the arithmetic provided in the bottom layer. On the third layer, scalar multiplication is realized, which uses the group operations of the previous layer. The top layer implements the actual protocol, e.g., ECDH or ECDSA. It is important to note that two entirely different finite algebraic structures are involved in an elliptic curve cryptosystem. There is a *finite field $GF(p)$* over which the curve is defined, and there is the *cyclic group* which is formed by the points on the curve.

In software, a highly optimized 256-bit ECC implementation on a 3-GHz, 64-bit CPU can take approximately 2 ms for one point multiplication. Slower throughputs due to smaller microprocessors or less optimized algorithms are common with performances in the range of 10 ms. For high-performance applications, e.g., for Internet servers that have to perform a large number of elliptic curve signatures per second, hardware implementations are desirable. The fastest implementations can compute a point multiplication in the range of 40 μs, while speeds of several 100 μs are more common.

On the other side of the performance spectrum, ECC is the most attractive public-key algorithm for lightweight applications such as RFID tags. Highly compact ECC engines are possible which need as little as 10,000 gate equivalences and run at a speed of several tens of milliseconds. Even though ECC engines are much larger than implementations of symmetric ciphers such as 3DES, they are considerably smaller than RSA implementations.

The computational complexity of ECC is cubic in the bit length of the prime used. This is due to the fact that modular multiplication, which is the main operation on the bottom layer, is quadratic in the bit length, and scalar multiplication (i.e., with the Double-and-Add algorithm) contributes another linear dimension, so that we have, in total, a cubic complexity. This implies that doubling the bit length of an ECC implementation results in performance degradation by a factor of roughly $2^3 = 8$. RSA and DL systems show the same cubic runtime behavior. The advantage of ECC over the other two popular public-key families is that the parameters have to be increased much more slowly to enhance the security level. For instance, doubling the effort of an attacker for a given ECC system requires an increase in the length of the parameter by 2 bits, whereas RSA or DL schemes require an increase of 20–30 bits. This behavior is due to the fact that only generic attacks (cf. Sect. 8.3.3) are known ECC cryptosystems, whereas more powerful algorithms are available for attacking RSA and DL schemes.

9.6 Discussion and Further Reading

History and General Remarks ECC was independently invented in 1987 by Neal Koblitz and in 1986 by Victor Miller. During the 1990s there was much speculation about the security and practicality of ECC, especially if compared to RSA. After a period of intensive research, they appear nowadays very secure, just like RSA and DL schemes. An important step for building confidence in ECC was the issuing of two ANSI banking standards for elliptic curve digital signature and key establishment in 1999 and 2001, respectively [6, 7]. Interestingly, in Suite B—a collection of crypto algorithms selected by the NSA for use in US government systems—only ECC schemes are allowed as asymmetric algorithms [130]. Elliptic curves are also widely used in commercial standards such as IPsec or Transport Layer Security (TLS).

At the time of writing, there still exist far more fielded RSA and DL applications than elliptic curve ones. This is mainly due to historical reasons and due to the quite complex patent situation of some ECC variants. Nevertheless, in many new applications with security needs, especially in embedded systems such as mobile devices, ECC is often the preferred public-key scheme. For instance, ECC is used in the most popular business handheld devices. Most likely, ECC will become more widespread in the years to come. Reference [100] describes the historical development of ECC with respect to scientific and commercial aspects, and makes excellent reading.

For readers interested in a deeper understanding of ECC, the books [25, 24, 90, 44] are recommended. The overview article [103], even though a bit dated now, provides a good state-of-the-art summary as of the year 2000. For more recent developments, the annual *Workshop on Elliptic Curve Cryptography (ECC)* is recommended as an excellent resource [166]. The workshop includes both theoretical and

applied topics related to ECC and related crypto schemes. There is also a rich literature that deals with the mathematics of elliptic curves [154, 101, 155], regardless
of their use in cryptography.

Implementation and Variants In the first few years after the invention of ECC,
these algorithms were believed to be computationally more complex than existing
public-key schemes, especially RSA. This assumption is somewhat ironic in hind-
sight, given that ECC tends to be often faster than most other public-key schemes.
During the 1990s, fast implementation techniques for ECC was intensively re-
searched, which resulted in considerable performance improvements.

In this chapter, elliptic curves over prime fields $GF(p)$ were introduced. These
are currently in practice somewhat more widely used than over other finite fields, but
curves over binary Galois fields $GF(2^m)$ are also popular. For efficient implementations, improvements are possible at the finite field arithmetic layer, at the group
operation layer and at the point multiplication layer. There is a wealth of techniques
and in the following is a summary of the most common acceleration techniques in
practice. For curves over $GF(p)$, generalized Mersenne primes are often used at the
arithmetic level. These are primes such as $p = 2^{192} - 2^{64} - 1$. Their major advantage
is that modulo reduction is extremely simple. If general primes are used, methods
similar to those described in Sect. 7.10 are applicable. With respect to ECC over
fields $GF(2^m)$, efficient arithmetic algorithms are described in [90]. On the group
operation layer, several optimizations are possible. A popular one is to switch from
the affine coordinates that were introduced here to projective coordinates, in which
each point is represented as a triple (x, y, z). Their advantage is that no inversion
is required within the group operation. The number of multiplications increases,
however. On the next layer, fast scalar multiplication techniques are applicable. Improved versions of the Double-and-Add algorithm which make use of the fact that
adding or subtracting a point come at almost identical costs are commonly being
applied. An excellent compilation of efficient computation techniques for ECC is
the book [90].

A special type of elliptic curve that allows for particularly fast point multiplication is the *Koblitz curve* [158]. These are curves over $GF(2^m)$ where the coefficients
have the values 0 or 1. There have also been numerous other suggestions for elliptic curves with good implementation properties. One such proposal involves elliptic
curves over optimum extension fields, i.e., fields of the form $GF(p^m)$, $p > 2$ [10].

As mentioned in Sect. 9.5, standardized curves are often being used in practice.
A widely used set of curves is provided in the FIPS Standard [126, Appendix D].
Alternatives are curves specified by the ECC Brainpool consortium or the Standards
for Efficient Cryptography Group (SECG) [34, 9].

Elliptic curves also allow for many variants and generalization. They are a special
case of hyperelliptic curves, which can also be used to build discrete logarithm cryptosystems [44]. A summary of implementation techniques for hyperelliptic curves is
given in [175]. A completely different type of public-key scheme which also makes
use of elliptic curves is identity-based cryptosystems [30], which have drawn much
attention over the last few years.

9.7 Lessons Learned

- Elliptic Curve Cryptography (ECC) is based on the discrete logarithm problem. It requires arithmetic modulo a prime or in a Galois field $GF(2^m)$.
- ECC can be used for key exchange, for digital signatures and for encryption.
- ECC provides the same level of security as RSA or discrete logarithm systems over $\mathbb{Z}_p^*$ with considerably shorter operands (approximately 160–256 bit vs. 1024–3072 bit), which results in shorter ciphertexts and signatures.
- In many cases ECC has performance advantages over other public-key algorithms. However, signature verification with short RSA keys is still considerably faster than ECC.
- ECC is slowly gaining popularity in applications, compared to other public-key schemes, i.e., many new applications, especially on embedded platforms, make use of elliptic curve cryptography.

Problems

9.1. Show that the condition $4a^3 + 27b^2 \neq 0 \bmod p$ is fulfilled for the curve

$$y^2 \equiv x^3 + 2x + 2 \bmod 17 \tag{9.3}$$

9.2. Perform the additions

1. $(2,7) + (5,2)$
2. $(3,6) + (3,6)$

in the group of the curve $y^2 \equiv x^3 + 2x + 2 \bmod 17$. Use only a pocket calculator.

9.3. In this chapter the elliptic curve $y^2 \equiv x^3 + 2x + 2 \bmod 17$ is given with $\#E = 19$. Verify Hasse's theorem for this curve.

9.4. Let us again consider the elliptic curve $y^2 \equiv x^3 + 2x + 2 \bmod 17$. Why are *all* points primitive elements?

Note: In general it is not true that all elements of an elliptic curve are primitive.

9.5. Let E be an elliptic curve defined over $\mathbb{Z}_7$:

$$E : y^2 = x^3 + 3x + 2.$$

1. Compute all points on E over $\mathbb{Z}_7$.
2. What is the order of the group? (Hint: Do not miss the neutral element $\mathcal{O}$.)
3. Given the element $\alpha = (0,3)$, determine the order of α. Is α a primitive element?

9.6. In practice, a and k are both in the range $p \approx 2^{150} \cdots 2^{250}$, and computing $T = a \cdot P$ and $y_0 = k \cdot P$ is done using the Double-and-Add algorithm as shown in Sect. 9.2.

1. Illustrate how the algorithm works for $a = 19$ and for $a = 160$. Do *not* perform elliptic curve operations, but keep P a variable.
2. How many (i) point additions and (ii) point doublings are required on average for one "multiplication"? Assume that all integers have $n = \lceil \log_2 p \rceil$ bit.
3. Assume that all integers have $n = 160$ bit, i.e., p is a 160-bit prime. Assume one group operation (addition or doubling) requires 20 μsec. What is the time for one double-and-add operation?

9.7. Given an elliptic curve E over $\mathbb{Z}_{29}$ and the base point $P = (8,10)$:

$$E : y^2 = x^3 + 4x + 20 \bmod 29.$$

Calculate the following point multiplication $k \cdot P$ using the Double-and-Add algorithm. Provide the intermediate results after each step.

1. $k = 9$
2. $k = 20$

9.8. Given is the same curve as in 9.7. The order of this curve is known to be #$E =$ 37. Furthermore, an additional point $Q = 15 \cdot P = (14, 23)$ on this curve is given. Determine the result of the following point multiplications by using as few group operations as possible, i.e., make smart use of the known point Q. Specify *how* you simplified the calculation each time.

Hint: In addition to using Q, use the fact that it is easy to compute $-P$.

1. $16 \cdot P$
2. $38 \cdot P$
3. $53 \cdot P$
4. $14 \cdot P + 4 \cdot Q$
5. $23 \cdot P + 11 \cdot Q$

You should be able to perform the scalar multiplications with considerably fewer steps than a straightforward application of the double-and-add algorithm would allow.

9.9. Your task is to compute a session key in a DHKE protocol based on elliptic curves. Your private key is $a = 6$. You receive Bob's public key $B = (5, 9)$. The elliptic curve being used is defined by

$$y^2 \equiv x^3 + x + 6 \bmod 11.$$

9.10. An example for an elliptic curve DHKE is given in Sect. 9.3. Verify the two scalar multiplications that Alice performs. Show the intermediate results within the group operation.

9.11. After the DHKE, Alice and Bob possess a mutual secret point $R = (x, y)$. The modulus of the used elliptic curve is a 64-bit prime. Now, we want to derive a session key for a 128-bit block cipher. The session key is calculated as follows:

$$K_{AB} = h(x||y)$$

Describe an *efficient* brute-force attack against the symmetric cipher. How many of the key bits are truly random in this case? (Hint: You do not need to describe the mathematical details. Provide a list of the necessary steps. Assume you have a function that computes square roots modulo p.)

9.12. Derive the formula for addition on elliptic curves. That is, given the coordinates for P and Q, find the coordinates for $R = (x_3, y_3)$.

Hint: First, find the equation of a line through the two points. Insert this equation in the elliptic curve equation. At some point you have to find the roots of a cubic polynomial $x^3 + a_2 x^2 + a_1 x + a_0$. If the three roots are denoted by x_0, x_1, x_2, you can use the fact that $x_0 + x_1 + x_2 = -a_2$.

Chapter 10
Digital Signatures

Digital signatures are one of the most important cryptographic tools they and are widely used today. Applications for digital signatures range from digital certificates for secure e-commerce to legal signing of contracts to secure software updates. Together with key establishment over insecure channels, they form the most important instance for public-key cryptography.

Digital signatures share some functionality with handwritten signatures. In particular, they provide a method to assure that a message is authentic to one user, i.e., it in fact originates from the person who claims to have generated the message. However, they actually provide much more functionality, as we'll learn in this chapter.

In this chapter you will learn:

- The principle of digital signatures
- Security services, that is, the specific objectives that can be achieved by a security system
- The RSA digital signature scheme
- The Elgamal digital signature scheme and two extensions of it, the digital signature algorithm (DSA) and the elliptic curve digital signature algorithm (ECDSA)

10.1 Introduction

In this section, we first provide a motivating example why digital signatures are needed and why they must be based on asymmetric cryptography. We then develop the principles of digital signatures. Actual signature algorithms are introduced in subsequent sections.

10.1.1 Odd Colors for Cars, or: Why Symmetric Cryptography Is Not Sufficient

The crypto schemes that we have encountered so far had two main goals: either to encrypt data (e.g., with AES, 3DES or RSA encryption) or to establish a shared key (e.g., with the Diffie–Hellman or elliptic curve key exchange). One might be tempted to think that we are now in a position to satisfy any security needs that arise in practice. However, there are many other security needs besides encryption and key exchange, which are in fact termed security services; these are discussed in detail in Sect. 10.1.3. We now discuss a setting in which symmetric cryptography fails to provide a desirable security function.

Assume we have two communicating parties, Alice and Bob, who share a secret key. Furthermore, the secret key is used for encryption with a block cipher. When Alice receives and decrypts a message which makes semantic sense, e.g., the decrypted message is an actual (English) text, she can in many cases conclude that the message was in fact generated by a person with whom he shares the secret key[1]. If only Alice and Bob know the key, they can be reasonably sure that an attacking third party has not changed the message in transit. So far we've always assumed that the bad guy is an external party that we often named Oscar. However, in practice it is often the case that Alice and Bob do want to communicate securely with each other, but at the same time they might be interested in cheating each other. It turns out that symmetric-key schemes do not protect the two parties *against each other*. Consider the following scenario:

Suppose that Alice owns a dealership for new cars where you can select and order cars online. We assume that Bob, the customer, and Alice, the dealer, have established a shared secret k_{AB}, e.g., by using the Diffie–Hellman key exchange. Bob now specifies the car that he likes, which includes a color choice of pink for the interior and an external color of orange — choices most people would not make. He sends the order form AES-encrypted to Alice. She decrypts the order and is happy to have sold another model for $25,000. Upon delivery of the car three weeks later, Bob has second thoughts about his choice, in part because his spouse is threatening

---

[1] One has to be a bit careful with such a conclusion, though. For instance, if Alice and Bob use a stream cipher an attacker can flip individual bits of the ciphertext, which results in bit flips in the received plaintext. Depending on the application, the attacker might be able to manipulate the message in a way that is semantically still correct. However, using block ciphers, especially in a chaining mode, makes it quite likely that ciphertext manipulations can be detected after decryption.

him with divorce after *seeing* the car. Unfortunately for Bob (and his family), Alice has a "no return" policy. Given that she is an experienced car dealer, she knows too well that it will not be easy to sell a pink and orange car, and she is thus set on not making any exceptions. Since Bob now claims that he never ordered the car, she has no other choice but to sue him. In front of the judge, Alice's lawyer presents Bob's digital car order together with the encrypted version of it. Obviously, the lawyer argues, Bob must have generated the order since he is in possession of k_{AB} with which the ciphertext was generated. However, if Bob's lawyer is worth his money, he will patiently explain to the judge that the car dealer, Alice, also knows k_{AB} and that Alice has, in fact, a high incentive to generate faked car orders. The judge, it turns out, has no way of knowing whether the plaintext–ciphertext pair was generated by Bob or Alice! Given the laws in most countries, Bob probably gets away with his dishonesty.

This might sound like a rather specific and somewhat artificially constructed scenario, but in fact it is not. There are many, many situations where it is important to prove to a neutral third party, i.e., a person acting as a judge, that one of two (or more) parties generated a message. By *proving* we mean that the judge can conclude without doubt who has generated the message, even if all parties are potentially dishonest. Why can't we use some (complicated) symmetric-key scheme to achieve this goal? The high-level explanation is simple: Exactly because we have a symmetric set-up, Alice and Bob have the same knowledge (namely of keys) and thus the same capabilities. Everything that Alice can do can be done by Bob, too. Thus, a neutral third party cannot distinguish whether a certain cryptographic operation was performed by Alice or by Bob or by both. Generally speaking, the solution to this problem lies in public-key cryptography. The asymmetric set-up that is inherent in public-key algorithms might potentially enable a judge to distinguish between actions that only one person can perform (namely the person in possession of the private key), and those that can be done by both (namely computations involving the public key). It turns out that digital signatures are public-key algorithms which have the properties that are needed to resolve a situation of cheating participants. In the e-commerce car scenario above, Bob would have been required to digitally sign his order using his private key.

10.1.2 Principles of Digital Signatures

The property of proving that a certain person generated a message is obviously also very important outside the digital domain. In the real, "analog" world, this is achieved by handwritten signatures on paper. For instance, if we sign a contract or sign a check, the receiver can prove to a judge that we actually signed the message. (Of course, one can try to forge signatures, but there are legal and social barriers that prevent most people from even attempting to do so.) As with conventional handwritten signatures, only the person who creates a digital message must be capable of generating a valid signature. In order to achieve this with cryptographic primitives, we have to apply public-key cryptography. The basic idea is that the person who signs the message uses a private key, and the receiving party uses the matching public key. The principle of a digital signature scheme is shown in Fig. 10.1.

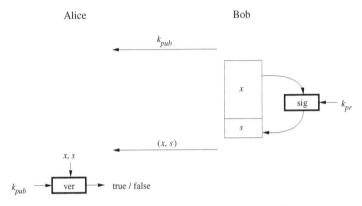

Fig. 10.1 Principle of digital signatures which involves signing and verifying a message

The process starts with Bob signing the message x. The signature algorithm is a function of Bob's private key, k_{pr}. Hence, assuming he in fact keeps his private key private, only Bob can sign a message x on his behalf. In order to relate a signature to the message, x is also an input to the signature algorithm. After signing the message, the signature s is appended to the message x and the pair (x, s) is sent to Alice. It is important to note that a digital signature by itself is of no use unless it is accompanied by the message. A digital signature without the message is the equivalent of a handwritten signature on a strip of paper without the contract or a check that is supposed to be signed.

The digital signature itself is merely a (large) integer value, for instance, a string of 2048 bits. The signature is only useful to Alice if she has means to *verify* whether the signature is valid or not. For this, a verification function is needed which takes both x and the signature s as inputs. In order to link the signature to Bob, the function also requires his public key. Even though the verification function has long inputs, its only output is the binary statement "true" or "false". If x was actually signed with the private key that belongs to the public verification key, the output is true, otherwise it is false.

From these general observations we can easily develop a generic digital signature protocol:

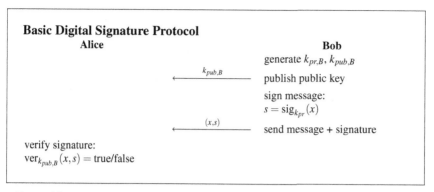

From this set-up, the core property of digital signatures follows: A signed message can unambiguously be traced back to its originator since a valid signature can only be computed with the unique signer's private key. Only the signer has the ability to generate a signature on his behalf. Hence, we can *prove* that the signing party has actually generated the message. Such a proof can even have legal meaning, for instance, as in the Electronic Signatures in Global and National Commerce Act (ESIGN) in the USA or in the *Signaturgesetz*, or Signature Law, in Germany. We note that the basic protocol above does not provide any confidentiality of the message since the message x is being sent in the clear. Of course, the message can be kept confidential by also encrypting it, e.g., with AES or 3DES.

Each of the three popular public-key algorithm families, namely integer factorization, discrete logarithms and elliptic curves, allows us to construct digital signatures. In the remainder of this chapter we learn about most signature schemes that are of practical relevance.

10.1.3 Security Services

It is very instructive to discuss in more detail the security functions we can achieve with digital signatures. In fact, at this point we will step for a moment away from digital signature and ask ourselves in general: What are possible *security objectives* that a security system might possess? More accurately the objectives of a security systems are called *security services*. There exist many security services, but the most important ones which are desirable in many applications are as follows:

1. **Confidentiality:** Information is kept secret from all but authorized parties.
2. **Integrity:** Messages have not been modified in transit.
3. **Message Authentication:** The sender of a message is authentic. An alternative term is *data origin authentication*.
4. **Nonrepudiation:** The sender of a message can not deny the creation of the message.

Different applications call for different sets of security services. For instance, for private e-mail the first three functions are desirable, whereas a corporate e-mail system might also require nonrepudiation. As another example, if we want to secure software updates for a cell phone, the chief objectives might be integrity and message authentication because the manufacturer primarily wants to assure that only original updates are loaded into the handheld device. We note that message authentication always implies data integrity; the opposite is not true.

The four security services can be achieved in a more or less straightforward manner with the schemes introduced in this book: For confidentiality one uses primarily symmetric ciphers and less frequently asymmetric encryption. Integrity and message authentication are provided by digital signatures and message authentication codes which, are introduced in Chap. 12. Nonrepudiation can be achieved with digital signatures as discussed above.

In addition to the four core security services there are several other ones:

5. **Identification/entity authentication:** Establish and verify the identity of an entity, e.g., a person, a computer or a credit card.
6. **Access control:** Restrict access to the resources to privileged entities.
7. **Availability:** Assures that the electronic system is reliably available.
8. **Auditing:** Provide evidence about security-relevant activities, e.g., by keeping logs about certain events.
9. **Physical security:** Provide protection against physical tampering and/or responses to physical tampering attempts.
10. **Anonymity:** Provide protection against discovery and misuse of identity.

Which security services are desired in a given system is heavily application-specific. For instance, anonymity might make no sense for an e-mail system since e-mails are supposed to have a clearly identifiable sender. On the other hand, car-to-car communication systems for collision avoidance (one of the many exciting new applications for cryptography we will see in the next ten years or so) have a strong need to keep cars and drivers anonymous in order to avoid tracking. As a further example, in order to secure an operating system, access control to certain parts of a computer system is often of paramount importance. Most but not all of these advanced services can be achieved with the crypto algorithms from this book. However, in some cases noncryptographic approaches need to be taken. For instance, availability is often achieved by using redundancy, e.g., running redundant computing or storage systems in parallel. Such solutions are only indirectly, if at all, related to cryptography.

10.2 The RSA Signature Scheme

The RSA signature scheme is based on RSA encryption introduced in Chap. 7. Its security relies on the difficulty of factoring a product of two large primes (the integer factorization problem). Since its first description in 1978 in [143], the RSA signature scheme has emerged as the most widely used digital signatures scheme in practice.

10.2.1 Schoolbook RSA Digital Signature

Suppose Bob wants to send a signed message x to Alice. He generates the same RSA keys that were used for RSA encryption as shown in Chap. 7. At the end of the set-up he has the following parameters:

RSA Keys
- Bob's private key: $k_{pr} = (d)$
- Bob's public key: $k_{pub} = (n, e)$

The actual signature protocol is shown in the following. The message x that is being signed is in the range $(1, 2, \ldots, n-1)$.

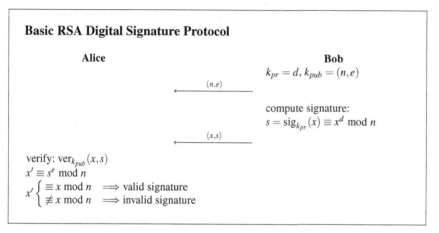

As can be seen from the protocol, Bob computes the signature s for a message x by RSA-encrypting x with his private key k_{pr}. Bob is the only party who can apply k_{pr}, and hence the ownership of k_{pr} authenticates him as the author of the signed message. Bob appends the signature s to the message x and sends both to Alice. Alice receives the signed message and RSA-decrypts s using Bob's public key k_{pub}, yielding x. If x and x' match, Alice knows two important things: First, the author of the message was in possession of Bob's secret key, and if only Bob has had access to the key, it was in fact Bob who signed the message. This is called message authentication. Second, the message has not been changed in transit, so that message integrity is given. We recall from the previous section that these are two of the fundamental security services which are often needed in practice.

Proof. We now prove that the scheme is correct, i.e., that the verification process yields a "true" statement if the message and signature have not been altered during transmission. We start from the verification operation $s^e \bmod n$:

$$s^e = (x^d)^e = x^{de} \equiv x \bmod n$$

Due to the mathematical relationship between the private and the public key, namely that

$$de \equiv 1 \bmod \phi(n),$$

raising any integer $x \in \mathbb{Z}_n$ to the (de)th power yields the integer itself again. The proof for this was given in Sect. 7.3. $\square$

The role of the public and the private keys are swapped compared to the RSA encryption scheme. Whereas RSA encryption applies the public key to the message x, the signature scheme applies the private key k_{pr}. On the other side of the communication channel, RSA encryption requires the use of the private key by the receiver, while the digital signature scheme applies the public key for verification.

Let's look at an example with small numbers.

Example 10.1. Suppose Bob wants to send a signed message $(x = 4)$ to Alice. The first steps are exactly the same as it is done for an RSA encryption: Bob computes his RSA parameters and sends the public key to Alice. In contrast to the encryption scheme, now the private key is used for signing while the public key is needed to verify the signature.

Alice		**Bob**
		1. choose $p = 3$ and $q = 11$
		2. $n = p \cdot q = 33$
		3. $\Phi(n) = (3-1)(11-1) = 20$
		4. choose $e = 3$
		5. $d \equiv e^{-1} \equiv 7 \bmod 20$
	$\underleftarrow{\quad (n,e)=(33,3) \quad}$	
		compute signature for message $x = 4$:
		$s = x^d \equiv 4^7 \equiv 16 \bmod 33$
	$\underleftarrow{\quad (x,s)=(4,16) \quad}$	

verify:
$x' = s^e \equiv 16^3 \equiv 4 \bmod 33$
$x' \equiv x \bmod 33 \Longrightarrow$ valid signature

Alice can conclude from the valid signature that Bob generated the message and that it was not altered in transit, i.e., message authentication and message integrity are given.

$\diamond$

It should be noted that we introduced a digital signature scheme only. In particular, the message itself is not encrypted and, thus, there is not confidentiality. If this security service is required, the message together with the signature should be encrypted, e.g., using a symmetric algorithm like AES.

10.2.2 Computational Aspects

First, we note that the signature is as long as the modulus n, i.e., roughly $\lceil \log_2 n \rceil$ bit. As discussed earlier, n is typically in the range from 1024 to 3072 bit. Even though such a signature length is not a problem in most Internet applications, it can be undesirable in systems that are bandwidth and/or energy constrained, e.g., mobile phones.

The key generation process is identical to the one we used for RSA encryption, which was discussed in detail in Chap. 7. To compute and verify the signature, the square-and-multiply algorithm introduced in Sect. 7.4 is used. The acceleration techniques for RSA introduced in Sect. 7.5 are also applicable to the digital signature scheme. Particularly interesting are short public keys e, for instance, the choice $e = 2^{16} + 1$. This makes verification a very fast operation. Since in many practical scenarios a message is signed only once but verified many times, the fact that verification is very fast is helpful. This is, e.g., the case in public-key infrastructures which use certificates. Certificates are signed only once but are verified over and over again every time a user uses his asymmetric keys (cf. Sect. 13.3.3).

10.2.3 Security

Like in every other asymmetric scheme, it must be assured that the public keys are authentic. This means that the verifying party in fact has the public key that is associated with the private signature key. If an attacker succeeds in providing the verifier with an incorrect public key that supposedly belongs to the signer, the attacker can obviously sign messages. In order to prevent the attack, certificates can be used, a topic which is discussed in Chap. 13.

Algorithmic Attacks

The first group of attacks attempts to break the underlying RSA scheme by computing the private key d. The most general of these attacks tries to factor the modulus n into the primes p and q. If an attacker succeeds with this, she can compute the private key d from e. In order to prevent factoring attacks the modulus must be sufficiently large, as discussed in Sect. 7.8. In practice, 1024 bit or more are recommended.

Existential Forgery

Another attack against the schoolbook RSA signature scheme allows an attacker to generate a valid signature for a *random* message x. The attack works as follows:

Existential Forgery Attack Against RSA Digital Signature

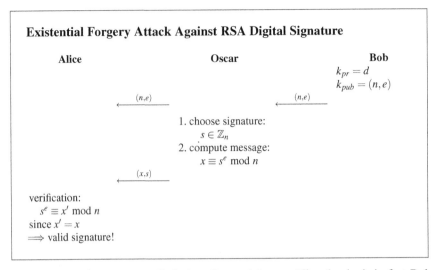

The attacker impersonates Bob, i.e., Oscar claims to Alice that he is in fact Bob. Because Alice performs exactly the same computations as Oscar, she will verify the signature as correct. However, by closely looking at Steps 1 and 2 that Oscar performs, one sees that the attack is somewhat odd. The attacker chooses the signature first and then *computes* the message. As a consequence, he cannot control the semantics of the message x. For instance, Oscar cannot generate a message such as "Transfer $1000 into Oscar's account". Nevertheless, the fact that an automated verification process does not recognize the forgery is certainly not a desirable feature. For this reason, schoolbook RSA signature is rarely used in practice, and padding schemes are applied in order to prevent this and other attacks.

RSA Padding: The Probabilistic Signature Standard (PSS)

The attack above can be prevented by allowing only certain message formats. Roughly speaking, formatting imposes a rule which allows the verifier, Alice in our examples, to distinguish between valid and invalid messages; this is called *padding*. For example, a simple formatting rule could specify that all messages x have 100 trailing bits with the value zero (or any other specific bit pattern). If Oscar chooses signature values s and computes the "message" $x \equiv s^e \bmod n$, it is extremely unlikely that x has this specific format. If we require a certain value for the 100 trailing bits, the chance that x has this format is 2^{-100}, which is considerably lower than winning any lottery.

We now look at a padding scheme which is widely used in practice. Note that a padding scheme for RSA encryption was already discussed in Sect. 7.7. The *probabilistic signature scheme* (RSA-PSS) is a signature scheme based on the RSA cryptosystem. It combines signature and verification with an encoding of the message.

Let's have a closer look at RSA-PSS. Almost always in practice, the message itself is not signed directly but rather the hashed version of it. Hash functions compute a digital fingerprint of messages. The fingerprint has a fixed length, say 160 or 256 bit, but accepts messages as inputs of arbitrary lengths. More about hash functions and the role the play in digital signatures is found in Chap. 11.

In order to be consistent with the terminology used in standards, we denote the message with M rather than with x. Figure 10.2 depicts the encoding procedure which is known as Encoding Method for Signature with Appendix (EMSA) Probabilistic Signature Scheme (PSS).

Encoding for the EMSA Probabilistic Signature Scheme

Let $|n|$ be the size of the RSA modulus in bits. The encoded message EM has a length $\lceil (|n| - 1)/8 \rceil$ bytes such that the bit length of EM is at most $|n| - 1$ bit.

1. Generate a random value *salt*.
2. Form a string M' by concatenating a fixed padding $padding_1$, the hash value $mHash = h(M)$ and *salt*.
3. Compute a hash value H of the string M'.
4. Concatenate a fixed padding $padding_2$ and the value *salt* to form a data block DB.
5. Apply a mask generation function MGF to the string M' to compute the mask value $dbMask$. In practice, a hash function such as SHA-1 is often used as MGF.
6. XOR the mask value $dbMask$ and the data block DB to compute $maskedDB$.
7. The encoded message EM is obtained by concatenating $maskedDB$, the hash value H and the fixed padding bc.

After the encoding, the actual signing operation is applied to the encoded message EM, e.g.,

$$s = \text{sig}_{k_{pr}}(x) \equiv EM^d \bmod n$$

The verification operation then proceeds in a similar way: recovery of the *salt* value and checking whether the EMSA-PSS encoding of the message is correct. Note that the receiver knows the values of $padding_1$ and $padding_2$ from the standard.

The value H in EM is in essence the hashed version of the message. By adding a random value *salt* prior to the second hashing, the encoded value becomes probabilistic. As a consequence, if we encode and sign the same message twice, we obtain different signatures, which is a desirable feature.

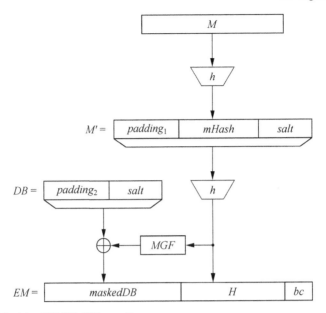

Fig. 10.2 Principle of EMSA-PSS encoding

10.3 The Elgamal Digital Signature Scheme

The Elgamal signature scheme, which was published in 1985, is based on the difficulty of computing discrete logarithms (cf. Chap. 8). Unlike RSA, where encryption and digital signature are almost identical operations, the Elgamal digital signature is quite different from the encryption scheme with the same name.

10.3.1 Schoolbook Elgamal Digital Signature

Key Generation

As with every public-key scheme, there is a set-up phase during which the keys are computed. We start by finding a large prime p and constructing a discrete logarithm problem as follows:

Key Generation for Elgamal Digital Signature

1. Choose a large prime p.
2. Choose a primitive element α of $\mathbb{Z}_p^*$ or a subgroup of $\mathbb{Z}_p^*$.
3. Choose a random integer $d \in \{2, 3, \ldots, p-2\}$.
4. Compute $\beta = \alpha^d \bmod p$.

The public key is now formed by $k_{pub} = (p, \alpha, \beta)$, and the private key by $k_{pr} = d$.

Signature and Verification

Using the private key and the parameters of the public key, the signature

$$\text{sig}_{k_{pr}}(x, k_E) = (r, s)$$

for a message x is computed during the signing process. Note that the signature consists of two integers r and s. The signing consists of two main steps: choosing a random value k_E, which forms an ephemeral private key, and computing the actual signature of x.

Elgamal Signature Generation

1. Choose a random ephemeral key $k_E \in \{0, 1, 2, \ldots, p-2\}$ such that $\gcd(k_E, p-1) = 1$.
2. Compute the signature parameters:

$$r \equiv \alpha^{k_E} \bmod p,$$
$$s \equiv (x - d \cdot r) k_E^{-1} \bmod p - 1.$$

On the receiving side, the signature is verified as $\text{ver}_{k_{pub}}(x, (r, s))$ using the public key, the signature and the message.

Elgamal Signature Verification

1. Compute the value
$$t \equiv \beta^r \cdot r^s \bmod p$$

2. The verification follows from:

$$t \begin{cases} \equiv \alpha^x \bmod p & \Longrightarrow \text{valid signature} \\ \not\equiv \alpha^x \bmod p & \Longrightarrow \text{invalid signature} \end{cases}$$

In short, the verifier accepts a signature (r,s) only if the relation $\beta^r \cdot r^s \equiv \alpha^x$ mod p is satisfied. Otherwise, the verification fails. In order to make sense of the rather arbitrary looking rules for computing the signature parameters r and s as well as the verification, it is helpful to study the following proof.

Proof. We'll prove the correctness of the Elgamal signature scheme. More specifically, we show that the verification process yields a "true" statement if the verifier uses the correct public key and the correct message, and if the signature parameters (r,s) were chosen as specified. We start with the verification equation:

$$\beta^r \cdot r^s \equiv (\alpha^d)^r (\alpha^{k_E})^s \text{ mod } p$$
$$\equiv \alpha^{dr+k_E s} \text{ mod } p.$$

We require that the signature is considered valid if this expression is identical to α^x:

$$\alpha^x \equiv \alpha^{dr+k_E s} \text{ mod } p. \tag{10.1}$$

According to Fermat's Little Theorem, the relationship (10.1) holds if the exponents on both sides of the expression are identical modulo $p-1$:

$$x \equiv dr + k_E s \text{ mod } p-1$$

from which the construction rule of the signature parameters s follows:

$$s \equiv (x - d \cdot r)k_E^{-1} \text{ mod } p-1.$$

$\square$

The condition that $\gcd(k_E, p-1) = 1$ is required since we have to invert the ephemeral key modulo $p-1$ when computing s.

Let's look at an example with small numbers.

Example 10.2. Again, Bob wants to send a message to Alice. This time, it should be signed with the Elgamal digital signature scheme. The signature and verification process is as follows:

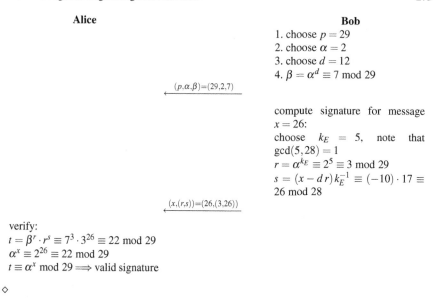

Alice   **Bob**

                                                   1. choose $p = 29$
                                                   2. choose $\alpha = 2$
                                                   3. choose $d = 12$
                                                   4. $\beta = \alpha^d \equiv 7 \bmod 29$

                $(p,\alpha,\beta)=(29,2,7)$
          $\longleftarrow$

                                                   compute signature for message
                                                   $x = 26$:
                                                   choose $\quad k_E \;=\; 5,\quad$ note that
                                                   $\gcd(5,28) = 1$
                                                   $r = \alpha^{k_E} \equiv 2^5 \equiv 3 \bmod 29$
                                                   $s = (x - dr)\,k_E^{-1} \equiv (-10)\cdot 17 \equiv$
                                                   $26 \bmod 28$

                $(x,(r,s))=(26,(3,26))$
          $\longleftarrow$

verify:
$t = \beta^r \cdot r^s \equiv 7^3 \cdot 3^{26} \equiv 22 \bmod 29$
$\alpha^x \equiv 2^{26} \equiv 22 \bmod 29$
$t \equiv \alpha^x \bmod 29 \Longrightarrow$ valid signature

$\diamond$

10.3.2 *Computational Aspects*

The key generation phase is identical to the set-up phase of Elgamal encryption, which we introduced in Sect. 8.5.2. Because the security of the signature scheme relies on the discrete logarithm problem, p needs to have the properties discussed in Sect. 8.3.3. In particular, it should have a length of at least 1024 bits. The prime can be generated using the prime-finding algorithms introduced in Sect 7.6. The private key should be generated by a true random number generator. The public key requires one exponentiation using the square-and-multiply algorithm.

The signature consists of the pair (r,s). Both have roughly the same bit length as p, so that the total length of the package $(x,(r,s))$ is about three times as long as only the message x. Computing r requires an exponentiation modulo p, which can be achieved with the square-and-multiply algorithm. The main operation when computing s is the inversion of k_E. This can be done using the extended Euclidean algorithm. A speed-up is possible through precomputing. The signer can generate the ephemeral key k_E and r in advance and store both values. When a message is to be signed, they can be retrieved and used to compute s. The verifier performs two exponentiations that are again computed with the square-and-multiply algorithm, and one multiplication.

10.3.3 Security

First, we must make sure that the verifier has the correct public key. Otherwise, the attack sketched in Sect. 10.2.3 is applicable. Other attacks are described in the following.

Computing Discrete Logarithms

The security of the signature scheme relies on the discrete logarithm problem (DLP). If Oscar is capable of computing discrete logarithms, he can compute the private key d from β as well as the ephemeral key k_E from r. With this knowledge, he can sign arbitrary messages on behalf of the signer. Hence the Elgamal parameters must be chosen such that the DLP is intractable. We refer to Sect. 8.3.3 for a discussion of possible discrete logarithm attacks. One of the key requirements is that the prime p should be at least 1024-bit long. We have also make sure that small subgroup attacks are not possible. In order to counter this attack, in practice primitive elements α are used to generate a subgroup of prime order. In such groups, all elements are primitive and small subgroups do not exist.

Reuse of the Ephemeral Key

If the signer reuses the ephemeral key k_E, an attacker can easily compute the private key a. This constitutes a complete break of the system. Here is how the attack works.

Oscar observes two digital signatures and messages of the form $(x, (r, s))$. If the two messages x_1 and x_2 have the same ephemeral key k_E, Oscar can detect this since the two r values are the same because they were constructed as $r_1 = r_2 = \alpha^{k_E}$. The two s values are different, and Oscar obtains the following two expressions:

$$s_1 \equiv (x_1 - d\,r)k_E^{-1} \bmod p - 1 \tag{10.2}$$
$$s_2 \equiv (x_2 - d\,r)k_E^{-1} \bmod p - 1 \tag{10.3}$$

This is an equation system with the two unknowns d, which is Bob's private key (!) and the ephemeral key k_E. By multiplying both equations by k_E it becomes a linear system of equations which can be solved easily. Oscar simply subtracts the second equation from the first one, yielding:

$$s_1 - s_2 \equiv (x_1 - x_2)k_E^{-1} \bmod p - 1$$

from which the ephemeral key follows as

$$k_E \equiv \frac{x_1 - x_2}{s_1 - s_2} \bmod p - 1.$$

If $\gcd(s_1 - s_2, p - 1) \neq 1$, the equation has multiple solutions for k_E, and Oscar has to verify which is the correct one. In any case, using k_E, Oscar can now also compute the private key through either Eq. (10.2) or Eq. (10.3):

$$d \equiv \frac{x_1 - s_1 k_E}{r} \mod p - 1.$$

With the knowledge of the private key d and the public key parameters, Oscar can now freely sign any documents on Bob's behalf. In order to avoid the attack, fresh ephemeral keys stemming from a random number generator should be used for every digital signature.

An attack with small numbers is given in the next example.

Example 10.3. Let's assume the situation where Oscar eavesdrops on the following two messages that were previously signed with Bob's private key and that use the same ephemeral key k_E:

1. $(x_1, (r, s_1)) = (26, (3, 26))$,
2. $(x_2, (r, s_2)) = (13, (3, 1))$.

Additionally, Oscar knows Bob's public key, which is given as

$$(p, \alpha, \beta) = (29, 2, 7).$$

With this information, Oscar is now able to compute the ephemeral key

$$
\begin{aligned}
k_E &\equiv \frac{x_1 - x_2}{s_1 - s_2} \mod p - 1 \\
&\equiv \frac{26 - 13}{26 - 1} \equiv 13 \cdot 9 \\
&\equiv 5 \mod 28
\end{aligned}
$$

and finally reveal Bob's private key d:

$$
\begin{aligned}
d &\equiv \frac{x_1 - s_1 \cdot k_E}{r} \mod p - 1 \\
&\equiv \frac{26 - 26 \cdot 5}{3} \equiv 8 \cdot 19 \\
&\equiv 12 \mod 28.
\end{aligned}
$$

◇

Existential Forgery Attack

Similar to the case of RSA digital signatures, it is also possible that an attacker generates a valid signature for a *random* message x. The attacker, Oscar, impersonates Bob, i.e., Oscar claims to Alice that he is in fact Bob. The attack works as follows:

Existential Forgery Attack Against Elgamal Digital Signature

Alice	Oscar	Bob
		$k_{pr} = d$
		$k_{pub} = (p, \alpha, \beta)$

$\xleftarrow{\quad (p,\alpha,\beta) \quad}$ $\xleftarrow{\quad (p,\alpha,\beta) \quad}$

1. select integers i, j
   where $\gcd(j, p-1) = 1$
2. compute signature:
   $r \equiv \alpha^i \beta^j \bmod p$
   $s \equiv -rj^{-1} \bmod p - 1$
3. compute message:
   $x \equiv si \bmod p - 1$

$\xleftarrow{\quad (x,(r,s)) \quad}$

verification:
$t \equiv \beta^r \cdot r^s \bmod p$
since $t \equiv \alpha^x \bmod p$:
   valid signature!

The verification yields a "true" statement because the following holds:

$$t \equiv \beta^r \cdot r^s \bmod p$$
$$\equiv \alpha^{dr} \cdot r^s \bmod p$$
$$\equiv \alpha^{dr} \cdot \alpha^{(i+dj)s} \bmod p$$
$$\equiv \alpha^{dr} \cdot \alpha^{(i+dj)(-rj^{-1})} \bmod p$$
$$\equiv \alpha^{dr-dr} \cdot \alpha^{-rij^{-1}} \bmod p$$
$$\equiv \alpha^{si} \bmod p$$

Since the message was constructed as $x \equiv si \bmod p - 1$, the last expression is equal to

$$\alpha^{si} \equiv \alpha^x \bmod p$$

which is exactly Alice's condition for accepting the signature as valid.

The attacker computes in Step 3 the message x, the semantics of which he cannot control. Thus, Oscar can only compute valid signatures for pseudorandom messages.

The attack is not possible if the message is hashed, which is, in practice, very often the case. Rather than using the message directly for computing the signature, one applies a hash function to the message prior to signing, i.e., the signing equation becomes:

$$s \equiv (h(x) - d \cdot r)k_E^{-1} \bmod p - 1.$$

10.4 The Digital Signature Algorithm (DSA)

The native Elgamal signature algorithm described in this section is rarely used in practice. Instead, a much more popular variant is used, known as the *Digital Signature Algorithm (DSA)*. It is a federal US government standard for digital signatures (DSS) and was proposed by the National Institute of Standards and Technology (NIST). Its main advantages over the Elgamal signature scheme are that the signature is only 320-bit long and that some of the attacks that can threaten the Elgamal scheme are not applicable.

10.4.1 The DSA Algorithm

We introduce here the DSA standard with a bit length of 1024 bit. Note that longer bit lengths are also possible in the standard.

Key Generation

The keys for DSA are computed as follows:

Key Generation for DSA

1. Generate a prime p with $2^{1023} < p < 2^{1024}$.
2. Find a prime divisor q of $p - 1$ with $2^{159} < q < 2^{160}$.
3. Find an element α with $\text{ord}(\alpha) = q$, i.e., α generates the subgroup with q elements.
4. Choose a random integer d with $0 < d < q$.
5. Compute $\beta \equiv \alpha^d \bmod p$.

The keys are now:

$k_{pub} = (p, q, \alpha, \beta)$

$k_{pr} = (d)$

The central idea of DSA is that there are two cyclic groups involved. One is the large cyclic group $\mathbb{Z}_p^*$, the order of which has bit length of 1024 bit. The second one is in the 160-bit subgroup of $\mathbb{Z}_p^*$. This set-up yields shorter signatures, as we see in the following.

In addition to the 1024-bit prime p and a 160-bit prime q, there are two other bit length combinations possible for the primes p and q. According to the latest version of the standard, the combinations shown in Table 10.1 are allowed.

If one of the other bit lengths is required, only Steps 1 and 2 of the key generation phase have to be adjusted accordingly. More about the issue of bit length will be said in Sect. 10.4.3 below.

Table 10.1 Bit lengths of important parameters of DSA

p	q	Signature
1024	160	320
2048	224	448
3072	256	512

Signature and Verification

As in the Elgamal signature scheme, the DSA signature consists of a pair of integers (r,s). Since each of the two parameters is only 160-bit long, the total signature length is 320 bit. Using the public and private key, the signature for a message x is computed as follows:

DSA Signature Generation

1. Choose an integer as random ephemeral key k_E with $0 < k_E < q$.
2. Compute $r \equiv (\alpha^{k_E} \bmod p) \bmod q$.
3. Compute $s \equiv (SHA(x) + d \cdot r) k_E^{-1} \bmod q$.

According to the standard, the message x has to be hashed using the hash function SHA-1 in order to compute s. Hash functions, including SHA-1, are described in Chap. 11. For now it is sufficient to know that SHA-1 compresses x and computes a 160-bit fingerprint. This fingerprint can be thought of as a representative of x.

The signature verification process is as follows:

DSA Signature Verification

1. Compute auxiliary value $w \equiv s^{-1} \bmod q$.
2. Compute auxiliary value $u_1 \equiv w \cdot SHA(x) \bmod q$.
3. Compute auxiliary value $u_2 \equiv w \cdot r \bmod q$.
4. Compute $v \equiv (\alpha^{u_1} \cdot \beta^{u_2} \bmod p) \bmod q$.
5. The verification $ver_{k_{pub}}(x,(r,s))$ follows from:

$$v \begin{cases} \equiv r \bmod q \implies \text{valid signature} \\ \not\equiv r \bmod q \implies \text{invalid signature} \end{cases}$$

The verifier accepts a signature (r,s) only if $v \equiv r \bmod q$ is satisfied. Otherwise, the verification fails. In this case, the message or the signature may have been modified or the verifier is not in possession of the correct public key. In any case, the signature should be considered invalid.

Proof. We show that a signature (r,s) satisfies the verification condition $v \equiv r \bmod q$. We'll start with the signature parameter s:

$$s \equiv (SHA(x) + dr)k_E^{-1} \bmod q$$

which is equivalent to:

$$k_E \equiv s^{-1} SHA(x) + d s^{-1} r \bmod q.$$

The right-hand side can be expressed in terms of the auxiliary values u_1 and u_2:

$$k_E \equiv u_1 + d u_2 \bmod q.$$

We can raise α to either side of the equation if we reduce modulo p:

$$\alpha^{k_E} \bmod p \equiv \alpha^{u_1 + d u_2} \bmod p.$$

Since the public key value β was computed as $\beta \equiv \alpha^d \bmod p$, we can write:

$$\alpha^{k_E} \bmod p \equiv \alpha^{u_1} \beta^{u_2} \bmod p.$$

We now reduce both sides of the equation modulo q:

$$(\alpha^{k_E} \bmod p) \bmod q \equiv (\alpha^{u_1} \beta^{u_2} \bmod p) \bmod q.$$

Since r was constructed as $r \equiv (\alpha^{k_E} \bmod p) \bmod q$ and $v \equiv (\alpha^{u_1} \beta^{u_2} \bmod p) \bmod q$, this expression is identical to the condition for verifying a signature as valid:

$$r \equiv v \bmod q.$$

□

Let's look at an example with small numbers.

Example 10.4. Bob wants to send a message x to Alice which is to be signed with the DSA algorithm. Suppose the hash value of x is $h(x) = 26$. Then the signature and verification process is as follows:

Alice	Bob
	1. choose $p = 59$
	2. choose $q = 29$
	3. choose $\alpha = 3$
	4. choose private key $d = 7$
	5. $\beta = \alpha^d \equiv 4 \bmod 59$

$$\xleftarrow{\quad (p,q,\alpha,\beta)=(59,29,3,4) \quad}$$

sign:
compute hash of message $h(x) = 26$
1. choose ephemeral key $k_E = 10$
2. $r = (3^{10} \bmod 59) \equiv 20 \bmod 29$
3. $s = (26 + 7 \cdot 20) \cdot 3 \equiv 5 \bmod 29$

$$\xleftarrow{\quad (x,(r,s))=(x,(20,5)) \quad}$$

verify:
1. $w = 5^{-1} \equiv 6 \bmod 29$
2. $u_1 = 6 \cdot 26 \equiv 11 \bmod 29$
3. $u_2 = 6 \cdot 20 \equiv 4 \bmod 29$
4. $v = (3^{11} \cdot 4^4 \bmod 59) \bmod 29 = 20$
5. $v \equiv r \bmod 29 \Longrightarrow$ valid signature

In this example, the subgroup has a prime order of $q = 29$, whereas the "large" cyclic group modulo p has 58 elements. We note that $58 = 2 \cdot 29$. We replaced the function $SHA(x)$ by $h(x)$ since the SHA hash function has an output of length 160 bit.

◇

10.4.2 Computational Aspects

We discuss now the computations involved in the DSA scheme. The most demanding part is the key-generation phase. However, this phase only has to be executed once at set-up time.

Key Generation

The challenge in the key-generation phase is to find a cyclic group $\mathbb{Z}_p^*$ with a bit length of 1024, and which has a prime subgroup in the range of 2^{160}. This condition is fulfilled if $p - 1$ has a prime factor q of 160 bit. The general approach to generating such parameters is to first find the 160-bit prime q and then to construct the larger prime p from it. Below is the prime-generating algorithm. Note that the NIST-specified scheme is slightly different.

Prime Generation for DSA
Output: two primes (p, q), where $2^{1023} < p < 2^{1024}$ and $2^{159} < q < 2^{160}$, such that $p - 1$ is a multiple of q.
Initialization: $i = 1$
Algorithm:

1   find prime q with $2^{159} < q < 2^{160}$ using the Miller–Rabin algorithm
2   FOR $i = 1$ TO 4096
2.1   generate random integer M with $2^{1023} < M < 2^{1024}$
2.2 $M_r \equiv M \mod 2q$
2.3 $p - 1 \equiv M - M_r$ (note that $p - 1$ is a multiple of $2q$.)
         IF p is prime   (use Miller–Rabin primality test)
 2.4   RETURN (p, q)
 2.5 $i = i + 1$
3   GOTO Step 1

The choice of $2q$ as modulus in step 2.3 assures that the prime candidates generated in step 2.3 are odd numbers. Since $p - 1$ is divisible by $2q$, it is also divisible by q. If p is a prime, $\mathbb{Z}_p^*$ thus has a subgroup of order q.

Signing

During signing we compute the parameters r and s. Computing r involves first evaluation $g^{k_E} \bmod p$ using the square-and-multiply algorithm. Since k_E has only 160 bit, about $1.5 \times 160 = 240$ squarings and multiplications are required on average, even though the arithmetic is done with 1024-bit numbers. The result, which has also a length of 1024 bit, is then reduced to 160 bit by the operation "$\bmod q$". Computing s involves only 160-bit numbers. The most costly step is the inversion of k_E.

Of these operations, the exponentiation is the most costly one in terms of computational complexity. Since the parameter r does not depend on the message, it can be precomputed so that the actual signing can be a relatively fast operation.

Verification

Computing the auxiliary parameters w, u_1 and u_2 only involves 160-bit operands, which makes them relatively fast.

10.4.3 Security

An interesting aspect of DSA is that we have to protect against two different discrete logarithm attacks. If an attacker wants to break DSA, he could attempt to compute the private key d by solving the discrete logarithm in the large cyclic group modulo p:

$$d = \log_\alpha \beta \bmod p.$$

The most powerful method for this is the index calculus attack, which was sketched in Sect. 8.3.3. In order to thwart this attack, p must be at least 1024-bit long. It is estimated that this provides a security level of 80 bit, i.e., an attack would need on the order of 2^{80} operations (cf. Table 6.1 in Chap. 6). For higher security levels, NIST allows primes with lengths of 2048 and 3072 bit.

The second discrete logarithm attack on DSA is to exploit the fact that α generates only a small subgroup of order q. Hence, it seems promising to attack only the subgroup, which has a size of about 2^{160}, rather than the large cyclic group with about 2^{1024} elements formed by p. However, it turns out that the powerful index-calculus attack is not applicable if Oscar wants to exploit the subgroup property. The best he can do is to perform one of the generic DLP attacks, i.e., either the baby-step giant-step method or Pollard's rho method (cf. Sect. 8.3.3). These are so-called square root attacks, and given that the subgroup has an order of approximately 2^{160}, these attacks provide a security level of $\sqrt{2^{160}} = 2^{80}$. It is not a coincidence that the index calculus attack and the square root attack have a comparable complexity, in fact the parameter sizes were deliberately chosen that way. One has to be careful, though, if the size of p is increased to 2048 or 3072 bit. This only increases the difficulty of the index-calculus attack, but the small subgroup attack would still have a complexity of 2^{80} if the subgroup stays the same size. For this reason q also must be increased if larger p values are chosen. Table 10.2 shows the NIST-specified lengths of the primes p and q together with the resulting security levels. The security level of the hash function must also match the one of the discrete logarithm problem. Since the cryptographic strength of a hash function is mainly determined by the bit length of the hash output, the minimum hash output is also given in the table. More about security of hash functions will be said in Chap. 11.

Table 10.2 Standardized parameter bit lengths and security levels for DSA

p	q	Hash output (min)	Security levels
1024	160	160	80
2048	224	224	112
3072	256	256	128

It should be stressed that the record for discrete logarithm calculations is 532 bit, so that the 1024-bit DSA variant is currently secure, and the 2048-bit and 3072-bit variants seem to provide good long-term security.

In addition to discrete logarithm attacks, DSA becomes vulnerable if the ephemeral key is reused. This attack is completely analogues to the case of Elgamal digital signature. Hence, it must be assured that a fresh randomly-genererated key k_E is used in every signing operation.

10.5 The Elliptic Curve Digital Signature Algorithm (ECDSA)

As discussed in Chap. 9, elliptic curves have several advantages over RSA and over DL schemes like Elgamal or DSA. In particular, in absence of strong attacks against elliptic curve cryptosystems (ECC), bit lengths in the range of 160–256 bit can be chosen which provide security equivalent to 1024–3072-bit RSA and DL schemes. The shorter bit length of ECC often results in shorter processing time and in shorter signatures. For these reasons, the Elliptic Curve Digital Signature Algorithm (ECDSA) was standardized in the US by the American National Standards Institute (ANSI) in 1998.

10.5.1 The ECDSA Algorithm

The steps in the ECDSA standard are conceptionally closely related to the DSA scheme. However, its discrete logarithm problem is constructed in the group of an elliptic curve. Thus, the arithmetic to be performed for actually computing an ECDSA signature is entirely different from that used for DSA.

The ECDSA standard is defined for elliptic curves over prime fields $\mathbb{Z}_p$ and Galois fields $GF(2^m)$. The former is often preferred in practice, and we will only introduce this one in what follows.

Key Generation

The keys for the ECDSA are computed as follows:

Key Generation for ECDSA

1. Use an elliptic curve E with

   - modulus p
   - coefficients a and b
   - a point A which generates a cyclic group of prime order q

2. Choose a random integer d with $0 < d < q$.
3. Compute $B = dA$.
The keys are now:
$k_{pub} = (p, a, b, q, A, B)$
$k_{pr} = (d)$

Note that we have set up a discrete logarithm problem where the integer d is the private key and the result of the scalar multiplication, point B, is the public key. Similar to DSA, the cyclic group has an order q which should have a size of at least 160 bit or more for higher security levels.

Signature and Verification

Like DSA, an ECDSA signature consists of a pair of integers (r, s). Each value has the same bit length as q, which makes for fairly compact signatures. Using the public and private key, the signature for a message x is computed as follows:

ECDSA Signature Generation

1. Choose an integer as random ephemeral key k_E with $0 < k_E < q$.
2. Compute $R = k_E A$.
3. Let $r = x_R$.
4. Compute $s \equiv (h(x) + d \cdot r) k_E^{-1} \bmod q$.

In step 3 the x-coordinate of the point R is assigned to the variable r. The message x has to be hashed using the function h in order to compute s. The hash function output length must be at least as long as q. More about the choice of the hash function will be said in Chap. 11. However, for now it is sufficient to know that the hash function compresses x and computes a fingerprint which can be viewed as a representative of x.

The signature verification process is as follows:

ECDSA Signature Verification

1. Compute auxiliary value $w \equiv s^{-1} \bmod q$.
2. Compute auxiliary value $u_1 \equiv w \cdot h(x) \bmod q$.
3. Compute auxiliary value $u_2 \equiv w \cdot r \bmod q$.
4. Compute $P = u_1 A + u_2 B$.
5. The verification $ver_{k_{pub}}(x, (r,s))$ follows from:

$$x_P \begin{cases} \equiv r \bmod q \Longrightarrow \text{valid signature} \\ \not\equiv r \bmod q \Longrightarrow \text{invalid signature} \end{cases}$$

In the last step, the notation x_P indicates the x-coordinate of the point P. The verifier accepts a signature (r,s) only if the x_P has the same value as the signature parameter r modulo q. Otherwise, the signature should be considered invalid.

Proof. We show that a signature (r,s) satisfies the verification condition $r \equiv x_P \bmod q$. We'll start with the signature parameter s:

$$s \equiv (h(x) + d\,r)\,k_E^{-1} \bmod q$$

which is equivalent to:

$$k_E \equiv s^{-1} h(x) + d\,s^{-1} r \bmod q.$$

The right-hand side can be expressed in terms of the auxiliary values u_1 and u_2:

$$k_E \equiv u_1 + d\,u_2 \bmod q.$$

Since the point A generates a cyclic group of order q, we can multiply both sides of the equation with A:

$$k_E A = (u_1 + d\,u_2)\,A.$$

Since the group operation is associative, we can write

$$k_E A = u_1 A + d\,u_2 A$$

and

$$k_E A = u_1 A + u_2 B.$$

What we showed so far is that the expression $u_1 A + u_2 B$ is equal to $k_E A$ if the correct signature and key (and message) have been used. But this is exactly the condition that we check in the verification process by comparing the x-coordinates of $P = u_1 A + u_2 B$ and $R = k_E A$.

□

Using the small elliptic curve from Chap. 9, we look at a simple ECDSA example.

Example 10.5. Bob wants to send a message to Alice that is to be signed with the ECDSA algorithm. The signature and verification process is as follows:

Alice   **Bob**

choose E with $p = 17, a = 2, b = 2$,
and $A = (5, 1)$ with $q = 19$
choose $d = 7$
compute $B = dA = 7 \cdot (5, 1) = (0, 6)$

$$\xleftarrow{\quad (p,a,b,q,A,B)= \quad}$$
$$(17,2,2,19,(5,1),(0,6))$$

sign:
compute hash of message $h(x) = 26$
choose ephemeral key $k_E = 10$
$R = 10 \cdot (5, 1) = (7, 11)$
$r = x_R = 7$
$s = (26 + 7 \cdot 7) \cdot 2 \equiv 17 \bmod 19$

$$\xleftarrow{\quad (x,(r,s))=(x,(7,17)) \quad}$$

verify:
$w = 17^{-1} \equiv 9 \bmod 19$
$u_1 = 9 \cdot 26 \equiv 6 \bmod 19$
$u_2 = 9 \cdot 7 \equiv 6 \bmod 19$
$P = 6 \cdot (5, 1) + 6 \cdot (0, 6) = (7, 11)$
$x_P \equiv r \bmod 19 \Longrightarrow$ valid signature

Note that we chose the elliptic curve

$$E : y^2 \equiv x^3 + 2x + 2 \bmod 17$$

which is discussed in Sect. 9.2. Because all points of the curve form a cyclic group of order 19, i.e., a prime, there are no subgroups and hence in this case $q = \#E = 19$.

◇

10.5.2 Computational Aspects

We discuss now the computations involved in the three stages of the ECDSA scheme.

Key Generation As discussed earlier, finding an elliptic curve with good crypto-graphic properties is a nontrivial task. In practice, standardized curves such as the ones proposed by NIST or the Brainpool consortium are often used. The remaining computation in the key generation phase is one point multiplication, which can be done using the double-and-add algorithm.

Signing During signing we first compute the point R, which requires one point multiplication, and from which r immediately follows. For the parameter s we have to invert the ephemeral key, which is done with the extended Euclidean algorithm. The other main operations are hashing of the message and one reduction modulo q.

  The point multiplication, which is in most cases by the far the most arithmetic-intensive operation, can be precomputed by choosing the ephemeral key ahead of time, e.g., during the idle time of a CPU. Thus, in situations where precomputation is an option, signing becomes a very fast operation.

Verification Computing the auxiliary parameters w, u_1 and u_2 involves straightfor-ward modular arithmetic. The main computational load occurs during the evaluation of $Pu_1 A + u_2 B$. This can be accomplished by two separate point multiplications. However, there are specialized methods for simultaneous exponentiations (remem-ber from Chap. 9 that point multiplication is closely related to exponentiation) which are faster than two individual point multiplications.

10.5.3 Security

Given that the elliptic curve parameters are chosen correctly, the main analytical at-tack against ECDSA attempts to solve the elliptic curve discrete logarithm problem. If an attacker were capable of doing this, he could compute the private key d and/or the ephemeral key. However, the best known ECC attacks have a complexity propor-tional to the square root of the size of the group in which the DL problem is defined, i.e., proportional to $\sqrt{q}$. The parameter length of ECDSA and the corresponding security levels are given in Table 10.3. We recall that the prime p is typically only slightly larger than q, so that all arithmetic for ECDSA is done with operands which have roughly the bit length of q.

  The security level of the hash function must also match that of the discrete loga-rithm problem. The cryptographic strength of a hash function is mainly determined by the length of its output. More about security of hash functions will be said in Chap. 11.

  The security levels of 128, 192 and 256 were chosen so that they match the security offered by AES with its three respective key sizes.

  More subtle attacks against ECDSA are also possible. For instance, at the begin-ning of verification it must be checked whether $r, s \in \{1, 2, \dots, q\}$ in order to prevent a certain attack. Also, protocol-based weaknesses, e.g., reusing the ephemeral key, must be prevented.

Table 10.3 Bit lengths and security levels of ECDSA

q	Hash output (min)	Security levels
192	192	96
224	224	112
256	256	128
384	384	192
512	512	256

10.6 Discussion and Further Reading

Digital Signature Algorithms The first practical realization of digital signatures was introduced in the original paper by Rivest, Shamir and Adleman [143]. RSA digital signatures have been standardized by several bodies for a long time, see, e.g., [95]. RSA signatures were, and in many cases still are, the de facto standard for many applications, especially for certificates on the Internet.

The Elgamal digital signature was published in 1985 in [73]. Many variants of this scheme are possible and have been proposed over the years. For a compact summary, see [120, Note 11.70].

The DSA algorithm was proposed in 1991 and became a US standard in 1994. There were two possible motivations for the government to create this standard as an alternative to RSA. First, RSA was patented at that time and having a free alternative was attractive for US industry. Second, an RSA digital signature implementation can also be used for encryption. This was not desirable (from the US government viewpoint) since there were still rather strict export restrictions for cryptography in the US at that time. In contrast, a DSA implementation can only be used for signing and not for encryption, and it was easier to export systems that only included signature functionality. Note that DSA refers to the digital signature *algorithm*, and the corresponding standard is referred to as DSS, the digital signature *standard*. Today, DSS includes not only the DSA algorithm but also ECDSA and RSA digital signatures [126].

In addition to the algorithms discussed in this chapter, there exist several other schemes for digital signatures. These include, e.g., the Rabin signature [140], the Fiat–Shamir signature [76], the Pointcheval–Stern signature [134] and the Schnorr signature [150].

Using Digital Signatures With digital signatures, the problem of authentic public keys is acute: How can Alice (or Bob) assure that they possess the correct public keys for each other? Or, phrased differently, how can Oscar be prevented from injecting faked public keys in order to perform an attack? We discuss this question in detail in Chap. 13, where certificates are introduced. Certificates are based on digital signatures and are one of the main applications of digital signatures. They bind an identity (e.g., Alice's e-mail address) to a public key.

One of the more interesting interactions between society and cryptography is digital signature laws. They basically assure that a cryptographic digital signature has a legally binding meaning. For instance, an electronic contract that was digitally signed can be enforced in the same way as a conventionally signed contract. Around the turn of the millennium, many nations introduced corresponding laws. This was at a time that the "brave new world" of the Internet had opened up seemingly endless opportunities for doing business online, and digital signature laws seemed to be crucial to allow trusted business transactions via the Internet. Examples of digital signature laws are the Electronic Signatures in Global and National Commerce Act (ESIGN) in the US [138], or the corresponding directive of the European Union [133]. A good online source for more information is the Digital Law Survey [167]. Even though much electronic commerce is today conducted without making use of signature laws, there will be without doubt more and more situations where those laws are actually needed.

One crucial issue when using digital signatures in the real world is that the private keys, especially if used in a setting with legal significance, have to be kept strictly confidential. This requires a secure way to store this delicate key material. One way to satisfy this requirement is to employ *smart cards* that can be used as secure containers for secret keys. A secret key never leaves the smart card, and signatures are performed within the CPU inside the smart card. For applications with high security requirements, so called *tamper-resistant* smart cards are protected against several types of hardware attacks. Reference [141] provides excellent insight into the various facets of the highly sophisticated smart card technology.

10.7 Lessons Learned

- Digital signatures provide message integrity, message authentication and nonrepudiation.
- One of the main application areas of digital signatures is certificates.
- RSA is currently the most widely used digital signature algorithm. Competitors are the Digital Signature Standard (DSA) and the Elliptic Curve Digital Signature Standard (ECDSA).
- The Elgamal signature scheme is the basis for DSA. In turn, ECDSA is a generalization of DSA to elliptic curves.
- RSA verification can be done with short public keys e. Hence, in practice, RSA verification is usually faster than signing.
- DSA and ECDSA have the advantage over RSA in that the signatures are much shorter.
- In order to prevent certain attacks, RSA should be used with padding.
- The modulus of DSA and the RSA signature schemes should be at least 1024-bits long. For true long-term security, a modulus of length 3072 bits should be chosen. In contrast, ECDSA achieves the same security levels with bit lengths in the range 160–256 bits.

Problems

10.1. In Sect. 10.1.3 we state that sender (or message) authentication always implies data integrity. Why? Is the opposite true too, i.e., does data integrity imply sender authentication? Justify both answers.

10.2. In this exercise, we want to consider some basic aspects of security services.

1. Does privacy always guarantee integrity? Justify your answer.
2. In which order should confidentiality and integrity be assured (should the entire message be encrypted first or last)? Give the rationale for your answer.

10.3. Design a security service that provides data integrity, data confidentiality and nonrepudiation using public-key cryptography in a two-party communication system over an insecure channel. Give a rationale that data integrity, confidentiality and nonrepudiation are achieved by your solution. (Recommendation: Consider the corresponding threats in your argumentation.)

10.4. A painter comes up with a new business idea: He wants to offer custom paintings from photos. Both the photos and paintings will be transmitted in digital form via the Internet. One concern that he has is discretion towards his customers, since potentially embarrassing photos, e.g., nude photos, might be sent to him. Hence, the photo data should not be accessible for third parties during transmission. The painter needs multiple weeks for the creation of a painting, and hence he wants to assure that he cannot be fooled by someone who sends in a photo assuming a false name. He also wants to be assured that the painting will definitely be accepted by the customer and that she cannot deny the order.

1. Choose the necessary security services for the transmission of the digitalized photos from the customers to the painter.
2. Which cryptographic elements (e.g., symmetric encryption) can be utilized to achieve the security services? Assume that several megabytes of data have to be transmitted for every photo.

10.5. Given an RSA signature scheme with the public key $(n = 9797, e = 131)$, which of the following signatures are valid?

1. $(x = 123, \text{sig}(x) = 6292)$
2. $(x = 4333, \text{sig}(x) = 4768)$
3. $(x = 4333, \text{sig}(x) = 1424)$

10.6. Given an RSA signature scheme with the public key $(n = 9797, e = 131)$, show how Oscar can perform an existential forgery attack by providing an example of such for the parameters of the RSA digital signature scheme.

10.7. In an RSA digital signature scheme, Bob signs messages x_i and sends them together with the signatures s_i and her public key to Alice. Bob's public key is the pair (n, e); her private key is d.

Oscar can perform man-in-the-middle attacks, i.e., he can replace Bob's public key by his own on the channel. His goal is to alter messages and provide these with a digital signature which will check out correctly on Alice's side. Show everything that Oscar must do for a successful attack.

10.8. Given is an RSA signature scheme with EMSA-PSS padding as shown in Sect. 10.2.3. Describe the verification process step-by-step that has to be performed by the receiver of a signature that was EMSA-PSS encoded.

10.9. One important aspect of digital signatures is the computational effort required to (i) sign a message, and (ii) to verify a signature. We study the computational complexity of the RSA algorithm used as a digital signature in this problem.

1. How many multiplications do we need, on average, to perform (i) signing of a message with a general exponent, and (ii) verification of a signature with the short exponent $e = 2^{16} + 1$? Assume that n has $l = \lceil \log_2 n \rceil$ bits. Assume the square-and-multiply algorithm is used for both signing and verification. Derive general expressions with l as a variable.
2. Which takes longer, signing or verification?
3. We now derive estimates for the speed of actual software implementation. Use the following timing model for multiplication: The computer operates with 32-bit data structures. Hence, each full-length variable, in particular n and x, is represented by an array with $m = \lceil l/32 \rceil$ elements (with x being the basis of the exponentiation operation). We assume that one multiplication or squaring of two of these variables modulo n takes m^2 time units (a time unit is the clock period times some constant larger than one which depends on the implementation). Note that you never multiply with the exponents d and e. That means, the bit length of the exponent does not influence the time it takes to perform an individual modular squaring or multiplication.

   How long does it take to compute a signature/verify a signature if the time unit on a certain computer is 100 nsec, and n has 512 bits? How long does it take if n has 1024 bit?
4. Smart cards are one very important platform for the use of digital signatures. Smart cards with an 8051 microprocessor kernel are popular in practice. The 8051 is an 8-bit processor. What time unit is required in order to perform one signature generation in 0.5 sec if n has (i) 512 bits and (ii) 1024 bits? Since these processors cannot be clocked at more than, say, 10 MHz, is the required time unit realistic?

10.10. We now consider the Elgamal signature scheme. You are given Bob's private key $K_{pr} = (d) = (67)$ and the corresponding public key $K_{pub} = (p, \alpha, \beta) = (97, 23, 15)$.

1. Calculate the Elgamal signature (r, s) and the corresponding verification for a message from Bob to Alice with the following messages x and ephemeral keys k_E:

   a. $x = 17$ and $k_E = 31$

   b. $x = 17$ and $k_E = 49$
   c. $x = 85$ and $k_E = 77$

2. You receive two alleged messages x_1, x_2 with their corresponding signatures (r_i, s_i) from Bob. Verify whether the messages $(x_1, r_1, s_1) = (22, 37, 33)$ and $(x_2, r_2, s_2) = (82, 13, 65)$ both originate from Bob.
3. Compare the RSA signature scheme with the Elgamal signature scheme. Where are their relative advantages and drawbacks?

10.11. Given is an Elgamal signature scheme with $p = 31$, $\alpha = 3$ and $\beta = 6$. You receive the message $x = 10$ twice with the signatures (r, s):

$$(i) \quad (17, 5)$$
$$(ii) \quad (13, 15)$$

1. Are both signatures valid?
2. How many valid signatures are there for each message x and the specific parameters chosen above?

10.12. Given is an Elgamal signature scheme with the public parameters $(p = 97, \alpha = 23, \beta = 15)$. Show how Oscar can perform an existential forgery attack by providing an example for a valid signature.

10.13. Given is an Elgamal signature scheme with the public parameters $p, \alpha \in \mathbb{Z}_p^*$ and an unknown private key d. Due to faulty implementation, the following dependency between two consecutive ephemeral keys is fulfilled:

$$k_{E_{i+1}} = k_{E_i} + 1.$$

Furthermore, two consecutive signatures to the plaintexts x_1 and x_2

$$(r_1, s_1)$$
$$\text{and} \quad (r_2, s_2)$$

are given. Explain how an attacker is able to calculate the private key with the given values.

10.14. The parameters of DSA are given by $p = 59, q = 29, \alpha = 3$, and Bob's private key is $d = 23$. Show the process of signing (Bob) and verification (Alice) for following hash values $h(x)$ and ephemeral keys k_E:

1. $h(x) = 17, k_E = 25$
2. $h(x) = 2, k_E = 13$
3. $h(x) = 21, k_E = 8$

10.15. Show how DSA can be attacked if the same ephemeral key is used to sign two different messages.

10.16. The parameters of ECDSA are given by the curve $E : y^2 = x^3 + 2x + 2 \bmod 17$, the point $A = (5,1)$ of order $q = 19$ and Bob's private $d = 10$. Show the process of signing (Bob) and verification (Alice) for following hash values $h(x)$ and ephemeral keys k_E:

1. $h(x) = 12, k_E = 11$
2. $h(x) = 4, k_E = 13$
3. $h(x) = 9, k_E = 8$

Chapter 11
Hash Functions

Hash functions are an important cryptographic primitive and are widely used in protocols. They compute a *digest* of a message which is a short, fixed-length bit-string. For a particular message, the message digest, or *hash value*, can be seen as the fingerprint of a message, i.e., a unique representation of a message. Unlike all other crypto algorithms introduced so far in this book, hash functions do not have a key. The use of hash functions in cryptography is manifold: Hash functions are an essential part of digital signature schemes and message authentication codes, as discussed in Chapter 12. Hash functions are also widely used for other cryptographic applications, e.g., for storing of password hashes or key derivation.

In this chapter you will learn:

- Why hash functions are required in digital signature schemes
- Important properties of hash functions
- A security analysis of hash functions, including an introduction to the birthday paradox
- An overview of different families of hash functions
- How the popular hash function SHA-1 works

11.1 Motivation: Signing Long Messages

Even though hash functions have many applications in modern cryptography, they are perhaps best known for the important role they play in the practical use of digital signatures. In the previous chapter, we have introduced signature schemes based on the asymmetric algorithms RSA and the discrete logarithm problem. For all schemes, the length of the plaintext is limited. For instance, in the case of RSA, the message cannot be larger than the modulus, which is in practice often between 1024 and 3072-bits long. Remember this translates into only 128–384 bytes; most emails are longer than that. Thus far, we have ignored the fact that in practice the plaintext x will often be (much) larger than those sizes. The question that arises at this point is simple: How are we going to efficiently compute signatures of large messages? An intuitive approach would be similar to the ECB mode for block ciphers: Divide the message x into blocks x_i of size less than the allowed input size of the signature algorithm, and sign each block separately, as depicted in Figure 11.1.

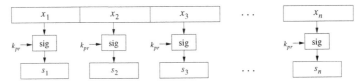

Fig. 11.1 Insecure approach to signing of long messages

However, this approach yields three serious problems:

Problem 1: High Computational Load Digital signatures are based on computationally intensive asymmetric operations such as modular exponentiations of large integers. Even if a single operation consumes a small amount of time (and energy, which is relevant in mobile applications), the signatures of large messages, e.g., email attachments or multimedia files, would take too long on current computers. Furthermore, not only does the signer have to compute the signature, but the verifier also has to spend a similar amount of time and energy to verify the signature.

Problem 2: Message Overhead Obviously, this naïve approach doubles the message overhead because not only must the message be sent but also the signature, which is of the same length in this case. For instance, a 1-MB file must yield an RSA signature of length 1 MB, so that a total of 2 MB must be transmitted.

Problem 3: Security Limitations This is the most serious problem if we attempt to sign a long message by signing a sequence of message blocks *individually*. The approach shown in Fig. 11.1 leads immediately to new attacks: For instance, Oscar could remove individual messages and the corresponding signatures, or he could reorder messages and signatures, or he could reassemble new messages and signatures out of fragments of previous messages and signatures, etc. Even though an attacker cannot perform manipulations *within* an individual block, we do not have protection for the whole message.

Hence, for performance as well as for security reasons we would like to have *one short signature* for a message of arbitrary length. The solution to this problem is hash functions. If we had a hash function that somehow computes a fingerprint of the message x, we could perform the signature operation as shown in Figure 11.2

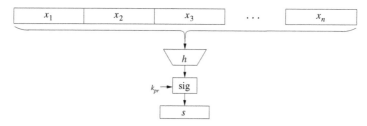

Fig. 11.2 Signing of long messages with a hash function

Assuming we possess such a hash function, we now describe a basic protocol for a digital signature scheme with a hash function. Bob wants to send a digitally signed message to Alice.

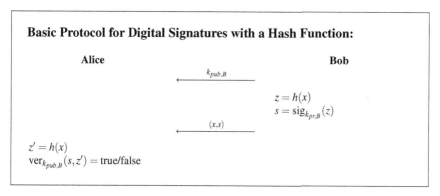

Bob computes the hash of the message x and signs the hash value z with his private key $k_{pr,B}$. On the receiving side, Alice computes the hash value z' of the received message x. She verifies the signature s with Bob's public key $k_{pub,B}$. We note that both the signature generation and the verification operate on the hash value z rather than on the message itself. Hence, the hash value represents the message. The hash is sometimes referred to as the *message digest* or the *fingerprint* of the message.

Before we discuss the security properties of hash functions in the next section, we can now get a rough feeling for a desirable input–output behavior of hash functions: We want to be able to apply a hash function to messages x of any size, and it is thus desirable that the function h is computationally efficient. Even if we hash large messages in the range of, say, hundreds of megabytes, it should be relatively fast to compute. Another desirable property is that the output of a hash function is of fixed length and independent of the input length. Practical hash functions have output lengths between 128–512 bits. Finally, the computed fingerprint should be highly sensitive to all input bits. That means even if we make minor modifications to the input x, the fingerprint should look very different. This behavior is similar to that of block ciphers. The properties which we just described are symbolized in Figure 11.3.

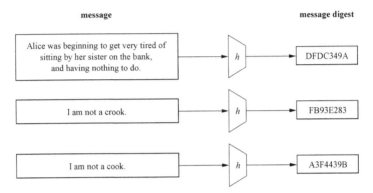

Fig. 11.3 Principal input–output behavior of hash functions

11.2 Security Requirements of Hash Functions

As mentioned in the introduction, unlike all other crypto algorithms we have dealt with so far, hash functions do not have keys. The question is now whether there are any special properties needed for a hash function to be "secure". In fact, we have to ask ourselves whether hash functions have any impact on the security of an application at all since they do not encrypt and they don't have keys. As is often the case in cryptography, things can be tricky and there are attacks which use weaknesses of hash functions. It turns out that there are three central properties which hash functions need to possess in order to be secure:

1. preimage resistance (or one-wayness)
2. second preimage resistance (or weak collision resistance)
3. collision resistance (or strong collision resistance)

These three properties are visualized in Figure 11.4. They are derived in the following.

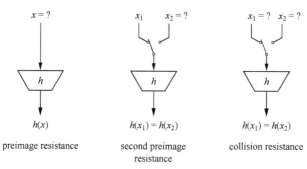

Fig. 11.4 The three security properties of hash functions

11.2.1 Preimage Resistance or One-Wayness

Hash functions need to be *one-way*: Given a hash output z it must be computationally infeasible to find an input message x such that $z = h(x)$. In other words, given a fingerprint, we cannot derive a matching message. We demonstrate now why preimage resistance is important by means of a fictive protocol in which Bob is encrypting the message but not the signature, i.e., he transmits the pair:

$$(e_k(x), \mathrm{sig}_{k_{pr,B}}(z)).$$

Here, $e_k()$ is a symmetric cipher, e.g., AES, with some symmetric key shared by Alice and Bob. Let's assume Bob uses an RSA digital signature, where the signature is computed as:

$$s = \mathrm{sig}_{k_{pr,B}}(z) \equiv z^d \bmod n$$

The attacker Oscar can use Bob's public key to compute

$$s^e \equiv z \bmod n.$$

If the hash function is *not* one-way, Oscar can now compute the message x from $h^{-1}(z) = x$. Thus, the symmetric encryption of x is circumvented by the signature, which leaks the plaintext. For this reason, $h(x)$ should be a one-way function.

In many other applications which make use of hash functions, for instance in key derivation, it is even more crucial that they are preimage resistant.

11.2.2 Second Preimage Resistance or Weak Collision Resistance

For digital signatures with hash it is essential that two different messages do not hash to the same value. This means it should be computationally infeasible to create two different messages $x_1 \neq x_2$ with equal hash values $z_1 = h(x_1) = h(x_2) = z_2$. We differentiate between two different types of such collisions. In the first case, x_1

is given and we try to find x_2. This is called second preimage resistance or weak collision resistance. The second case is given if an attacker is free to choose both x_1 and x_2. This is referred to as strong collision resistance and is dealt with in the subsequent section.

It is easy to see why second preimage resistance is important for the basic signature with hash scheme that we introduced above. Assume Bob hashes and signs a message x_1. If Oscar is capable of finding a second message x_2 such that $h(x_1) = h(x_2)$, he can run the following substitution attack:

Alice	**Oscar**		**Bob**
		$\xleftarrow{\quad k_{pub,B} \quad}$	
			$z = h(x_1)$
			$s = \text{sig}_{k_{pr,B}}(z)$
$\xleftarrow{\quad (x_2,s) \quad}$	$\not\equiv$ substitute	$\xleftarrow{\quad (x_1,s) \quad}$	

$z = h(x_2)$
$\text{ver}_{k_{pub,B}}(s,z) = \text{true}$

As we can see, Alice would accept x_2 as a correct message since the verification gives her the statement "true". How can this happen? From a more abstract viewpoint, this attack is possible because both signing (by Bob) and verifying (by Alice) do not happen with the actual message itself, but rather with the hashed version of it. Hence, if an attacker manages to find a second message with the same fingerprint (i.e., hash output), signing and verifying are the same for this second message.

The question now is how we can prevent Oscar from finding x_2. Ideally, we would like to have a hash function for which weak collisions do not exist. This is, unfortunately, impossible due to the *pigeonhole principle*, a more impressive term for which is *Dirichlet's drawer principle*. The pigeonhole principle uses a counting argument in situations like the following: If you are the owner of 100 pigeons but in your pigeon loop are only 99 holes, at least one pigeonhole will be occupied by 2 birds. Since the output of every hash function has a fixed bit length, say n bit, there are "only" 2^n possible output values. At the same time, the number of inputs to the hash functions is infinite so that multiple inputs must hash to the same output value. In practice, each output value is equally likely for a random input, so that weak collisions exist for all output values.

Since weak collisions exist in theory, the next best thing we can do is to assure that they cannot be found in practice. A strong hash function should be designed such that given x_1 and $h(x_1)$ it is impossible to *construct* x_2 such that $h(x_1) = h(x_2)$. This means there is no analytical attack. However, Oscar can always randomly pick x_2 values, compute their hash values and check whether they are equal to $h(x_1)$. This is similar to an exhaustive key search for a symmetric cipher. In order to prevent this attack given today's computers, an output length of $n = 80$ bit is sufficient. However, we see in the next section that more powerful attacks exist which force us to use even longer output bit lengths.

11.2.3 Collision Resistance and the Birthday Attack

We call a hash function collision resistant or strong collision resistant if it is computationally infeasible to find two different inputs $x_1 \neq x_2$ with $h(x_1) = h(x_2)$. This property is harder to achieve than weak collision resistance since an attacker has two degrees of freedom: Both messages can be altered to achieve similar hash values. We show now how Oscar could turn his ability to find collisions into an attack. He starts with two messages, for instance:

$$x_1 = \texttt{Transfer \$10 into Oscar's account}$$
$$x_2 = \texttt{Transfer \$10,000 into Oscar's account}$$

He now alters x_1 and x_2 at "nonvisible" locations, e.g., he replaces spaces by tabs, adds spaces or return signs at the end of the message, etc. This way, the semantics of the message is unchanged (e.g., for a bank), but the hash value changes for every version of the message. Oscar continues until the condition $h(x_1) = h(x_2)$ is fulfilled. Note that if an attacker has, e.g., 64 locations that he can alter or not, this yields 2^{64} versions of the same message with 2^{64} different hash values. With the two messages, he can launch the following attack:

Alice		**Oscar**		**Bob**
			$\xleftarrow{\quad k_{pub,B} \quad}$	
			$\xrightarrow{\quad x_1 \quad}$	
				$z = h(x_1)$
				$s = sig_{k_{pr,B}}(z)$
	$\xleftarrow{\quad (x_2,s) \quad}$	$\cancel{z}$ substitute	$\xleftarrow{\quad (x_1,s) \quad}$	
$z = h(x_2)$				
$ver_{k_{pub,B}}(s,z) = $ true				

This attack assumes that Oscar can trick Bob into signing the message x_1. This is, of course, not possible in every situation, but one can imagine scenarios where Oscar can pose as an innocent party, e.g., an e-commerce vendor on the Internet, and x_1 is the purchase order that is generated by Oscar.

As we saw earlier, due to the pigeonhole principle, collisions always exist. The question is how difficult it is to find them. Our first guess is probably that this is as difficult as finding second preimages, i.e., if the hash function has an output length of 80 bits, we have to check about 2^{80} messages. However, it turns out that an attacker needs only about 2^{40} messages! This is a quite surprising result which is due to the *birthday attack*. This attack is based on the *birthday paradox*, which is a powerful tool that is often used in cryptanalysis.

It turns out that the following real-world question is closely related to finding collisions for hash functions: How many people are needed at a party such that there is a reasonable chance that at least two people have the same birthday? By birthday we mean any of the 365 days of the year. Our intuition might lead us to assume that we need around 183 people (i.e., about half the number of days in a year) for a collision to occur. However, it turns out that we need far fewer people. The piecewise approach to solve this problem is to first compute the probability of two people *not* having the same birthday, i.e., having no collision of their birthdays. For one person, the probability of no collision is 1, which is trivial since a single birthday cannot collide with anyone else's. For the second person, the probability of no collision is 364 over 365, since there is only one day, the birthday of the first person, to collide with:

$$P(\text{no collision among 2 people}) = \left(1 - \frac{1}{365}\right)$$

If a third person joins the party, he or she can collide with both of the people already there, hence:

$$P(\text{no collision among 3 people}) = \left(1 - \frac{1}{365}\right) \cdot \left(1 - \frac{2}{365}\right)$$

Consequently, the probability for t people having no birthday collision is given by:

$$P(\text{no collision among } t \text{ people}) = \left(1 - \frac{1}{365}\right) \cdot \left(1 - \frac{2}{365}\right) \cdots \left(1 - \frac{t-1}{365}\right)$$

For $t = 366$ people we will have a collision with probability 1 since a year has only 365 days. We return now to our initial question: how many people are needed to have a 50% chance of two colliding birthdays? Surprisingly—following from the equations above—it only requires 23 people to obtain a probability of about 0.5 for a birthday collision since:

$$P(\text{at least one collision}) = 1 - P(\text{no collision})$$
$$= 1 - \left(1 - \frac{1}{365}\right) \cdots \left(1 - \frac{23-1}{365}\right)$$
$$= 0.507 \approx 50\%.$$

Note that for 40 people the probability is about 90%. Due to the surprising outcome of this gedankenexperiment, it is often referred to as the birthday paradox.

Collision search for a hash function $h()$ is exactly the same problem as finding birthday collisions among party attendees. For a hash function there are not 365 values each element can take but 2^n, where n is the output width of $h()$. In fact, it turns out that n is the crucial security parameter for hash functions. The question is how many messages $(x_1, x_2, \ldots, x_t)$ does Oscar need to hash until he has a reasonable chance that $h(x_i) = h(x_j)$ for some x_i and x_j that he picked. The probability for no collisions among t hash values is:

$$P(\text{no collision}) = \left(1 - \frac{1}{2^n}\right)\left(1 - \frac{2}{2^n}\right)\cdots\left(1 - \frac{t-1}{2^n}\right)$$

$$= \prod_{i=1}^{t-1}\left(1 - \frac{i}{2^n}\right)$$

We recall from our calculus courses that the approximation

$$e^{-x} \approx 1 - x,$$

holds[1] since $i/2^n \ll 1$. We can approximate the probability as:

$$P(\text{no collision}) \approx \prod_{i=1}^{t-1} e^{-\frac{i}{2^n}}$$

$$\approx e^{-\frac{1+2+3+\cdots+t-1}{2^n}}$$

The arithmetic series

$$1 + 2 + \cdots + t - 1 = t(t-1)/2,$$

is in the exponent, which allows us to write the probability approximation as

$$P(\text{no collision}) \approx e^{-\frac{t(t-1)}{2 \cdot 2^n}}.$$

Recall that our goal is to find out how many messages $(x_1, x_2, \ldots, x_t)$ are needed to find a collision. Hence, we solve the equation now for t. If we denote the probability of at least one collision by $\lambda = 1 - P(\text{no collision})$, then

$$\lambda \approx 1 - e^{-\frac{t(t-1)}{2^{n+1}}}$$

$$\ln(1 - \lambda) \approx -\frac{t(t-1)}{2^{n+1}}$$

$$t(t-1) \approx 2^{n+1} \ln\left(\frac{1}{1-\lambda}\right).$$

Since in practice $t \gg 1$, it holds that $t^2 \approx t(t-1)$ and thus:

$$t \approx \sqrt{2^{n+1} \ln\left(\frac{1}{1-\lambda}\right)}$$

$$t \approx 2^{(n+1)/2} \sqrt{\ln\left(\frac{1}{1-\lambda}\right)}. \tag{11.1}$$

---

[1] This follows from the Taylor series representation of the exponential function: $e^{-x} = 1 - x + x^2/2! - x^3/3! + \cdots$ for $x \ll 1$.

Equation (11.1) is extremely important: it describes the relationship between the number of hashed messages t needed for a collision as a function of the hash output length n and the collision probability λ. **The most important consequence of the birthday attack is that the number of messages we need to hash to find a collision is roughly equal to the square root of the number of possible output values, i.e., about $\sqrt{2^n} = 2^{n/2}$.** Hence, for a security level (cf. Section 6.2.4) of x bit, the hash function needs to have an output length of $2x$ bit. As an example, assume we want to find a collision for a hypothetical hash function with 80-bit output. For a success probability of 50%, we expect to hash about:

$$t = 2^{81/2} \sqrt{\ln(1/(1-0.5))} \approx 2^{40.2}$$

input values. Computing around 2^{40} hashes and checking for collisions can be done with current laptops! In order to thwart collision attacks based on the birthday paradox, the output length of a hash function must be about twice as long as an output length which protects merely against a second preimage attack. For this reason, all hash functions have an output length of at least 128 bit, where most modern ones are much longer. Table 11.1 shows the number of hash computations needed for a birthday-paradox collision for output lengths found in current hash functions. Interestingly, the desired likelihood of a collision does not influence the attack complexity very much, as is evidenced by the small difference between the success probabilities $\lambda = 0.5$ and $\lambda = 0.9$. It should be stressed that the birthday attack is a

Table 11.1 Number of hash values needed for a collision for different hash function output lengths and for two different collision likelihoods

λ	Hash output length				
	128 bit	160 bit	256 bit	384 bit	512 bit
0.5	2^{65}	2^{81}	2^{129}	2^{193}	2^{257}
0.9	2^{67}	2^{82}	2^{130}	2^{194}	2^{258}

generic attack. This means it is applicable against any hash function. On the other hand, it is not guaranteed that it is the most powerful attack available for a given hash function. As we will see in the next section, for some of the most popular hash functions, in particular MD5 and SHA-1, mathematical collision attacks exist which are faster than the birthday attack.

It should be stressed that there are many applications for hash functions, e.g., storage of passwords, which only require preimage resistance. Thus, a hash function with a relatively short output, say 80 bit, might be sufficient since collision attacks do not pose a threat.

At the end of this section we summarize all important properties of hash functions $h(x)$. Note that the first three are practical requirements, whereas the last three relate to the security of hash functions.

Properties of Hash Functions
1. **Arbitrary message size** $h(x)$ can be applied to messages x of any size.
2. **Fixed output length** $h(x)$ produces a hash value z of fixed length.
3. **Efficiency** $h(x)$ is relatively easy to compute.
4. **Preimage resistance** For a given output z, it is impossible to find any input x such that $h(x) = z$, i.e, $h(x)$ is one-way.
5. **Second preimage resistance** Given x_1, and thus $h(x_1)$, it is computationally infeasible to find any x_2 such that $h(x_1) = h(x_2)$.
6. **Collision resistance** It is computationally infeasible to find any pairs $x_1 \neq x_2$ such that $h(x_1) = h(x_2)$.

11.3 Overview of Hash Algorithms

So far we only discussed the requirements for hash functions. We now introduce how to actually built them. There are two general types of hash functions:

1. **Dedicated hash functions** These are algorithms that are specifically designed to serve as hash functions.
2. **Block cipher-based hash functions** It is also possible to use block ciphers such as AES to construct hash functions.

As we saw in the previous section, hash functions can process an arbitrary-length message and produce a fixed-length output. In practice, this is achieved by segmenting the input into a series of blocks of equal size. These blocks are processed sequentially by the hash function, which has a compression function at its heart. This iterated design is known as *Merkle–Damgård construction*. The hash value of the input message is then defined as the output of the last iteration of the compression function (Fig. 11.5).

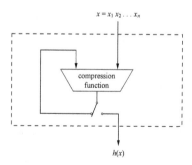

Fig. 11.5 Merkle–Damgård hash function construction

11.3.1 Dedicated Hash Functions: The MD4 Family

Dedicated hash functions are algorithms that have been custom designed. A large number of such constructions have been proposed over the last two decades. In practice, by far the most popular ones have been the hash functions of what is called the MD4 family. MD5, the SHA family and RIPEMD are all based on the principles of MD4. MD4 is a message digest algorithm developed by Ronald Rivest. MD4 was an innovative idea because it was especially designed to allow very efficient software implementation. It uses 32-bit variables, and all operations are bitwise Boolean functions such as logical AND, OR, XOR and negation. All subsequent hash functions in the MD4 family are based on the same software-friendly principles.

A strengthened version of MD4, named MD5, was proposed by Rivest in 1991. Both hash functions compute a 128-bit output, i.e., they possess a collision resistance of about 2^{64}. MD5 became extremely widely used, e.g., in Internet security protocols, for computing checksums of files or for storing of password hashes. There were, however, early signs of potential weaknesses. Thus, the US NIST published a new message digest standard, which was coined the Secure Hash Algorithm (SHA), in 1993. This is the first member of the SHA family and is officially called SHA, even though it is nowadays commonly referred to as SHA-0. In 1995, SHA-0 was modified to SHA-1. The difference between the SHA-0 and SHA-1 algorithms lies in the schedule of the compression function to improve its cryptographic security. Both algorithms have an output length of 160 bit. In 1996, a partial attack against MD5 by Hans Dobbertin led to more and more experts recommending SHA-1 as a replacement for the widely used MD5. Since then, SHA-1 has gained wide adoption in numerous products and standards.

In the absence of analytical attacks, the maximum collision resistance of SHA-0 and SHA-1 is about 2^{80}, which is not a good fit if they are used in protocols together with algorithms such as AES, which has a security level of 128–256 bits. Similarly, most public-key schemes can offer higher security levels, for instance, elliptic curves can have security levels of 128 bits if 256 bits curves are used. Thus, in 2001 NIST introduced three more variants of SHA-1: SHA-256, SHA-384 and SHA-512, with message digest lengths of 256, 384 and 512 bits, respectively. A further modification, SHA-224, was introduced in 2004 in order to fit the security level of 3DES. These four hash functions are often referred to as SHA-2.

In 2004, collision-finding attacks against MD5 and SHA-0 where announced by Xiaoyun Wang. One year later it was claimed that the attack could be extended to SHA-1 and it was claimed that a collision search would take 2^{63} steps, which is considerably less than the 2^{80} achieved by the birthday attack. Table 11.2 gives an overview of the main parameters of the MD4 family.

In Section 11.4 we will learn about the internal functioning of SHA-1, which is to date—despite its potential weakness—the most widely deployed hash function.

At this point we would like to note that finding a collision does not necessarily mean that the hash function is insecure in every situation. There are many applications for hash functions, e.g., key derivation or storage of passwords, where only

Table 11.2 The MD4 family of hash functions

Algorithm		Output [bit]	Input [bit]	No. of rounds	Collisions found
MD5		128	512	64	yes
SHA-1		160	512	80	not yet
SHA-2	**SHA-224**	224	512	64	no
	SHA-256	256	512	64	no
	SHA-384	384	1024	80	no
	SHA-512	512	1024	80	no

preimage and second preimage resistance are required. For such applications, MD5 is still sufficient.

11.3.2 Hash Functions from Block Ciphers

Hash functions can also be constructed using block cipher chaining techniques. As in the case of dedicated hash functions like SHA-1, we divide the message x into blocks x_i of a fixed size. Figure 11.6 shows a construction of such a hash function: The message chunks x_i are encrypted with a block cipher e of block size b. As m-bit key input to the cipher, we use a mapping g from the previous output H_{i-1}, which is a b-to-m-bit mapping. In the case of $b = m$, which is, for instance, given if AES with a 128-bit key is being used, the function g can be the identity mapping. After the encryption of the message block x_i, we XOR the result to the original message block. The last output value computed is the hash of the whole message $x_1, x_2, \ldots, x_n$, i.e., $H_n = h(x)$.

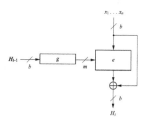

Fig. 11.6 The Matyas–Meyer–Oseas hash function construction from block ciphers

The function can be expressed as:

$$H_i = e_{g(H_{i-1})}(x_i) \oplus x_i$$

This construction, which is named after its inventors, is called the Matyas–Meyer–Oseas hash function.

There exist several other variants of block cipher based realizations of hash functions. Two popular ones are shown in Figure 11.7.

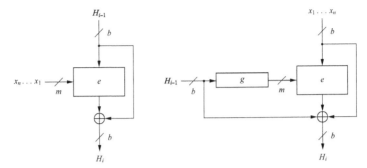

Fig. 11.7 Davies–Meyer (left) and Miyaguchi–Preneel hash function constructions from block ciphers

The expressions for the two hash functions are:

$$H_i = H_{i-1} \oplus e_{x_i}(H_{i-1}) \qquad \text{(Davies–Meyer)}$$
$$H_i = H_{i-1} \oplus x_i \oplus e_{g(H_{i-1})}(x_i) \qquad \text{(Miyaguchi–Preneel)}$$

All three hash functions need to have initial values assigned to H_0. These can be public values, e.g., the all-zero vector. All schemes have in common that the bit size of the hash output is equal to the block width of the cipher used. In situations where only preimage and second preimage resistance is required, block ciphers like AES with 128-bit block width can be used, because they provide a security level of 128 bit against those attacks. For application which require collision resistance, the 128-bit length provided by most modern block ciphers is not sufficient. The birthday attack reduces the security level to mere 64 bit, which is a computational complexity that is within reach of PC clusters and certainly is doable for attackers with large budgets.

One solution to this problem is to use Rijndael with a block width of 192 or 256 bit. These bit lengths provide a security level of 96 and 128 bit, respectively, against birthday attacks, which is sufficient for most applications. We recall from Section 4.1 that Rijndael is the cipher that became AES but allows block sizes of 128, 192 and 256 bit.

Another way of obtaining larger message digests is to use constructions which are composed of several instances of a block cipher and which yield twice the width of the block length b. Figure 11.8 shows such a construction for the case that a cipher e is being employed whose key length is twice the block length. This is in particular the case for AES with a 256-bit key. The message digest output are the $2b$ bit $(H_{n,L} || H_{n,R})$. If AES is being used, this output is $2b = 256$ bit long, which provides a high level of security against collision attacks. As can be seen from the figure, the previous output of the left cipher $H_{i-1,L}$ is fed back as input to both block ciphers. The concatenation of the previous output of the right cipher, $H_{i-1,R}$, with the next message block x_i, forms the key for both ciphers. For security reasons a constant c has to be XORed to the input of the right block cipher. c can have any value other than the all-zero vector. As in the other three constructions described above, initial values have to be assigned to the first hash values ($H_{0,L}$ and $H_{0,R}$).

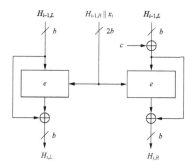

Fig. 11.8 Hirose construction for a hash function with twice the block width

We introduce here the Hirose construction for the case that the key length be twice the block width. There are many other ciphers that satisfy this condition in addition to AES, e.g., the block ciphers Blowfish, Mars, RC6 and Serpent. If a hash function for resource-constrained applications is needed, the lightweight block cipher PRESENT (cf. Section 3.7) allows an extremely compact hardware implementation. With a key size of 128-bit and a block size of 64 bit, the construction computes a 128-bit hash output. This message digest size resists preimage and second preimage attacks, but offers only marginal security against birthday attacks.

11.4 The Secure Hash Algorithm SHA-1

The Secure Hash Algorithm (SHA-1) is the most widely used message digest function of the MD4 family. Even though new attacks have been proposed against the algorithm, it is very instructive to look at its details because the stronger versions in the SHA-2 family show a very similar internal structure. SHA-1 is based on a Merkle–Damgård construction, as can be seen in Figure 11.9.

An interesting interpretation of the SHA-1 algorithm is that the compression function works like a block cipher, where the input is the previous hash value H_{i-1} and the key is formed by the message block x_i. As we will see below, the actual rounds of SHA-1 are in fact quite similar to a Feistel block cipher.

SHA-1 produces a 160-bit output of a message with a maximum length of 2^{64} bit. Before the hash computation, the algorithm has to preprocess the message. During the actual computation, the compression function processes the message in 512-bit

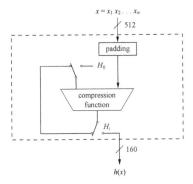

Fig. 11.9 High-level diagram of SHA-1

chunks. The compression function consists of 80 rounds which are divided into four
stages of 20 rounds each.

11.4.1 Preprocessing

Before the actual hash computation, the message x has to be padded to fit a size of
a multiple of 512 bit. For the internal processing, the padded message must then be
divided into blocks. Also, the initial value H_0 is set to a predefined constant.

Padding   Assume that we have a message x with a length of l bit. To obtain an
overall message size of a multiple of 512 bits, we append a single "1" followed
by k zero bits and the binary 64-bit representation of l. Consequently, the number
of required zeros k is given by

$$k \equiv 512 - 64 - 1 - l$$
$$= 448 - (l+1) \bmod 512.$$

Figure 11.10 illustrates the padding of a message x.

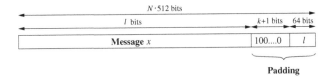

Fig. 11.10 Padding of a message in SHA-1

Example 11.1. Given is the message "abc" consisting of three 8-bit ASCII characters with a total length of $l = 24$ bits:

$$\underbrace{01100001}_{a} \quad \underbrace{01100010}_{b} \quad \underbrace{01100011}_{c}.$$

We append a "1" followed by $k = 423$ zero bits, where k is determined by

$$k \equiv 448 - (l+1) = 448 - 25 = 423 \bmod 512.$$

Finally, we append the 64-bit value which contains the binary representation of the length $l = 24_{10} = 11000_2$. The padded message is then given by

$$\underbrace{01100001}_{a} \quad \underbrace{01100010}_{b} \quad \underbrace{01100011}_{c} \quad 1 \quad \underbrace{00...0}_{423\ zeros} \quad \underbrace{00...011000}_{l=24}.$$

◇

Dividing the padded message Prior to applying the compression function, we need to divide the message into 512-bit blocks $x_1, x_2, \ldots, x_n$. Each 512-bit block can be subdivided into 16 words of size of 32 bits. For instance, the ith block of the message x is split into:

$$x_i = \left(x_i^{(0)}\, x_i^{(1)} \ldots x_i^{(15)}\right)$$

where $x_i^{(k)}$ are words of size of 32 bits.

Initial value H_0 A 160-bit buffer is used to hold the initial hash value for the first iteration. The five 32-bit words are fixed and given in hexadecimal notation as:

$$A = H_0^{(0)} = \texttt{67452301}$$
$$B = H_0^{(1)} = \texttt{EFCDAB89}$$
$$C = H_0^{(2)} = \texttt{98BADCFE}$$
$$D = H_0^{(3)} = \texttt{10325476}$$
$$E = H_0^{(4)} = \texttt{C3D2E1F0}.$$

11.4.2 Hash Computation

Each message block x_i is processed in four stages with 20 rounds each as shown in Figure 11.11. The algorithm uses

- a message schedule which computes a 32-bit word $W_0, W_1, \ldots, W_{79}$ for each of the 80 rounds. The words W_j are derived from the 512-bit message block as follows:

$$W_j = \begin{cases} x_i^{(j)} & 0 \le j \le 15 \\ (W_{j-16} \oplus W_{j-14} \oplus W_{j-8} \oplus W_{j-3})_{\lll 1} & 16 \le j \le 79, \end{cases}$$

where $X_{\lll n}$ indicates a circular left shift of the word X by n bit positions.

- five working registers of size of 32 bits A, B, C, D, E
- a hash value H_i consisting of five 32-bit words $H_i^{(0)}, H_i^{(1)}, H_i^{(2)}, H_i^{(3)}, H_i^{(4)}$. In the beginning, the hash value holds the initial value H_0, which is replaced by a new hash value after the processing of each single message block. The final hash value H_n is equal to the output $h(x)$ of SHA-1.

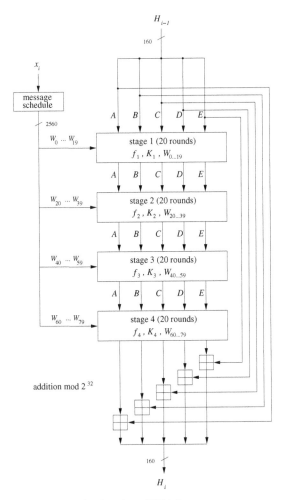

Fig. 11.11 Eighty-round compression function of SHA-1

The four SHA-1 stages have a similar structure but use different internal functions f_t and constants K_t, where $1 \le t \le 4$. Each stage is composed of 20 rounds, where parts of the message block are processed by the function f_t together with some stage-dependent constant K_t. The output after 80 rounds is added to the input value H_{i-1} modulo 2^{32} in word-wise fashion.

The operation within round j in stage t is given by

$$A, B, C, D, E = (E + f_t(B, C, D) + (A)_{\lll 5} + W_j + K_t), A, (B)_{\lll 30}, C, D$$

and is depicted in Figure 11.12. The internal functions f_t and constants K_t change

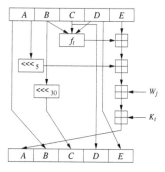

Fig. 11.12 Round j in stage t of SHA-1

depending on the stage according to Table 11.3, i.e., every 20 rounds a new function and a new constant are being used. The function only uses bitwise Boolean operations, namely logical AND ($\wedge$), OR ($\vee$), NOT (top bar) and XOR. These operation are applied to 32-bit variables and are very fast to implement on modern PCs.

A SHA-1 round as shown in Figure 11.12 has some resemblance to the round of a Feistel network. Such structures are sometimes referred to as generalized Feistel networks. Feistel networks are generally characterized by the fact the first part of the input is copied directly to the output. The second part of the input is encrypted using the first part, where the first part is sent through some function, e.g., the f-function in the case of DES. In the SHA-1 round, the inputs A, B, C and D are passed to the output with no change (A, C, D), or only minimal change (rotation of B). However, the input word E is "encrypted" by adding values derived from the other four input words. The message-derived value W_i and the round constant play the role of subkeys.

Table 11.3 Round functions and round constants for the SHA rounds

Stage t	Round j	Constant K_t	Function f_t
1	$0 \ldots 19$	$K_1 = \text{5A827999}$	$f_1(B, C, D) = (B \wedge C) \vee (\bar{B} \wedge D)$
2	$20 \ldots 39$	$K_2 = \text{6ED9EBA1}$	$f_2(B, C, D) = B \oplus C \oplus D$
3	$40 \ldots 59$	$K_3 = \text{8F1BBCDC}$	$f_3(B, C, D) = (B \wedge C) \vee (B \wedge D) \vee (C \wedge D)$
4	$60 \ldots 79$	$K_4 = \text{CA62C1D6}$	$f_4(B, C, D) = B \oplus C \oplus D$

11.4.3 Implementation

SHA-1 was designed to be especially amenable to software implementations. Each round requires only bitwise Boolean operation with 32-bit registers. Somewhat countering this effect is the large number of rounds. Nevertheless, optimized implementations on modern 64-bit microprocessors can achieve throughputs of 1 Gbit/sec or beyond. These are highly optimized assembly code software and typical implementations are most likely considerably slower. Generally speaking, one drawback of SHA-1 and other MD4 family algorithms is that they are difficult to parallelize. It is hard to execute many of the Boolean operations that constitute a round in parallel.

With respect to hardware, SHA-1 is certainly not a truly large algorithm but there are several factors which cause it to be larger than one might expect. Recent hardware implementations on conventional FPGAs can reach a few Gbit/sec which is not that groundbreaking compared to PC-based implementations. One reason is that the function f_t depends on the stage number t. Another reason is the many registers that are required to store the 512 bit intermediate results. Hence, block ciphers like AES are typically smaller and faster in hardware. Also in some applications, hash functions built from block ciphers as described in Section 11.3.2 are sometimes desirable for hardware implementations.

11.5 Discussion and Further Reading

MD4 family and General Remarks It is instructive to have a look at the attack history of the MD4 family. A predecessor of MD4 was Rivest's MD2 hash function, which did not appear to become widely used. It is doubtful that the algorithm would withstand today's attacks. The first attacks against reduced versions of MD4 (the first or the last rounds were missing) were developed by Boer and Bosselaers in 1992 [53]. In 1995, Dobbertin showed how collisions for the full MD4 can be constructed in less than a minute on conventional PCs [61]. Later Dobbertin showed that a variant of MD4 (a round was not executed) does not have the one-wayness property. In 1994, Boer and Bosselaer found collisions in MD5 [54]. In 1995, Dobbertin was able to find collisions for the compression function of MD5 [62]. In order to construct a collision for the popular SHA-1 algorithm, about 2^{63} computations have to be executed. This is still a formidable task. In 2007, a distributed hash collision search over the Internet was organized by Rechberger at the Technical University of Graz in Austria. At the time of writing, about two years into the search, no collisions have been found.

RIPEMD-160 plays a somewhat special role in the MD4 family of hash functions. Unlike all SHA-1 and SHA-2 algorithms, it is the only one that was not designed by NIST and NSA, but rather by a team of European researchers. Even though there is no indication that any of the SHA algorithms are artificially weakened or contain backdoors (introduced by the US government, that is), RIPEMD-160 might appeal to some people who heavily distrust governments. Currently, no successful attacks against the hash functions are known. On the other hand, due to its more limited deployment, there has been less scrutiny by the research community with respect to RIPEMD-160.

It is important to point out that in addition to the MD4 family, numerous other algorithms have been proposed over the years including, for instance, Whirlpool [12], which is related to AES. Most of them did not gain widespread adoption, however. Entirely different from the MD4 family are hash functions which are based on algebraic structures such as MASH-1 and MASH-2 [96]. Many of these algorithms were found to be insecure.

SHA-3 Due to the serious attacks against SHA-1, NIST held two public workshops to assess the status of SHA and to solicit public input on its cryptographic hash function policy and standard. As a consequence, NIST decided to develop additional hash functions, to be named SHA-3, through a public competition. This approach is quite similar to the selection process of AES. In the fall of 2008, 64 algorithms had been submitted to NIST. At the time of writing, 33 of those hash functions are still in the competition. The final decision is expected in 2012. In the meantime the SHA-2 algorithm, against which no attacks are known to date, appears to be the safest choice when selecting a hash function.

Hash Functions from Block Ciphers The four block cipher based hash functions introduced in the chapter are all provable secure. This means, the best possible preimage and second preimage attacks have a complexity of 2^b, where b is the message digest length, and the best possible collision attack requires $2^{b/2}$ steps. The security proof only holds if the block cipher is being treated as a black box, i.e, no (possible) specific weaknesses of the cipher are being exploited. In addition to the four methods of building hash functions from block ciphers introduced in this chapter, there are several other constructions [136]. In Problem 11.3, 12 variants are treated in more detail.

The Hirose construction is relatively new [92]. It can also be realized with AES with a 192-bit key and message blocks x_i of length 64 bit. However, the efficiency is roughly half of that of the construction presented in this chapter (AES256 with 128-bit message blocks). There are also various other methods to build hash functions with twice the output size of the block ciphers used. A prominent one is MDC-2, which was originally designed for DES but works with any block cipher [137]. MDC-2 is standardized in ISO/IEC 10118-2.

11.6 Lessons Learned

- Hash functions are keyless. The two most important applications of hash functions are their use in digital signatures and in message authentication codes such as HMAC.
- The three security requirements for hash functions are one-wayness, second preimage resistance and collision resistance.

■ Hash functions should have at least 160-bit output length in order to withstand collision attacks; 256 bit or more is desirable for long-term security.

■ MD5, which was widely used, is insecure. Serious security weaknesses have been found in SHA-1, and the hash function should be phased out. The SHA-2 algorithms all appear to be secure.

■ The ongoing SHA-3 competition will result in new standardized hash functions in a few years.

Problems

11.1. Compute the output of the first round of stage 1 of SHA-1 for a 512-bit input block of

1. $x = \{0...00\}$
2. $x = \{0...01\}$ (i.e., bit 512 is one).

Ignore the initial hash value H_0 for this problem (i.e., $A_0 = B_0 = ... = 00000000_{hex}$).

11.2. One of the earlier applications of cryptographic hash functions was the storage of passwords for user authentication in computer systems. With this method, a password is hashed after its input and is compared to the stored (hashed) reference password. People realized early that it is sufficient to only store the hashed versions of the passwords.

1. Assume you are a hacker and you got access to the hashed password list. Of course, you would like to recover the passwords from the list in order to impersonate some of the users. Discuss which of the three attacks below allow this. Exactly describe the consequences of each of the attacks:

   ■ Attack A: You can break the one-way property of h.
   ■ Attack B: You can find second preimages for h.
   ■ Attack C: You can find collisions for h.

2. Why is this technique of storing hashed passwords often extended by the use of a so-called *salt*? (A *salt* is a random value appended to the password before hashing. Together with the hash, the value of the *salt* is stored in the list of hashed passwords.) Are the attacks above affected by this technique?
3. Is a hash function with an output length of 80 bit sufficient for this application?

11.3. Draw a block digram for the following hash functions built from a block cipher $e()$:

1. $e(H_{i-1}, x_i) \oplus x_i$
2. $e(H_{i-1}, x_i \oplus H_{i-1}) \oplus x_i \oplus H_{i-1}$
3. $e(H_{i-1}, x_i) \oplus x_i \oplus H_{i-1}$
4. $e(H_{i-1}, x_i \oplus H_{i-1}) \oplus x_i$
5. $e(x_i, H_{i-1}) \oplus H_{i-1}$
6. $e(x_i, x_i \oplus H_{i-1}) \oplus x_i \oplus H_{i-1}$
7. $e(x_i, H_{i-1}) \oplus x_i \oplus H_{i-1}$
8. $e(x_i, x_i \oplus H_{i-1}) \oplus H_{i-1}$
9. $e(x_i \oplus H_{i-1}, x_i) \oplus x_i$
10. $e(x_i \oplus H_{i-1}, H_{i-1}) \oplus H_{i-1}$
11. $e(x_i \oplus H_{i-1}, x_i) \oplus H_{i-1}$
12. $e(x_i \oplus H_{i-1}, H_{i-1}) \oplus x_i$

11.4. We define the rate of a block cipher-based hash function as follows: A block cipher-based hash function that processes u input bits at a time, produces v output bits and performs w block cipher encryptions per input block has a rate of

$$v/(u \cdot w).$$

What is the rate of the four block cipher constructions introduced in Section 11.3.2?

11.5. We consider three different hash functions which produce outputs of lengths 64, 128 and 160 bit. After how many random inputs do we have a probability of $\varepsilon = 0.5$ for a collision? After how many random inputs do we have a probability of $\varepsilon = 0.1$ for a collision?

11.6. Describe how exactly you would perform a collision search to find a pair x_1, x_2, such that $h(x_1) = h(x_2)$ for a given hash function h. What are the memory requirements for this type of search if the hash function has an output length of n bits?

11.7. Assume the block cipher PRESENT (block length 64 bits, 128-bit key) is used in a Hirose hash function construction. The algorithm is used to store the hashes of passwords in a computer system. For each user i with password PW_i, the system stores:

$$h(PW_i) = y_i$$

where the passwords (or passphrases) have an arbitrary length. Within the computer system only the values y_i are actually used for identifying users and giving them access.

Unfortunately, the password file that contains all hash values falls into your hands and you are widely known as a very dangerous hacker. This in itself should not pose a serious problem as it should be impossible to recover the passwords from the hashes due to the one-wayness of the hash function. However, you discovered a small but momentous implementation flaw in the software: The constant c in the hash scheme is assigned the value $c = 0$. Assume you also know the initial values $(H_{0,L}$ and $H_{0,R})$.

1. What is the size of each entry y_i?
2. Assume you want to log in as user U (you might be the CEO of the organization). Provide a detailed description that shows that finding a value PW_{hack} for which

$$PW_{hack} = y_U$$

   takes only about 2^{64} steps.
3. Which of the three general attacks against hash functions do you perform?
4. Why is the attack not possible if $c \neq 0$?

11.8. In this problem, we will examine why techniques that work nicely for error correction codes are not suited as cryptographic hash functions. We look at a hash function that computes an 8-bit hash value by applying the following equation:

$$C_i = b_{i1} \oplus b_{i2} \oplus b_{i3} \oplus b_{i4} \oplus b_{i5} \oplus b_{i6} \oplus b_{i7} \oplus b_{i8} \tag{11.2}$$

Every block of 8 bits constitutes an ASCII-encoded character.

1. Encode the string CRYPTO to its binary or hexadecimal representation.
2. Calculate the (6-bit long) hash value of the character string using the previously defined equation.
3. "Break" the hash function by pointing out how it is possible to find (meaningful) character strings which result in the same hash value. Provide an appropriate example.
4. Which cruical property of hash functions is missing in this case?

Chapter 12
Message Authentication Codes (MACs)

A *Message Authentication Code* (MAC), also known as a *cryptographic checksum* or a *keyed hash function*, is widely used in practice. In terms of security functionality, MACs share some properties with digital signatures, since they also provide message integrity and message authentication. However, unlike digital signatures, MACs are symmetric-key schemes and they do not provide nonrepudiation. One advantage of MACs is that they are much faster than digital signatures since they are based on either block ciphers or hash functions.

In this chapter you will learn:

■ The principle behind MACs
■ The security properties that can be achieved with MACs
■ How MACs can be realized with hash functions and with block ciphers

12.1 Principles of Message Authentication Codes

Similar to digital signatures, MACs append an authentication tag to a message. The crucial difference between MACs and digital signatures is that MACs use a symmetric key k for both generating the authentication tag and verifying it. A MAC is a function of the symmetric key k and the message x. We will use the notation

$$m = \text{MAC}_k(x)$$

for this in the following. The principle of the MAC calculation and verification is shown in Figure 12.1.

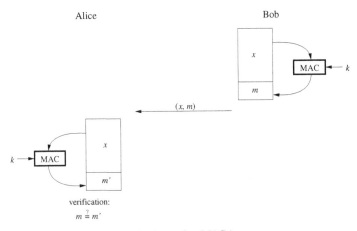

Fig. 12.1 Principle of message authentication codes (MACs)

The motivation for using MACs is typically that Alice and Bob want to be assured that any manipulations of a message x in transit are detected. For this, Bob computes the MAC as a function of the message and the shared secret key k. He sends both the message and the authentication tag m to Alice. Upon receiving the message and m, Alice verifies both. Since this is a symmetric set-up, she simply repeats the steps that Bob conducted when sending the message: She merely recomputes the authentication tag with the received message and the symmetric key.

The underlying assumption of this system is that the MAC computation will yield an incorrect result if the message x was altered in transit. Hence, *message integrity* is provided as a security service. Furthermore, Alice is now assured that Bob was the originator of the message since only the two parties with the same secret key k have the possibility to compute the MAC. If an adversary, Oscar, changes the message during transmission, he cannot simply compute a valid MAC since he lacks the secret key. Any malicious or accidental (e.g., due to transmission errors) forgery of the message will be detected by the receiver due to a failed verification of the MAC.

That means, from Alice's perspective, Bob must have generated the MAC. In terms of security services, message authentication is provided.

In practice, a messages x is often much larger than the corresponding MAC. Hence, similar to hash functions, the output of a MAC computation is a fixed-length authentication tag which is independent of the length of the input.

Together with earlier discussed characteristics of MACs, we can summarize all their important properties:

Properties of Message Authentication Codes

1. **Cryptographic checksum** A MAC generates a cryptographically secure authentication tag for a given message.
2. **Symmetric** MACs are based on secret symmetric keys. The signing and verifying parties must share a secret key.
3. **Arbitrary message size** MACs accept messages of arbitrary length.
4. **Fixed output length** MACs generate fixed-size authentication tags.
5. **Message integrity** MACs provide message integrity: Any manipulations of a message during transit will be detected by the receiver.
6. **Message authentication** The receiving party is assured of the origin of the message.
7. **No nonrepudiation** Since MACs are based on symmetric principles, they do not provide nonrepudiation.

The last point is important to keep in mind: MACs do not provide nonrepudiation. Since the two communicating parties share the same key, there is no possibility to prove towards a neutral third party, e.g., a judge, whether a message and its MAC originated from Alice or Bob. Thus, MACs offer no protection in scenarios where either Alice or Bob is dishonest, like the car-buying example we described in Section 10.1.1. A symmetric secret key is not tied to a certain person but rather to two parties, and hence a judge cannot distinguish between Alice and Bob in case of a dispute.

In practice, message authentication codes are constructed in essentially two different ways from block ciphers or from hash functions. In the subsequent sections of this chapter we will introduce both options for realizing MACs.

12.2 MACs from Hash Functions: HMAC

An option for realizing MACs is to use cryptographic hash functions such as SHA-1 as a building block. One possible construction, named HMAC, has become very popular in practice over the last decade. For instance, it is used in both the Transport Layer Security (TLS) protocol (indicated by the little lock symbol in your Web browser) as well as in the IPsec protocol suite. One reason for the widespread use of the HMAC construction is that it can be proven to be secure if certain assumptions are made.

The basic idea behind all hash-based message authentication codes is that the key is hashed together with the message. Two obvious constructions are possible. The first one:

$$m = \text{MAC}_k(x) = h(k||x)$$

is called *secret prefix MAC*, and the second one:

$$m = \text{MAC}_k(x) = h(x||k)$$

is known as *secret suffix MAC*. The symbol "$||$" denotes concatenation. Intuitively, due to the one-wayness and the good "scrambling properties" of modern hash functions, both approaches should result in strong cryptographic checksums. However, as is often the case in cryptography, assessing the security of a scheme can be trickier than it seems at first glance. We now demonstrate weaknesses in both constructions.

Attacks Against Secret Prefix MACs

We consider MACs realized as $m = h(k||x)$. For the attack we assume that the cryptographic checksum m is computed using a hash construction as shown in Figure 11.5. This iterated approach is used in the majority of today's hash functions. The message x that Bob wants to sign is a sequence of blocks $x = (x_1, x_2, \ldots, x_n)$, where the block length matches the input width of the hash function. Bob computes an authentication tag as:

$$m = \text{MAC}_K(x) = h(k||x_1, x_2, \ldots, x_n)$$

The problem is that the MAC for the message $x = (x_1, x_2, \ldots, x_n, x_{n+1})$, where x_{n+1} is an arbitrary additional block, can be constructed from m without knowing the secret key. The attack is shown in the protocol below.

Attack Against Secret Prefix MACs

Alice	Oscar	Bob
		$x = (x_1, \ldots, x_n)$
		$m = h(k||x_1, \ldots, x_n)$
	↯ intercept	$\xleftarrow{\quad (x,m) \quad}$
	$x_O = (x_1, \ldots, x_n, x_{n+1})$	
	$m_O = h(m||x_{n+1})$	
$\xleftarrow{\quad (x_O, m_O) \quad}$		
$m' \quad\quad =$		
$h(k||x_1, \ldots, x_n, x_{n+1})$		
since $m' = m_O$		
$\Rightarrow$ valid signature!		

Note that Alice will accept the message $(x_1, \ldots, x_n, x_{n+1})$ as valid, even though Bob only authenticated $(x_1, \ldots, x_n)$. The last block x_{n+1} could, for instance, be an appendix to an electronic contract, a situation that could have serious consequences.

The attack is possible since the MAC of the additional message block only needs the previous hash output, which is equal to Bob's m, and x_{n+1} as input but not the key k.

Attacks Against Secret Suffix MACs

After studying the attack above, it seems to be safe to use the other basic construction method, namely $m = h(x||k)$. However, a different weakness occurs here. Assume Oscar is capable of constructing a collision in the hash function, i.e., he can find x and x_O such that:

$$h(x) = h(x_O).$$

The two messages x and x_O can be, for instance, two versions of a contract which are different in some crucial aspect, e.g., the agreed upon payment. If Bob signs x with a message authentication code

$$m = h(x||k)$$

m is also a valid checksum for x_O, i.e.,

$$m = h(x||k) = h(x_O||k)$$

The reason for this is again given by the iterative nature of the MAC computation.

Whether this attack presents Oscar with an advantage depends on the parameters used in the construction. As a practical example, let's consider a secret suffix MAC which uses SHA-1 as hash function, which has an output length of 160 bits, and a 128-bit key. One would expect that this hash offers a security level of 128 bits, i.e., an attacker cannot do better than brute-forcing the entire key space to forge a message. However, if an attacker exploits the birthday paradox (cf. Section 11.2.3), he can forge a signature with about $\sqrt{2^{160}} = 2^{80}$ computations. There are indications that SHA-1 collisions can be constructed with even fewer steps, so that an actual attack might be even easier. In summary, we conclude that the secret suffix method also does not provide the security one would like to have from a MAC construction.

HMAC

A hash-based message authentication code which does not show the security weakness described above is the HMAC construction proposed by Mihir Bellare, Ran Canetti and Hugo Krawczyk in 1996. The scheme consists of an inner and outer hash and is visualized in Figure 12.2.

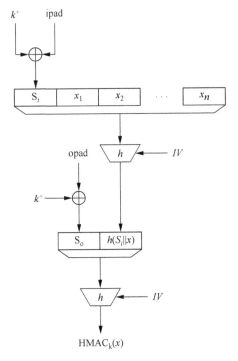

Fig. 12.2 HMAC construction

The MAC computation starts with expanding the symmetric key k with zeros on the left such that the result k^+ is b bits in length, where b is the input block width of the hash function. The expanded key is XORed with the inner pad, which consists of the repetition of the bit pattern:

$$\text{ipad} = 0011\,0110, 0011\,0110, \ldots, 0011\,0110$$

so that a length of b bit is achieved. The output of the XOR forms the first input block to the hash function. The subsequent input blocks are the message blocks $(x_1, x_2, \ldots, x_n)$.

The second, outer hash is computed with the padded key together with the output of the first hash. Here, the key is again expanded with zeros and then XORed with the outer pad:

$$\text{opad} = 0101\,1100, 0101\,1100, \ldots, 0101\,1100.$$

The result of the XOR operation forms the first input block for the outer hash. The other input is the output of the inner hash. After the outer hash has been computed, its output is the message authentication code of x. The HMAC construction can be expressed as:

$$\text{HMAC}_k(x) = h\left[(k^+ \oplus \text{opad}) \| h\left[(k^+ \oplus \text{ipad}) \| x\right]\right].$$

The hash output length l is in practice longer than the width b of an input block. For instance, SHA-1 has an $l = 160$ bit output but accepts $b = 512$ bit inputs. It does not pose a problem that the inner hash function output does not match the input size of outer hash because hash functions have preprocessing steps to match the input string to the block width. As an example, Section 11.4.1 described the preprocessing for SHA-1.

In terms of computational efficiency, it should be noted that the message x, which can be very long, is only hashed once in the inner hash function. The outer hash consists of merely two blocks, namely the padded key and the inner hash output. Thus, the computational overhead introduced through the HMAC construction is very low.

In addition to its computational efficiency, a major advantage of the HMAC construction is that there exists a *proof of security*. As for all schemes which are provable secure, HMAC is not secure per se, but its security is related to the security of some other building block. In the case of the HMAC construction it can be shown that if an attacker, Oscar, can break the HMAC, he can also break the hash function used in the scheme. Breaking HMAC means that even though Oscar does not know the key, he can construct valid authentication tags for messages. Breaking the hash function means that he can either find collisions or that he can compute a hash function output even though he does not know the initial value IV (which was the value H_0 in the case of SHA-1).

12.3 MACs from Block Ciphers: CBC-MAC

In the preceding section we saw that hash functions can be used to realize MACs. An alternative method is to construct MACs from block ciphers. The most popular approach in practice is to use a block cipher such as AES in cipher block chaining (CBC) mode, as discussed in Section 5.1.2.

Figure 12.3 depicts the complete setting for the application of a MAC on basis of a block cipher in CBC mode. The left side shows the sender, the right side the receiver. This scheme is also referred to as CBC-MAC.

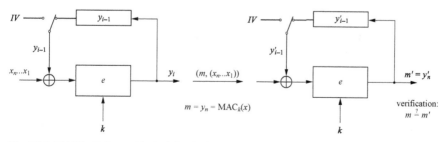

Fig. 12.3 MAC built from a block cipher in CBC mode

MAC Generation

For the generation of a MAC, we have to divide the message x into blocks $x_i, i = 1, ..., n$. With the secret key k and an initial value IV, we can compute the first iteration of the MAC algorithm as

$$y_1 = e_k(x_1 \oplus IV),$$

where the IV can be a public but random value. For subsequent message blocks we use the XOR of the block x_i and the previous output y_{i-1} as input to the encryption algorithm:

$$y_i = e_k(x_i \oplus y_{i-1}).$$

Finally, the MAC of the message $x = x_1 x_2 x_3 ... x_n$ is the output y_n of the last round:

$$m = \text{MAC}_k(x) = y_n$$

In contrast to CBC encryption, the values $y_1, y_2, y_3, ..., y_{n-1}$ are *not* transmitted. They are merely internal values which are used for computing the final MAC value $m = y_n$.

MAC Verification

As with every MAC, verification involves simply repeating the operation that were used for the MAC generation. For the actual verification decision we have to compare the computed MAC m' with the received MAC value m. In case $m' = m$, the message is verified as correct. In case $m' \neq m$, the message and/or the MAC value m have been altered during transmission. We note that the MAC verification is different from CBC decryption, which actually reverses the encryption operation.

The output length of the MAC is determined by the block size of the cipher used. Historically, DES was widely used, e.g., for banking applications. More recently, AES is often used; it yields a MAC of length 128 bit.

12.4 Galois Counter Message Authentication Code (GMAC)

GMAC is a variant of the Galois Counter Mode (GCM) introduced in Section 5.1.6. GMAC is specified in [160] and is a mode of operation for an underlying symmetric key block cipher. In contrast to the GCM mode, GMAC does not encrypt data but only computes a message authentication code. GMAC is easily parallelizable, which is attractive for high-speed applications. The use of GMAC in IPsec Encapsulating Security Payload (ESP) and Authentication Header (AH) is described in the RFC 4543 [119]. The RFC describes how to use AES in GMAC to provide data origin authentication without confidentiality within the IPsec ESP and AH. GMAC can be efficiently implemented in hardware and can reach a speed of 10 Gbit/sec and above.

12.5 Discussion and Further Reading

Block Cipher-Based MACs Historically, block cipher-based MACs have been the dominant method for constructing message authentication codes. As early as in 1977, i.e., only a couple of years after the announcement of the Data Encryption Standard (DES), it was suggested that DES could be used to compute cryptographic checksums [39]. In the following years, block cipher-based MACs were standardized in the US and became popular for assuring the integrity of financial transactions, see, e.g., the ANSI X9.17 standard [3]. Much more recently, the NIST recommendation [65] specifies a message authentication code algorithm based on a symmetric key block cipher (*CMAC*), which is similar to CBC-MAC. The AES-CMAC algorithm is specified in RFC 4493 [159].

In this chapter the CBC-MAC was introduced. In addition to the CBC-MAC, there are the OMAC and PMAC, which are both constructed with block ciphers. Counter with CBC-MAC (*CCM*) is a mode for authenticated encryption and is defined for use with a 128-bit block cipher [173]. It is described in the NIST recommendation [64]. The GMAC construction is standardized in IPSec [119] and in the NIST recommendation for Block Cipher Modes of Operation [66].

Hash Function-Based MACs The HMAC construction was originally proposed at the Crypto 1996 conference [14]. A very accessible treatment of the scheme can be found in [15]. HMAC was turned into an Internet RFC, and was quickly adopted in many Internet security protocols, including TLS and IPsec. In both cases it protects the integrity of a message during transmission. It is widely used with the hash functions SHA-1 and MD5, and its use with RIPEMD-160 has also been often discussed. It seems likely that the switch to more modern hash functions such as SHA-2 and SHA-3 will result in more and more HMAC constructions with these hash functions.

Other MAC Constructions Another type of message authentication code is based on *universal hashing* and is called *UMAC*. UMAC is backed by a formal security analysis, and the only internal cryptographic component is a block cipher used to generate the pseudorandom pads and internal key material. The universal hash function is used to produce a short hash value of fixed length. This hash is then XORed with a key-derived pseudorandom pad. The universal hash function is designed to be very fast in software (e.g., as low as one cycle per byte on contemporary processors) and is mainly based on additions of 32-bit and 64-bit numbers and multiplication of 32-bit numbers. Based on the original idea by Wegman and Carter [40], numerous schemes have been proposed, e.g., the schemes Multilinear-Modular-Hashing (MMH) and UMAC [89, 23].

12.6 Lessons Learned

- MACs provide two security services, *message integrity* and *message authentication*, using symmetric techniques. MACs are widely used in protocols.
- Both of these services are also provided by digital signatures, but MACs are much faster.
- MACs do not provide nonrepudiation.
- In practice, MACs are either based on block ciphers or on hash functions.
- HMAC is a popular MAC used in many practical protocols such as TLS.

Problems

12.1. As we have seen, MACs can be used to authenticate messages. With this problem, we want to show the difference between two protocols—one with a MAC, one with a digital signature. In the two protocols, the sending party performs the following operation:

1. Protocol A:

$$y = e_{k_1}[x || h(k_2 || x)]$$

where x is the message, $h()$ is a hash function such as SHA-1, e is a private-key encryption algorithm, "$||$" denotes simple concatenation, and k_1, k_2 are secret keys which are only known to the sender and the receiver.

2. Protocol B:

$$y = e_k[x || sig_{k_{pr}}(h(x))]$$

Provide a step-by-step description (e.g., with an itemized list) of what the receiver does upon receipt of y. You may want to draw a block diagram for the process on the receiver's side, but that's optional.

12.2. For hash functions it is crucial to have a sufficiently large number of output bits, with, e.g., 160 bits, in order to thwart attacks based on the birthday paradox. Why are much shorter output lengths of, e.g., 80 bits, sufficient for MACs?

For your answer, assume a message x that is sent in clear together with its MAC over the channel: $(x, MAC_k(x))$. Exactly clarify what Oscar has to do to attack this system.

12.3. We study two methods for integrity protection with encryption.

1. Assume we apply a technique for combined encryption and integrity protection in which a ciphertext c is computed as

$$c = e_k(x || h(x))$$

where $h()$ is a hash function. This technique is not suited for encryption with stream ciphers if the attacker knows the whole plaintext x. Explain *exactly* how an active attacker can now *replace* x by an arbitrary x' of his/her choosing and compute c' such that the receiver will verify the message correctly. Assume that x and x' are of equal length. Will this attack work too if the encryption is done with a one-time pad?

2. Is the attack still applicable if the checksum is computed using a keyed hash function such as a MAC:

$$c = e_{k_1}(x || MAC_{k_2}(x))$$

Assume that $e()$ is a stream cipher as above.

12.4. We will now discuss some issues when constructing an efficient MAC.

1. The messages X to be authenticated consists of z independent blocks, so that $X = x_1||x_2||\ldots||x_z$, where every x_i consists of $|x_i| = 8$ bits. The input blocks are consecutively put into the compression function

$$c_i = h(c_{i-1}, x_i) = c_{i-1} \oplus x_i$$

At the end, the MAC value

$$MAC_k(X) = c_z + k \mod 2^8$$

is calculated, where k is a 64-bit long shared key. Describe how exactly the (effective part of the) key k can be calculated with only one known message X.
2. Perform this attack for the following parameters and determine the key k:

$$X = \texttt{HELLO ALICE!}$$
$$c_0 = 11111111_2$$
$$MAC_k(X) = 10011101_2$$

3. What is the effective key length of k?
4. Although two different operations ($[\oplus, 2^8]$ and $[+, 2^8]$) are utilized in this MAC, this MAC-based signature possesses significant weaknesses. To which property of the design can these be ascribed, and where should one take care when constructing a cryptographic system? This essential property also applies for block ciphers and hash functions!

12.5. MACs are, in principle, also vulnerable against collision attacks. We discuss the issue in the following.

1. Assume Oscar found a collision between two messages, i.e.,

$$MAC_k(x_1) = MAC_k(x_2)$$

Show a simple protocol with an attack that is based on a collision.
2. Even though the birthday paradox can still be used for constructing collisions, why is it in a practical setting much harder to construct them for MACs than for hash functions? Since this is the case: what security is provided by a MAC with 80-bit output compared to a hash function with 80-bit output?

Chapter 13
Key Establishment

With the cryptographic mechanisms that we have learned so far, in particular symmetric and asymmetric encryption, digital signatures and message authentication codes (MACs), one can relatively easily achieve the basic security services (cf. Sect. 10.1.3):

■ Confidentiality (with encryption algorithms)
■ Integrity (with MACs or digital signatures)
■ Message authentication (with MACs or digital signatures)
■ Non-repudiation (with digital signatures)

Similarly, identification can be accomplished through protocols which make use of standard cryptographic primitives.

However, all cryptographic mechanisms that we have introduced so far assume that keys are properly distributed between the parties involved, e.g., between Alice and Bob. The task of key establishment is in practice one of the most important and often also most difficult parts of a security system. We already learned some ways of distributing keys, in particular Diffie–Hellman key exchange. In this chapter we will learn many more methods for establishing keys between remote parties. You will learn about the following important issues:

■ How keys can be established using symmetric cryptosystems
■ How keys can be established using public-key cryptosystems
■ Why public-key techniques still have shortcomings for key distribution
■ What certificates are and how they are used
■ The role that public-key infrastructures play

13.1 Introduction

In this section we introduce some terminology, some thoughts on key freshness and a very basic key distribution scheme. The latter is helpful for motivating the more advanced methods which will follow in this chapter.

13.1.1 Some Terminology

Roughly speaking, key establishment deals with establishing a shared secret between two or more parties. Methods for this can be classified into *key transport* and *key agreement* methods, as shown in Fig. 13.1. A key transport protocol is a technique where one party securely transfers a secret value to others. In a key agreement protocol two (or more) parties derive the shared secret where all parties contribute to the secret. Ideally, none of the parties can control what the final joint value will be.

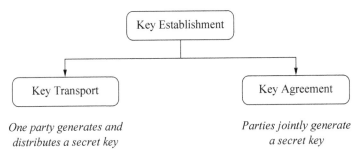

Fig. 13.1 Classification of key establishment schemes

Key establishment itself is strongly related to identification. For instance, you may think of attacks by unauthorized users who join the key establishment protocol with the aim of masquerading as either Alice or Bob with the goal of establishing a secret key with the other party. To prevent such attacks, each party must be assured of the identity of the other entity. All of these issues are addressed in this chapter.

13.1.2 Key Freshness and Key Derivation

In many (but not all) security systems it is desirable to use cryptographic keys which are only valid for a limited time, e.g., for one Internet connection. Such keys are called *session keys* or *ephemeral keys*. Limiting the period in which a cryptographic key is used has several advantages. A major one is that there is less damage if the key is exposed. Also, an attacker has less ciphertext available that was generated under one key, which can make cryptographic attacks much more difficult. Moreover, an attacker is forced to recover several keys if he is interested in decrypting larger parts of plaintext. Real-world examples where session keys are frequently generated include voice encryption in GSM cell phones and video encryption in pay-TV satellite systems; in both cases new keys are generated within a matter of minutes or sometimes even seconds.

The security advantages of *key freshness* are fairly obvious. However, the question now is, how can key updates be realized? The first approach is to simply execute the key establishment protocols shown in this chapter over and over again. However, as we see later, there are always certain costs associated with key establishment, typically with respect to additional communication connections and computations. The latter holds especially in the case of public-key algorithms which are very computationally intensive.

The second approach to key update uses an already established joint secret key to *derive* fresh session keys. The principal idea is to use a key derivation function (KDF) as shown in Fig. 13.2. Typically, a non-secret parameter r is processed together with the joint secret k_{AB} between the users Alice and Bob.

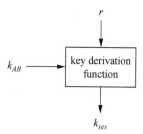

Fig. 13.2 Principle of key derivation

An important characteristic of the key derivation function is that it should be a one-way function. The one-way property prevents an attacker from deducing k_{AB} should any of the session keys become compromised, which in turn would allow the attacker to compute all other session keys.

One possible way of realizing the key derivation function is that one party sends a nonce, i.e., a numerical value that is used only once, to the other party. Both users encrypt the nonce using the shared secret key k_{AB} by means of a symmetric cipher such as AES. The corresponding protocol is shown below.

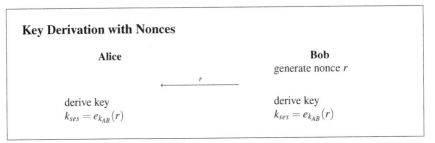

An alternative to encrypting the nonce is hashing it together with k_{AB}. One way of achieving this is that both parties perform a HMAC computation with the nonce serving as the "message":

$$k_{ses} = HMAC_{k_{AB}}(r)$$

Rather than sending a nonce, Alice and Bob can also simply encrypt a counter *cnt* periodically, where the ciphertext again forms the session key:

$$k_{ses} = e_{k_{AB}}(cnt)$$

or compute the HMAC of the counter:

$$k_{ses} = HMAC_{k_{AB}}(cnt)$$

Using a counter can save Alice and Bob one communication session because, unlike the case of the nonce-based key derivation, no value needs to be transmitted. However, this holds only if both parties know exactly when the next key derivation needs to take place. Otherwise, a counter synchronization message might be required.

13.1.3 The n^2 Key Distribution Problem

Until now we mainly assumed that the necessary keys for symmetric algorithms are distributed via a "secure channel", as depicted in the beginning of this book in Fig. 1.5. Distributing keys this way is sometimes referred to as *key predistribution* or *out-of-band transmission* since it typically involves a different mode (or band) of communication, e.g., the key is transmitted via a phone line or in a letter. Even though this seems somewhat clumsy, it can be a useful approach in certain practical situations, especially if the number of communicating parties is not too large. However, key predistribution quickly reaches its limits even if the number of entities in a network is only moderately large. This leads to the well-known n^2 key distribution problem.

We assume a network with n users, where every party is capable of communicating with every other one in a secure fashion, i.e., if Alice wants to communicate with Bob, these two share a secret key k_{AB} which is only known to them but not to any of the other $n-2$ parties. This situation is shown for the case of a network with $n = 4$ participants in Fig. 13.3.

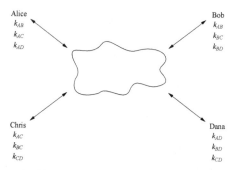

Fig. 13.3 Keys in a network with $n = 4$ users

We can extrapolate several features of this simple scheme for the case of n users:

■ Each user must store $n - 1$ keys.
■ There is a total of $n(n - 1) \approx n^2$ keys in the network.
■ A total of $n(n-1)/2 = \binom{n}{2}$ symmetric key pairs are in the network.
■ If a new user joins the network, a secure channel must be established with every other user in order to upload new keys.

The consequences of these observations are not very favorable if the number of users increases. The first drawback is that the number of keys in the system is roughly n^2. Even for moderately sized networks, this number becomes quite large. All these keys must be generated securely at one location, which is typically some type of trusted authority. The other drawback, which is often more serious in practice, is that adding one new user to the system requires updating the keys at all existing users. Since each update requires a secure channel, this is very burdensome.

Example 13.1. A mid-size company with 750 employees wants to set up secure e-mail communication with symmetric keys. For this purpose, $750 \times 749/2 = 280,875$ symmetric key pairs must be generated, and $750 \times 749 = 561,750$ keys must be distributed via secure channels. Moreover, if employee number 751 joins the company, all 750 other users must receive a key update. This means that 751 secure channels (to the 750 existing employees and to the new one) must be established.
    ◇

Obviously, this approach does not work for large networks. However, there are many cases in practice where the number of users is (i) small and (ii) does not change frequently. An example could be a company with a small number of branches which all need to communicate with each other securely. Adding a new branch does not happen too often, and if this happens it can be tolerated that one new key is uploaded to any of the existing branches.

13.2 Key Establishment Using Symmetric-Key Techniques

Symmetric ciphers can be used to establish secret (session) keys. This is somewhat surprising because we assumed for most of the book that symmetric ciphers themselves need a secure channel for establishing their keys. However, it turns out that it is in many cases sufficient to have a secure channel only when a new user joins the network. This is in practice often achievable for computer networks because at set-up time a (trusted) system administrator might be needed in person anyway who can install a secret key manually. In the case of embedded devices, such as cell phones, a secure channel is often given during manufacture, i.e., a secret key can be loaded into the device "in the factory".

The protocols introduced in the following all perform key transport and not key agreement.

13.2.1 Key Establishment with a Key Distribution Center

The protocols developed in the following rely on a *Key Distribution Center (KDC)*. This is a server that is fully trusted by all users and that shares a secret key with each user. This key, which is named the *Key Encryption Key* (KEK), is used to securely transmit session keys to users.

Basic Protocol

A necessary prerequisite is that each user U shares a unique secret key KEK k_U with the key distribution center which predistributed through a secure channel. Let's look what happens if one party requests a secure session from the KDC, e.g., Alice wants to communicate with Bob. The interesting part of this approach is that the KDC **encrypts the session key** that will eventually be used by Alice and Bob. In a basic protocol, the KDC generates two messages, y_A and y_B, for Alice and Bob, respectively:

$$y_A = e_{k_A}(k_{ses})$$
$$y_B = e_{k_B}(k_{ses})$$

Each message contains the session key encrypted with one of the two KEKs. The protocol looks like this:

Basic Key Establishment Using a Key Distribution Center

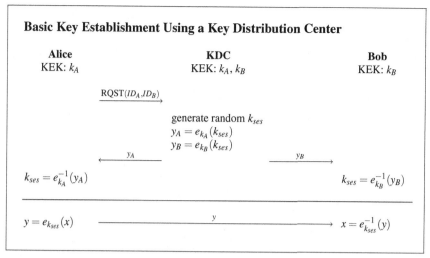

The protocol begins with a request message $RQST(ID_A, ID_B)$, where ID_A and ID_B simply indicate the users involved in the session. The actual key establishment protocol is executed subsequently in the upper part of the drawing. Below the solid line is, as an example, shown how Alice and Bob can now communicate with each other securely using the session key.

It is important to note that two types of keys are involved in the protocol. The KEKs k_A and k_B are long-term keys that do not change. The session key k_{ses} is an ephemeral key that changes frequently, ideally for every communication session. **In order to understand this protocol more intuitively, one can view the predistributed KEKs as forming a secret channel between the KDC and each user.** With this interpretation, the protocol is straightforward: The KDC simply sends a session key to Alice and Bob via the two respective secret channels.

Since the KEKs are long-term keys, whereas the session keys have typically a much shorter lifetime, in practice sometimes different encryption algorithms are used with both. Let's consider the following example. In a pay-TV system AES might be used with the long-term KEKs k_U for distributing session keys k_{ses}. The session keys might only have a lifetime of, say, one minute. The session keys are used to encrypt the actual plaintext (the digital TV signal in this example) with a fast stream cipher. A stream cipher might be required to assure real-time decryption. The advantage of this arrangement is that even if a session key becomes compromised, only one minute's worth of multimedia data can be decrypted by an adversary. Thus, the cipher that is used with the session key does not necessarily need to have the same cryptographic strength as the algorithm which is used for distributing the session keys. On the other hand, if one of the KEKs becomes compromised, all prior and future traffic can be decrypted by an eavesdropper.

It is easy to modify the above protocol such that we save one communication session. This is shown in the following:

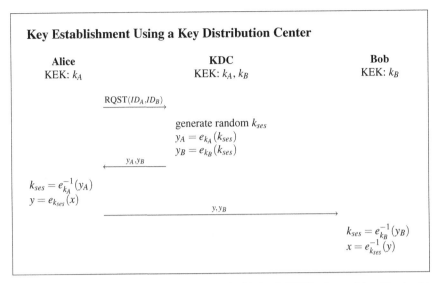

Key Establishment Using a Key Distribution Center

Alice receives the session key encrypted with both KEKs, k_A and k_B. She is able to compute the session key k_{ses} from y_A and can use it subsequently to encrypt the actual message she wants to send to Bob. The interesting part of the protocol is that Bob receives both the encrypted message y as well as y_B. He needs to decrypt the latter one in order to recover the session key which is needed for computing x.

Both of the KDC-based protocols have the advantage that there are only n long-term symmetric key pairs in the system, unlike the first naïve scheme that we encountered, where about $n^2/2$ key pairs were required. The n long-term KEKS only need to be stored by the KDC, while each user only stores his or her own KEK. Most importantly, if a new user Noah joins the network, a secure channel only needs to be established once between the KDC and Noah to distribute the KEK k_N.

Security

Even though the two protocols protect against a passive attacker, i.e, an adversary that can only eavesdrop, there are attacks if an adversary can actively manipulate messages and create faked ones.

Replay Attack One weakness is that a *replay attack* is possible. This attack makes use of the fact that neither Alice nor Bob know whether the encrypted session key they receive is actually a new one. If an old one is reused, key freshness is violated. This can be a particularly serious issue if an old session key has become compromised. This could happen if an old key is leaked, e.g., through a hacker, or if the encryption algorithm used with an old key has become insecure due to cryptanalytical advances.

If Oscar gets hold of a previous session key, he can impersonate the KDC and resend old messages y_A and y_B to Alice and Bob. Since Oscar knows the session key, he can decipher the plaintext that will be encrypted by Alice or Bob.

Key Confirmation Attack Another weakness of the above protocol is that Alice is not assured that the key material she receives from the KDC is actually for a session between her and Bob. This attack assumes that Oscar is also a legitimate (but malicious) user. By changing the session-request message Oscar can trick the KDC and Alice to set up session between him and Alice as opposed to between Alice and Bob. Here is the attack:

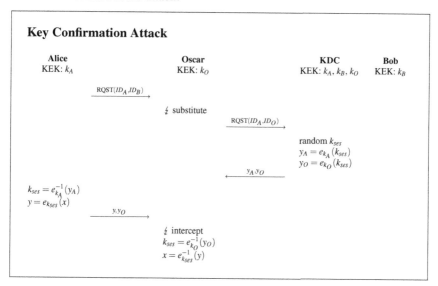

The gist of the attack is that the KDC believes Alice requests a key for a session between Alice and Oscar, whereas she really wants to communicate with Bob. Alice assumes that the encrypted key "y_O" is "y_B", i.e., the session key encrypted under Bob's KEK k_B. (Note that if the KDC puts a header ID_O in front of y_O which associates it with Oscar, Oscar might simply change the header to ID_B.) In other words, Alice has no way of knowing that the KDC prepared a session with her and Oscar; instead she still thinks she is setting up a session with Bob. Alice continues with the protocol and encrypts her actual message as y. If Oscar intercepts y, he can decrypt it.

The underlying problem for this attack is that there is *no* key confirmation. If key confirmation were given, Alice would be assured that Bob and no other user knows the session key.

13.2.2 Kerberos

A more advanced protocol that protects against both replay and key confirmation attacks is Kerberos. It is, in fact, more than a mere key distribution protocol; its main purpose is to provide user authentication in computer networks. Kerberos was standardized as an RFC 1510 in 1993 and is in widespread use. It is also based on a KDC, which is named the "authentication sever" in Kerberos terminology. Let's first look at a simplified version of the protocol.

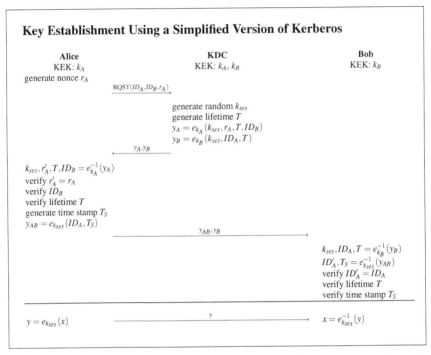

Key Establishment Using a Simplified Version of Kerberos

Kerberos assures the *timeliness* of the protocol through two measures. First, the KDC specifies a lifetime T for the session key. The lifetime is encrypted with both session keys, i.e., it is included in y_A and y_B. Hence, both Alice and Bob are aware of the period during which they can use the session key. Second, Alice uses a time stamp T_S, through which Bob can be assured that Alice's messages are recent and are not the result of a replay attack. For this, Alice's and Bob's system clocks must be synchronized, but not with a very high accuracy. Typical values are in the range of a few minutes. The usage of the lifetime parameter T and the time stamp T_S prevent replay attacks by Oscar.

Equally important is that Kerberos provides key confirmation and user authentication. In the beginning, Alice sends a random nonce r_A to the KDC. This can be considered as a *challenge* because she challenges the KDC to encrypt it with their joint KEK k_A. If the returned challenge r_A' matches the sent one, Alice is assured that the message y_A was actually sent by the KDC. This method to authenticate users is known as *challenge-response protocol* and is widely used, e.g., for authentication of smart cards.

Through the inclusion of Bob's identity ID_B in y_A Alice is assured that the session key is actually meant for a session between herself and Bob. With the inclusion of Alice's identity ID_A in both y_B and y_{AB}, Bob can verify that (i) the KDC included a session key for a connection between him and Alice and (ii) that he is currently actually talking to Alice.

13.2.3 Remaining Problems with Symmetric-Key Distribution

Even though Kerberos provides strong assurance that the correct keys are being used and that users are authenticated, there are still drawbacks to the protocols discussed so far. We now describe remaining general problems that exist for KDC-based schemes.

Communication requirements One problem in practice is that the KDC needs to be contacted if a new secure session is to be initiated between any two parties in the network. Even though this is a performance rather than a security problem, it can be a serious hindrance in a system with very many users. In Kerberos, one can alleviate this potential problem by increasing the lifetime T of the key. In practice, Kerberos can run with tens of thousands of users. However, it would be a problem to scale such an approach to "all" Internet users.

Secure channel during initialization As discussed earlier, all KDC-based protocols require a secure channel at the time a new user joins the network for transmitting that user's key encryption key.

Single point of failure All KDC-based protocols, including Kerberos, have the security drawback that they have a *single point of failure*, namely the database that contains the key encryption keys, the KEKs. If the KDC becomes compromised, all KEKs in the entire system become invalid and have to be re-established using secure channels between the KDC and each user.

No perfect forward secrecy If any of the KEKs becomes compromised, e.g., through a hacker or Trojan software running on a user's computer, the consequences are serious. First, all future communication can be decrypted by the attacker who eavesdrops. For instance, if Oscar got a hold of Alice's KEK k_A, he can recover the session key from all messages y_A that the KDC sends out. **Even more dramatic is the fact that Oscar can also decrypt past communications if he stored old messages y_A and y.** Even if Alice immediately realizes that her KEK has been compromised and she stops using it right away, there is nothing she can do to prevent Oscar from decrypting her *past* communication. Whether a system is vulnerable if long-term keys are compromised is an important feature of a security system and there is a special terminology used:

Definition 13.1. A cryptographic protocol has *perfect forward secrecy* (PFS) if the compromise of long-term keys does not allow an attacker to obtain past session keys.

Neither Kerberos nor the simpler protocols shown earlier offer PFS. The main mechanism to assure PFS is to employ public-key techniques, which we study in the following sections.

13.3 Key Establishment Using Asymmetric Techniques

Public-key algorithms are especially suited for key establishment protocols since they don't share most of the drawbacks that symmetric key approaches have. In fact, next to digital signatures, key establishment is the other major application domain of public-key schemes. They can be used for both key transport and key agreement. For the former, Diffie–Hellman key exchange, elliptic curve Diffie–Hellman or related protocols are often used. For key transport, any of the public-key encryption schemes, e.g., RSA or Elgamal, is often used. We recall at this point that public-key primitives are quite slow, and that for this reason actual data encryption is usually done with symmetric primitives like AES or 3DES, after a key has been established using asymmetric techniques.

At this moment it looks as though public-key schemes solve all key establishment problems. It turns out, however, that they all require what is termed an *authenticated channel* to distribute the public keys. The remainder of this chapter is chiefly devoted to solving the problem of authenticated public key distribution.

13.3.1 Man-in-the-Middle Attack

The *man-in-the-middle attack*[1] is a serious attack against public-key algorithms. The basic idea of the attack is that the adversary, Oscar, replaces the public keys sent out by the participants with his own keys. This is possible whenever public keys are not authenticated. The man-in-the-middle (MIM) attack has far-reaching consequences for asymmetric cryptography. For didactical reasons we will study the MIM attack against the Diffie–Hellman key exchange (DHKE). However, it is extremely important to bear in mind that the attack is applicable against any asymmetric scheme unless the public-keys are protected, e.g., through certificates, a topic that is discussed in Sect. 13.3.2.

We recall that the DHKE allows two parties who never met before to agree on a shared secret by exchanging messages over an insecure channel. For convenience, we restate the DHKE protocol here:

---

[1] The "man-in-the-middle attack" should not be confused with the similarly sounding but in fact entirely different "meet-in-the-middle attack" against block ciphers which was introduced in Sect. 5.3.1.

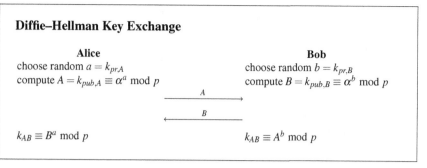

Diffie–Hellman Key Exchange

As we discussed in Sect. 8.4, if the parameters are chosen carefully, which includes especially a prime p with a length of 1024 or more bit, the DHKE is secure against eavesdropping, i.e., passive attacks. We consider now the case that an adversary is not restricted to only listening to the channel. Rather, Oscar can also actively take part in the message exchange by intercepting, changing and generating messages. The underlying idea of the MIM attack is that Oscar replaces both Alice's and Bob's public key by his own. The attack is shown here:

Man-in-the-Middle Attack Against the DHKE

Alice	Oscar	Bob
choose $a = k_{pr,A}$		choose $b = k_{pr,B}$
$A = k_{pub,A} \equiv \alpha^a \bmod p$		$B = k_{pub,B} \equiv \alpha^b \bmod p$
$\xrightarrow{\quad A \quad}$	$\frac{1}{2}$ substitute $\tilde{A} \equiv \alpha^o$ $\xrightarrow{\quad \tilde{A} \quad}$	
$\xleftarrow{\quad \tilde{B} \quad}$	$\frac{1}{2}$ substitute $\tilde{B} \equiv \alpha^o$ $\xleftarrow{\quad B \quad}$	
$k_{AO} \equiv (\tilde{B})^a \bmod p$	$k_{AO} \equiv A^o \bmod p$	$k_{BO} \equiv (\tilde{A})^b \bmod p$
	$k_{BO} \equiv B^o \bmod p$	

Let's look at the keys that are being computed by the three players, Alice, Bob and Oscar. The key Alice computes is:

$$k_{AO} = (\tilde{B})^a \equiv (\alpha^o)^a \equiv \alpha^{oa} \bmod p$$

which is identical to the key that Oscar computes as $k_{AO} = A^o \equiv (\alpha^a)^o \equiv \alpha^{ao} \bmod p$. At the same time Bob computes:

$$k_{BO} = (\tilde{A})^b \equiv (\alpha^o)^b \equiv \alpha^{ob} \bmod p$$

which is identical to Oscar's key $k_{BO} = B^o \equiv (\alpha^b)^o \equiv \alpha^{bo} \bmod p$. Note that the two malicious keys that Oscar sends out, $\tilde{A}$ and $\tilde{B}$, are in fact the same values. With use different names here merely to stress the fact that Alice and Bob assume that they have received each other's public keys.

What happens in this attack is that two DHKEs are being performed simultaneously, one between Alice and Oscar and another one between Bob and Oscar. As a result, Oscar has established a joined key with Alice, which we termed k_{AO}, and another one with Bob, which we named k_{BO}. **However, neither Alice nor Bob is aware of the fact that they share a key with Oscar and not with each other!** Both assume that they have computed a joint key k_{AB}.

From here on, Oscar has much control over encrypted traffic between Alice and Bob. As an example, here is how he can read encrypted messages in a way that goes unnoticed by Alice and Bob:

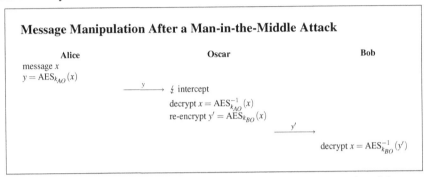

Message Manipulation After a Man-in-the-Middle Attack

For illustrative purposes, we assumed that AES is used for the encryption. Of course, any other symmetric cipher can be used as well. Please note that Oscar can not only read the plaintext x but can also alter it prior to re-encrypting it with k_{BO}. This can have serious consequences, e.g., if the message x describes a financial transaction.

13.3.2 Certificates

The underlying problem of the man-in-the-middle attack is that public keys are not authenticated. We recall from Sect. 10.1.3 that message authentication ensures that the sender of a message is authentic. However, in the scenario at hand Bob receives a public key which is supposedly Alice's, but he has no way of knowing whether that is in fact the case. To make this point clear, let's examine how a key of a user Alice would look in practice:

$$k_A = (k_{pub,A}, ID_A),$$

where ID_A is identifying information, e.g., Alice's IP address or her name together with date of birth. The actual public key $k_{pub,A}$, however, is a mere binary string, e.g., 2048 bit. If Oscar performs a MIM attack, he would change the key to:

$$k_A = (k_{pub,O}, ID_A).$$

Since everything is unchanged except the anonymous actual bit string, the receiver will not be able to detect that it is in fact Oscar's. This observation has far-reaching consequences which can be summarized in the following statement:

> **Even though public-key schemes do not require a secure channel, they require authenticated channels for the distribution of the public keys.**

We would like to stress here again that the MIM attack is not restricted to the DHKE, but is in fact applicable to any asymmetric crypto scheme. The attack always proceeds the same way: Oscar intercepts the public key that is being sent and replaces it with his own.

The problem of trusted distribution of private keys is central in modern public-key cryptography. There are several ways to address the problem of key authentication. The main mechanism is the use of *certificates*. The idea behind certificates is quite easy: Since the authenticity of the message $(k_{pub,A}, ID_A)$ is violated by an active attack, we apply a cryptographic mechanism that provides authentication. More specifically, we use digital signatures.[2] Thus, a certificate for a user Alice in its most basic form is the following structure:

$$\text{Cert}_A = [(k_{pub,A}, ID_A), \text{sig}_{k_{pr}}(k_{pub,A}, ID_A)]$$

The idea is that the receiver of a certificate verifies the signature prior to using the public key. We recall from Chap. 10 that the signature protects the signed message — which is the structure $(k_{pub,A}, ID_A)$ in this case — against manipulation. If Oscar attempts to replace $k_{pub,A}$ by $k_{pub,O}$ it will be detected. Thus, it is said that **certificates bind the identity of a user to their public key.**

Certificates require that the receiver has the correct verification key, which is a public key. If we were to use Alice's public key for this, we would have the same problem that we are actually trying to solve. Instead, the signatures for certificates are provided by a mutually trusted third party. This party is called the *Certification Authority* commonly abbreviated as *CA*. It is the task of the CA to generate and issue certificates for all users in the system. For certificate generation, we can distinguish between two main cases. In the first case, the user computes her own asymmetric key pair and merely requests the CA to sign the public key, as shown in the following simple protocol for a user named Alice:

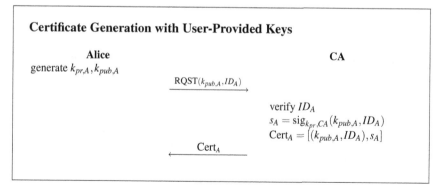

Certificate Generation with User-Provided Keys

Alice — generate $k_{pr,A}, k_{pub,A}$

CA

$\xrightarrow{\text{RQST}(k_{pub,A}, ID_A)}$

verify ID_A
$s_A = \text{sig}_{k_{pr,CA}}(k_{pub,A}, ID_A)$
$\text{Cert}_A = [(k_{pub,A}, ID_A), s_A]$

$\xleftarrow{\text{Cert}_A}$

---

[2] MACs also provide authentication and could, in principle, also be used for authenticating public keys. However, because MACs themselves are symmetric algorithms, we would again need a secure channel for distributing the MAC keys with all the associated drawbacks.

From a security point of view, the first transaction is crucial. It must be assured that Alice's message $(k_{pub,A}, ID_A)$ is sent via an authenticated channel. Otherwise, Oscar could request a certificate in Alice's name.

In practice it is often advantageous that the CA not only signs the public keys but also generates the public–private key pairs for each user. In this case, a basic protocol looks like this:

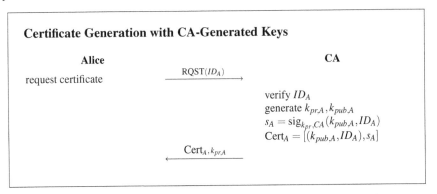

Certificate Generation with CA-Generated Keys

Alice		CA
request certificate	$\xrightarrow{\quad RQST(ID_A) \quad}$	
		verify ID_A
		generate $k_{pr,A}, k_{pub,A}$
		$s_A = \text{sig}_{k_{pr},CA}(k_{pub,A}, ID_A)$
		$\text{Cert}_A = [(k_{pub,A}, ID_A), s_A]$
	$\xleftarrow{\quad \text{Cert}_A, k_{pr,A} \quad}$	

For the first transmission, an authenticated channel is needed. In other words: The CA must be assured that it is really Alice who is requesting a certificate, and not Oscar who is requesting a certificate in Alice's name. Even more sensitive is the second transmission consisting of $(\text{Cert}_A, k_{pr,A})$. Because the private key is being sent here, not only an authenticated but a secure channel is required. In practice, this could be a certificate delivered by mail on a CD-ROM.

Before we discuss CAs in more detail, let's have a look at the DHKE which is protected with certificates:

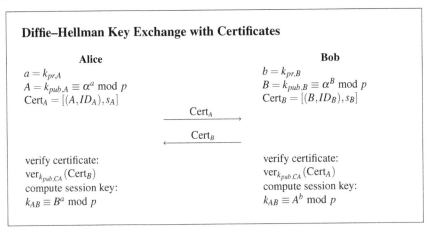

Diffie–Hellman Key Exchange with Certificates

Alice		Bob
$a = k_{pr,A}$		$b = k_{pr,B}$
$A = k_{pub,A} \equiv \alpha^a \bmod p$		$B = k_{pub,B} \equiv \alpha^B \bmod p$
$\text{Cert}_A = [(A, ID_A), s_A]$		$\text{Cert}_B = [(B, ID_B), s_B]$
	$\xrightarrow{\quad \text{Cert}_A \quad}$	
	$\xleftarrow{\quad \text{Cert}_B \quad}$	
verify certificate:		verify certificate:
$\text{ver}_{k_{pub,CA}}(\text{Cert}_B)$		$\text{ver}_{k_{pub,CA}}(\text{Cert}_A)$
compute session key:		compute session key:
$k_{AB} \equiv B^a \bmod p$		$k_{AB} \equiv A^b \bmod p$

One very crucial point here is the verification of the certificates. Obviously, without verification, the signatures within the certificates would be of no use. As can be seen in the protocol, verification requires the public key of the CA. This key must be transmitted via an authenticated channel, otherwise Oscar could perform MIM attacks again. It looks like we haven't gained much from the introduction of certificates since we again require an authenticated channel! **However, the difference from the former situation is that we need the authenticated channel only once, at set-up time.** For instance, public verification keys are nowadays often included in PC software such as Web browsers or Microsoft software products. The authenticated channel is here assumed to be given through the installation of original software which has not been manipulated. What's happening here from a more abstract point of view is extremely interesting, namely a **transfer of trust**. We saw in the earlier example of DHKE without certificates, that Alice and Bob have to trust each other's public keys directly. With the introduction of certificates, they only have to trust the CA's public key $k_{pub,CA}$. If the CA signs other public keys, Alice and Bob know that they can also trust those. This is called a *chain of trust*.

13.3.3 Public-Key Infrastructures (PKI) and CAs

The entire system that is formed by CAs together with the necessary support mechanisms is called a *public-key infrastructure*, usually referred to as *PKI*. As the reader can perhaps start to imagine, setting up and running a PKI in the real world is a complex task. Issues such as identifying users for certificate issuing and trusted distribution of CA keys have to be solved. There are also many other real-world issues; among the most complex are the existence of many different CAs and revocation of certificates. We discuss some aspects of using certificate systems in practice in the following.

X.509 Certificates

In practice, certificates not only include the ID and the public key of a user, they tend to be quite complex structures with many additional fields. As an example, we look at the a X.509 certificate in Fig. 13.4. X.509 is an important standard for network authentication services, and the corresponding certificates are widely used for Internet communication, i.e., in S/MIME, IPsec and SSL/TLS.

Discussing the fields defined in a X.509 certificate gives us some insight into many aspects of PKIs in the real world. We discuss the most relevant ones in the following:

1. *Certificate Algorithm*: Here it is specified which signature algorithm is being used, e.g., RSA with SHA-1 or ECDSA with SHA-2, and with which parameters, e.g., the bit lengths.
2. *Issuer*: There are many companies and organizations that issue certificates. This field specifies who generated the one at hand.
3. *Period of Validity*: In most cases, a public key is not certified indefinitely but rather for a limited time, e.g., for one or two years. One reason for doing this is that private keys which belong to the certificate may become compromised.

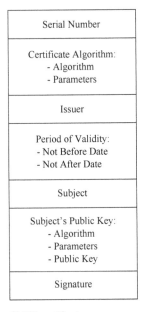

Fig. 13.4 Detailed structure of an X.509 certificate

By limiting the validity period, there is only a certain time span during which an attacker can maliciously use the private key. Another reason for a restricted lifetime is that, especially for certificates for companies, it can happen that the user ceases to exist. If the certificates, and thus the public keys, are only valid for limited time, the damage can be controlled.

4. *Subject*: This field contains what was called ID_A or ID_B in our earlier examples. It contains identifying information such as names of people or organizations. Note that not only actual people but also entities like companies can obtain certificates.

5. *Subject's Public Key*: The public key that is to be protected by the certificate is here. In addition to the binary string which is the public key, the algorithm (e.g., Diffie–Hellman) and the algorithm parameters, e.g., the modulus p and the primitive element α, are stored.

6. *Signature*: The signature over all other fields of the certificate.

We note that for every signature two public key algorithms are involved: the one whose public key is protected by the certificate and the algorithm with which the certificate is signed. These can be entirely different algorithms and parameter sets. For instance, the certificate might be signed with an RSA 2048-bit algorithm, while the public key within the certificate could belong to a 160-bit elliptic curve scheme.

Chain of Certificate Authorities (CAs)

In an ideal world, there would be one CA which issues certificates for, say, all Internet users on planet Earth. Unfortunately, that is not the case. There are many different entities that act as CAs. First of all, many countries have their own "official" CA, often for certificates that are used for applications that involve government business. Second, certificates for websites are currently issued by more than 50 mostly commercial entities. (Most Web browsers have the public key of those CAs preinstalled.) Third, many corporations issue certificate for their own employees and external entities who do business with them. It would be virtually impossible for a user to have the private keys of all these different CAs at hand. What is done instead is that CAs certify each other.

Let's look at an example where Alice's certificate is issued by CA1 and Bob's by CA2. At the moment, Alice is only in possession of the public key of "her" CA1, and Bob has only $k_{pub,CA2}$. If Bob sends his certificate to Alice, she cannot verify Bob's public key. This situation looks like this:

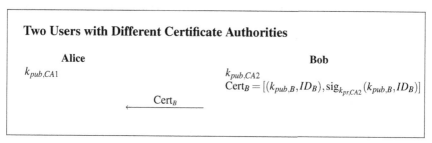

Alice can now request CA2's public key, which is itself contained in a certificate that was signed by Alice's CA1:

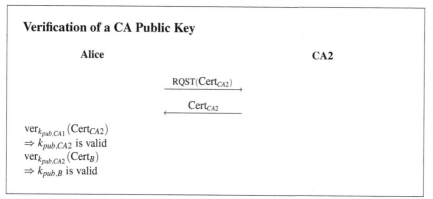

The structure $Cert_{CA2}$ contains the public key of CA2 signed by CA1, which looks like this:

$$Cert_{CA2} = [(k_{pub,CA2}, ID_{CA2}), sig_{k_{pr,CA1}}(k_{pub,CA2}, ID_{CA2})]$$

The important outcome of the process is that Alice can now verify Bob's key.

What's happening here is that a certificate chain is being established. CA1 trusts CA2 which is expressed by CA1 signing the public key $k_{pub,CA2}$. Now Alice can trust Bob's public key since it was signed by CA1. This situation is called a *chain of trust*, and it is said that *trust is delegated*.

In practice, CAs can be arranged hierarchically, where each CA signes the public key of the certificate authorities one level below. Alternatively, CAs can cross-certify each other without a strict hierarchical relationship.

Certificate Revocation Lists

One major issue in practice is that it must be possible to revoke certificates. A common reason is that a certificate is stored on a smart card which is lost. Another reason could be that a person left an organization and one wants to make sure that she is not using the public key that was given to her. The solution in these situations seems easy: Just publish a list with all certificates that are currently invalid. Such a list is called a *certificate revocation list*, or *CRL*. Typically, the serial numbers of certificates are used to identify the revoked certificates. Of course, a CRL must be signed by the CA since otherwise attacks are possible.

The problem with CLRs is how to transmit them to the users. The most straightforward way is that every user contacts the issuing CA every time a certificate of another user is received. The major drawback is that now the CA is involved in every session set-up. This was one major drawback of KDC-based, i.e., symmetrickey, approaches. The promise of certificate-based communication was that no online contact to a central authority was needed.

An alternative is that CRLs are sent out periodically. The problem with this approach is that there is always a period during which a certificate is invalid but users have not yet been informed. For instance, if the CRL is sent out at 3:00 am every morning (a time with relatively little network traffic otherwise), a dishonest person could have almost a whole day where a revoked certificate is still valid. To counter this, the CRL update period can be shortened, say to one hour. However, this would be a tremendous burden on the bandwidth of the network. This is an instructive example for the tradeoff between costs in the form of network traffic on one hand, and security on the other hand. In practice, a reasonable compromise must be found.

In order to keep the size of CRLs moderate, often only the changes from the last CRL broadcast are sent out. These update-only CRLs are referred to as *delta CRLs*.

13.4 Discussion and Further Reading

Key Establishment Protocols In most modern network security protocols, public-key approaches are used for establishing keys. In this book, we introduced the Diffie–Hellman key exchange and described a basic key transport protocol in Chap. 6 (cf. Fig. 6.5). In practice, often considerably more advanced asymmetric protocols are used. However, most of them are based on either the Diffie–Hellman or a key transport protocol. A comprehensive overview on this area is given in [33].

We now give a few examples of generic cryptographic protocols that are often preferred over the basic Diffie–Hellman key exchange. The *MTI* (Matsumoto–Takashima–Imai) protocols are an ensemble of authenticated Diffie–Hellman key exchanges which were already published in 1986. Good descriptions can be found in [33] and [120]. Another popular Diffie–Hellman extension is the station-to-station (STS) protocol. It uses certificates and provides both user and key authentication. A discussion about STS variants can be found in [60]. A more recent protocol for authenticated Diffie–Hellman is the MQV protocol which is discussed in [108]. It is typically used with elliptic curves.

A prominent practical example for a key establishment protocol is the Internet Key Exchange (IKE) protocol. IKE provides key material for IPsec, which is the "official" security mechanism for Internet traffic. IKE is quite complex and offers many options. At its core, however, is a Diffie–Hellman key agreement followed by an authentication. The latter can either be achieved with certificates or with pre-shared keys. A good starting point for more information on IPsec and IKE is the RFC [128] and, more accessibly, reference [161, Chapter 16].

Certificates and Alternatives During the second half of the 1990s there was a belief that essentially every Internet user would need a certificate in order to communicate securely, e.g., for doing ebusiness transactions. "PKI" was a buzzword for some time, and many companies were formed that provided certificates and PKI services. However, it turned out that there are major technical and practical hurdles to a PKI that truly encompasses all or most Internet users. What has happened instead is that nowadays many servers are authenticated with certificates, for instance Internet retailers, whereas most individual users are not. The needed CA verification keys are often preinstalled in users' Web browsers. This asymmetric set-up — the server is authenticated but the user is not — is acceptable since the user is typically the one who provides crucial information such as her credit card number. A comprehensive introduction to the large field of PKI and certificates is given in the book [2]. An interesting and entertaining discussion about the alleged shortcomings of PKI is given in [74], and an equally instructive rebuttal is online at [107].

We introduced certificates and a public-key infrastructure as the main method for authenticating public keys. Such hierarchical organized certificates are only one possible approach, though this is the most widely used one. Another concept is the *web of trust* that relies entirely on trust relationships between parties. The idea is as follows: If Alice trusts Bob, it is assumed that she also wants to trust all other users whom Bob trusts. This means that every party in such a web of trust implicitly trusts parties whom it does not know (or has never met before). The most popular example for such a system are *Pretty Good Privacy (PGP)* and *Gnu Privacy Guard (GPG)*, which are widely used for signing and encrypting emails.

13.5 Lessons Learned

- A key transport protocol securely transfers a secret key to other parties.
- In a key agreement protocol, two or more parties negotiate a common secret key.
- In most common symmetric protocols, the key exchange is coordinated by a trusted third party. A secure channel between the third party and each user is only required at set-up time.
- Symmetric key establishment protocols do not scale well to networks with large numbers of users and they provide typically no perfect forward secrecy.
- The most widely used asymmetric key establishment protocol is the Diffie–Hellman key exchange.
- All asymmetric protocols require that the public keys are authenticated, e.g., with certificates. Otherwise man-in-the-middle attacks are possible.

Problems

13.1. In this exercise, we want to analyze some variants of key derivation. In practice, one *masterkey* k_{MK} is exchanged in a secure way (e.g. certificate-based DHKE) between the involved parties. Afterwards, the session keys are regularly updated by use of key derivation. For this purpose, three different methods are at our disposal:

(1) $k_0 = k_{MK}; k_{i+1} = k_i + 1$
(2) $k_0 = h(k_{MK}); k_{i+1} = h(k_i)$
(3) $k_0 = h(k_{MK}); k_{i+1} = h(k_{MK}||i||k_i)$

where $h()$ marks a (secure) hash function, and k_i is the ith session key.

1. What are the main differences between these three methods?
2. Which method provides *Perfect Forward Secrecy*?
3. Assume Oscar obtains the nth session key (e.g., via brute-force). Which sessions can he now decrypt (depending on the chosen method)?
4. Which method remains secure if the masterkey k_{MK} is compromised? Give a rationale!

13.2. Imagine a peer-to-peer network where 1000 users want to communicate in an authenticated and confidential way without a central Trusted Third Party (TTP).

1. How many keys are collectively needed, if symmetric algorithms are deployed?
2. How are these numbers changed, if we bring in a central instance (Key Distribution Center, KDC)?
3. What is the main advantage of a KDC against the scenario without a KDC?
4. How many keys are necessary if we make use of asymmetric algorithms?

Also differentiate between keys which *every* user has to store and keys which are collectively necessary.

13.3. You have to choose the cryptographic algorithms for a KDC where two different classes of encryption occur:

■ $e_{k_{U,KDC}}()$, where U denotes an arbitrary network node (user),
■ $e_{k_{ses}}()$ for the communication between two users.

You have the choice between two different algorithms, DES and 3DES (Triple-DES), and you are advised to use distinct algorithms for both encryption classes. Which algorithm do you use for which class? Justify your answer including aspects of security as well as celerity.

13.4. This exercise considers the security of key establishment with the aid of a KDC. Assume that a hacker performs a successful attack against the KDC at the point of time t_x, where all keys are compromised. The attack is detected.

1. Which (practical) measures have to be taken in order to prevent decryption of future communication between the network nodes?

2. Which steps did the attacker have to take in order to decipher data transmissions which occurred at an earlier time $(t < t_x)$? Does such a KDC system provide Perfect Forward Secrecy (PFS) or not?

13.5. We will now analyze an improved KDC system. In contrast to the previous problem, all keys $e_{k_{U,KDC}}()$ are now refreshed in relatively short intervals:

■ The KDC generates a new (random) key: $k_{U,KDC}^{(i+1)}$
■ The KDC transmits the new key to user U, encrypted with the old one:

$$e_{k_{U,KDC}^{(i)}}(k_{U,KDC}^{(i+1)})$$

Which decryptions are possible, if a staff member of the KDC is corruptible and "sells" all recent keys $e_{k_{U,KDC}^{(i)}}$ of the KDC at the point of time t_x? We assume that this circumstance is not detected until the point of time t_y which could be much later, e.g., one year.

13.6. Show a key confirmation attack against the basic KDC protocol introduced in Sect. 13.2.1. Describe each step of the attack. Your drawing should look similar to the one showing a key confirmation attack against the second (modified) KDC-based protocol.

13.7. Show that PFS is in fact not given in the simplified Kerberos protocol. Show how Oscar can decrypt past and future communications if:

1. Alice's KEK k_A becomes compromised
2. Bob's KEK k_B becomes compromised

13.8. Extend the Kerberos protocol such that a mutual authentication between Alice and Bob is performed. Give a rationale that your solution is secure.

13.9. People at your new job are deeply impressed that you worked through this book. As the first job assignment you are asked to design a digital pay-TV system which uses encryption to prevent service theft through wire tapping. As key exchange protocol, a strong Diffie–Hellman with, e.g., 2048-bit modulus is being used. However, since your company wants to use cheap legacy hardware, only DES is available for data encryption algorithm. You decide to use the following key derivation approach:

$$K^{(i)} = f(K_{AB} \| i). \tag{13.1}$$

where f is an irreversible function.

1. First we have to determine whether the attacker can store an entire movie with reasonable effort (in particular, cost). Assume the data rate for the TV link is 1 Mbit/s, and that the longest movies we want to protect are 2 hours long. How many Gbytes (where $1M = 10^6$ and $1G = 10^9$) of data must be stored for a 2-hour film (don't mix up bit and byte here)? Is this realistic?

2. We assume that an attacker will be able to find a DES key in 10 minutes using a brute-force attack. Note that this is a somewhat optimistic assumption from an attacker's point of view, but we want to provide some medium-term security by assuming increasingly faster key searches in the future.

How frequently must a key be derived if the goal is to prevent an offline decryption of a 2-hour movie in less than 30 days?

13.10. We consider a system in which a key k_{AB} is established using the Diffie–Hellman key exchange protocol, and the encryption keys $k^{(i)}$ are then derived by computing:

$$k^{(i)} = h(k_{AB} \| i) \tag{13.2}$$

where i is just an integer counter, represented as a 32-bit variable. The values of i are public (e.g., the encrypting party always indicates which value for i was used in a header that precedes each ciphertext block). The derived keys are used for the actual data encryption with a symmetric algorithm. New keys are derived every 60 sec during the communication session.

1. Assume the Diffie–Hellman key exchange is done with a 512-bit prime, and the encryption algorithm is AES. Why doesn't it make cryptographic sense to use the key derivation protocol described above? Describe the attack that would require the least computational effort from Oscar.
2. Assume now that the Diffie–Hellman key exchange is done with a 2048-bit prime, and the encryption algorithm is DES. Describe in detail what the advantages are that the key derivation scheme offers compared to a system that just uses the Diffie–Hellman key for DES.

13.11. We reconsider the Diffie–Hellman key exchange protocol. Assume now that Oscar runs an active man-in-the-middle attack against the key exchange as explained in Sect. 13.3.1. For the Diffie–Hellman key exchange, use the parameters $p = 467$, $\alpha = 2$, and $a = 228$, $b = 57$ for Alice and Bob, respectively. Oscar uses the value $o = 16$. Compute the key pairs k_{AO} and k_{BO} (i) the way Oscar computes them, and (ii) the way Alice and Bob compute them.

13.12. We consider the Diffie–Hellman key exchange scheme with certificates. We have a system with the three users Alice, Bob and Charley. The Diffie–Hellman algorithm uses $p = 61$ and $\alpha = 18$. The three secret keys are $a = 11$, $b = 22$ and $c = 33$. The three IDs are ID(A)=1, ID(B)=2 and ID(C)=3.

For signature generation, the Elgamal signature scheme is used. We apply the system parameters $p' = 467$, $d' = 127$, $\alpha' = 2$ and β. The CA uses the ephemeral keys $k_E = 213$, 215 and 217 for Alice's, Bob's and Charley's signatures, respectively. (In practice, the CA should use a better pseudorandom generator to obtain the k_E values.)

To obtain the certificates, the CA computes $x_i = 4 \times b_i + \text{ID}(i)$ and uses this value as input for the signature algorithm. (Given x_i, ID(i) follows then from ID(i) $\equiv$ $x_i \bmod 4$.)

1. Compute three certificates $Cert_A$, $Cert_B$ and $Cert_C$.

2. Verify all three certificates.
3. Compute the three session keys k_{AB}, k_{AC} and k_{BC}.

13.13. Assume Oscar attempts to use an active (substitution) attack against the Diffie–Hellman key exchange with certificates in the following ways:

1. Alice wants to communicate with Bob. When Alice obtains C(B) from Bob, Oscar replaces it with (a valid!) C(O). How will this forgery be detected?
2. Same scenario: Oscar tries now to replace only Bob's public key b_B with his own public key b_O. How will this forgery be detected?

13.14. We consider certificate generation with CA-generated keys. Assume the second transmission of $(\text{Cert}_A, k_{pr,A})$ takes place over an authenticated but insecure channel, i.e., Oscar can read this message.

1. Show how he can decrypt traffic which is encrypted by means of a Diffie–Hellman key that Alice and Bob generated.
2. Can he also impersonate Alice such that he computes a DH key with Bob without Bob noticing?

13.15. Given is a user domain in which users share the Diffie–Hellman parameters α and p. Each user's public Diffie–Hellman key is certified by a CA. Users communicate securely by performing a Diffie–Hellman key exchange and then encrypting/decrypting messages with a symmetric algorithm such as AES.

Assume Oscar gets hold of the CA's signature algorithm (and especially its private key), which was used to generate certificates. Can he now decrypt old ciphertexts which were exchanged between two users before the CA signature algorithm was compromised, and which Oscar had stored? Explain your answer.

13.16. Another problem in certificate systems is the authenticated distribution of the CA's public key which is needed for certificate verification. Assume Oscar has full control over all of Bob's communications, that is, he can alter all messages to and from Bob. Oscar now replaces the CA's public key with his own (note that Bob has no means to authenticate the key that he receives, so he thinks that he received the CA public key.)

1. (Certificate issuing) Bob requests a certificate by sending a request containing (1) Bob's ID $ID(B)$ and (2) Bob's public key B from the CA. Describe exactly what Oscar has to do so that Bob doesn't find out that he has the wrong public CA key.
2. (Protocol execution) Describe what Oscar has to do to establish a session key with Bob using the authenticated Diffie–Hellman key exchange, such that Bob thinks he is executing the protocol with Alice.

13.17. Draw a diagram that shows a key transport protocol shown in Fig. 6.5 from Sect. 6.1, in which RSA encryption is used.

13.18. We consider RSA encryption with certificates in which Bob has the RSA keys. Oscar manages to send Alice a verification key $k_{pr,CA}$ which is, in fact, Oscar's key. Show an active attack in which he can decipher encrypted messages that Alice sends to Bob. Should Oscar run a MIM attack or should he set up a session only between himself and Alice?

13.19. Pretty Good Privacy (PGP) is a widespread scheme for electronic mail security to provide authentication and confidentiality. PGP does not necessarily require the use of certificate authorities. Describe the trust model of PGP and how the public-key management works in practice.

References

1. Michel Abdalla, Mihir Bellare, and Phillip Rogaway. DHAES: An encryption scheme based on the Diffie–Hellman problem. Available at `citeseer.ist.psu.edu/abdalla99dhaes.html`, 1999.
2. Carlisle Adams and Steve Lloyd. *Understanding PKI: Concepts, Standards, and Deployment Considerations*. Addison-Wesley Longman Publishing, Boston, MA, USA, 2002.
3. ANSI X9.17-1985. American National Standard X9.17: Financial Institution Key Management, 1985.
4. ANSI X9.31-1998. American National Standard X9.31, Appendix A.2.4: Public Key Cryptography Using Reversible Algorithms for the Financial Services Industry (rDSA). Technical report, Accredited Standards Committee X9, Available at `http://www.x9.org`, 2001.
5. ANSI X9.42-2003. Public Key Cryptography for the Financial Services Industry: Agreement of Symmetric Keys Using Discrete Logarithm Cryptography. Technical report, American Bankers Association, 2003.
6. ANSI X9.62-1999. The Elliptic Curve Digital Signature Algorithm (ECDSA). Technical report, American Bankers Association, 1999.
7. ANSI X9.62-2001. Elliptic Curve Key Agreement and Key Transport Protocols. Technical report, American Bankers Association, 2001.
8. Frederik Armknecht. *Algebraic attacks on certain stream ciphers*. PhD thesis, Department of Mathematics, University of Mannheim, Germany, December 2006. `http://madoc.bib.uni-mannheim.de/madoc/volltexte/2006/1352/`.
9. Standards for Efficient Cryptography — SEC 1: Elliptic Curve Cryptography, September 2000. Version 1.0.
10. Daniel V. Bailey and Christof Paar. Efficient arithmetic in finite field extensions with application in elliptic curve cryptography. *Journal of Cryptology*, 14, 2001.
11. Elad Barkan, Eli Biham, and Nathan Keller. Instant Ciphertext-Only Cryptanalysis of GSM Encrypted Communication. *Journal of Cryptology*, 21(3):392–429, 2008.
12. P. S. L. M. Barreto and V. Rijmen. The whirlpool hashing function, September 2999. (revised May 2003), `http://paginas.terra.com.br/informatica/paulobarreto/WhirlpoolPage.html`.
13. F. L. Bauer. *Decrypted Secrets: Methods and Maxims of Cryptology*. Springer, 4th edition, 2007.
14. Mihir Bellare, Ran Canetti, and Hugo Krawczyk. Keying Hash Functions for Message Authentication. In *CRYPTO '96: Proceedings of the 16th Annual International Cryptology Conference, Advances in Cryptology*, pages 1–15. Springer, 1996.
15. Mihir Bellare, Ran Canetti, and Hugo Krawczyk. Message Authentication using Hash Functions—The HMAC Construction. *CRYPTOBYTES*, 2, 1996.
16. C.H. Bennett, E. Bernstein, G. Brassard, and U. Vazirani. The strengths and weaknesses of quantum computation. *SIAM Journal on Computing*, 26:1510–1523, 1997.

17. Daniel J. Bernstein. Multidigit multiplication for mathematicians. URL: http://cr.yp.to/papers.html.

18. Daniel J. Bernstein, Johannes Buchmann, and Erik Dahmen. *Post-Quantum Cryptography*. Springer, 2009.

19. N. Biggs. *Discrete Mathematics*. Oxford University Press, New York, 2nd edition, 2002.

20. E. Biham. A fast new DES implementation in software. In *Fourth International Workshop on Fast Software Encryption*, volume 1267 of *LNCS*, pages 260–272. Springer, 1997.

21. Eli Biham and Adi Shamir. *Differential Cryptanalysis of the Data Encryption Standard*. Springer, 1993.

22. Alex Biryukov, Adi Shamir, and David Wagner. Real time cryptanalysis of A5/1 on a PC. In *FSE: Fast Software Encryption*, pages 1–18. Springer, 2000.

23. J. Black, S. Halevi, H. Krawczyk, T. Krovetz, and P. Rogaway. UMAC: Fast and secure message authentication. In *CRYPTO '99: Proceedings of the 19th Annual International Cryptology Conference, Advances in Cryptology*, volume 99, pages 216–233. Springer, 1999.

24. I. Blake, G. Seroussi, N. Smart, and J. W. S. Cassels. *Advances in Elliptic Curve Cryptography (London Mathematical Society Lecture Note Series)*. Cambridge University Press, New York, NY, USA, 2005.

25. Ian F. Blake, G. Seroussi, and N. P. Smart. *Elliptic Curves in Cryptography*. Cambridge University Press, New York, NY, USA, 1999.

26. Daniel Bleichenbacher, Wieb Bosma, and Arjen K. Lenstra. Some remarks on Lucas-based cryptosystems. In *CRYPTO '95: Proceedings of the 15th Annual International Cryptology Conference, Advances in Cryptology*, pages 386–396. Springer, 1995.

27. L Blum, M Blum, and M Shub. A simple unpredictable pseudorandom number generator. *SIAM J. Comput.*, 15(2):364–383, 1986.

28. Manuel Blum and Shafi Goldwasser. An efficient probabilistic public-key encryption scheme which hides all partial information. In *CRYPTO '84: Proceedings of the 4th Annual International Cryptology Conference, Advances in Cryptology*, pages 289–302, 1984.

29. Andrey Bogdanov, Gregor Leander, Lars R. Knudsen, Christof Paar, Axel Poschmann, Matthew J.B. Robshaw, Yannick Seurin, and Charlotte Vikkelsoe. PRESENT—An Ultra-Lightweight Block Cipher. In *CHES '07: Proceedings of the 9th International Workshop on Cryptographic Hardware and Embedded Systems*, number 4727 in LNCS, pages 450–466. Springer, 2007.

30. Dan Boneh and Matthew Franklin. Identity-based encryption from the Weil pairing. *SIAM J. Comput.*, 32(3):586–615, 2003.

31. Dan Boneh and Richard J. Lipton. Algorithms for black-box fields and their application to cryptography (extended abstract). In *CRYPTO '96: Proceedings of the 16th Annual International Cryptology Conference, Advances in Cryptology*, pages 283–297. Springer, 1996.

32. Dan Boneh, Ron Rivest, Adi Shamir, and Len Adleman. Twenty Years of Attacks on the RSA Cryptosystem. *Notices of the AMS*, 46:203–213, 1999.

33. Colin A. Boyd and Anish Mathuria. *Protocols for Key Establishment and Authentication*. Springer, 2003.

34. ECC Brainpool. ECC Brainpool Standard Curves and Curve Generation, 2005. http://www.ecc-brainpool.org/ecc-standard.htm.

35. Johannes Buchmann and Jintai Ding, editors. *Post-Quantum Cryptography, Second International Workshop, PQCrypto 2008, Proceedings*, volume 5299 of *LNCS*. Springer, 2008.

36. Johannes Buchmann and Jintai Ding, editors. *PQCrypto 2006: International Workshop on Post-Quantum Cryptography*, LNCS. Springer, 2008.

37. German Federal Office for Information Security (BSI). http://www.bsi.de/english/publications/bsi_standards/index.htm.

38. Mike Burmester and Yvo Desmedt. A secure and efficient conference key distribution system (extended abstract). In *Advances in Cryptology — EUROCRYPT'94*, pages 275–286, 1994.

39. C. M. Campbell. Design and specification of cryptographic capabilities. *NBS Special Publication 500-27: Computer Security and the Data Encryption Standard, U.S. Department of Commerce, National Bureau of Standards*, pages 54–66, 1977.

40. J.L. Carter and M.N. Wegman. New hash functions and their use in authentication and set equality. *Journal of Computer and System Sciences*, 22(3):265–277, 1981.
41. Çetin Kaya Koç, Tolga Acar, and Burton S. Kaliski. Analyzing and comparing Montgomery multiplication algorithms. *IEEE Micro*, 16(3):26–33, 1996.
42. P. Chodowiec and K. Gaj. Very compact FPGA implementation of the AES algorithm. In C. D. Walter, Ç. K. Koç, and C. Paar, editors, *CHES '03: Proceedings of the 5th International Workshop on Cryptographic Hardware and Embedded Systems*, volume 2779 of *LNCS*, pages 319–333. Springer, 2003.
43. C. Cid, S. Murphy, and M. Robshaw. *Algebraic Aspects of the Advanced Encryption Standard*. Springer, 2006.
44. H. Cohen, G. Frey, and R. Avanzi. *Handbook of Elliptic and Hyperelliptic Curve Cryptography*. Discrete Mathematics and Its Applications. Chapman and Hall/CRC, September 2005.
45. T. Collins, D. Hopkins, S. Langford, and M. Sabin. Public key cryptographic apparatus and method, 1997. United States Patent US 5,848,159. Jan. 1997.
46. Common Criteria for Information Technology Security Evaluation. http://www.commoncriteriaportal.org/.
47. COPACOBANA—A Cost-Optimized Parallel Code Breaker. http://www.copacobana.org/.
48. Sony Corporation. Clefia – new block cipher algorithm based on state-of-the-art design technologies, 2007. http://www.sony.net/SonyInfo/News/Press/200703/07-028E/index.html.
49. Ronald Cramer and Victor Shoup. A practical public key cryptosystem provably secure against adaptive chosen ciphertext attack. *CRYPTO '98: Proceedings of the 18th Annual International Cryptology Conference, Advances in Cryptology*, 1462:13–25, 1998.
50. Cryptool — Educational Tool for Cryptography and Cryptanalysis. https://www.cryptool.org/.
51. J. Daemen and V. Rijmen. AES Proposal: Rijndael. In *First Advanced Encryption Standard (AES) Conference*, Ventura, California, USA, 1998.
52. Joan Daemen and Vincent Rijmen. *The Design of Rijndael*. Springer, 2002.
53. B. den Boer and A. Bosselaers. An attack on the last two rounds of MD4. In *CRYPTO '91: Proceedings of the 11th Annual International Cryptology Conference, Advances in Cryptology*, LNCS, pages 194–203. Springer, 1992.
54. B. den Boer and A. Bosselaers. Collisions for the compression function of MD5. In *Advances in Cryptology - EUROCRYPT'93*, LNCS, pages 293–304. Springer, 1994.
55. Alexander W. Dent. A brief history of provably-secure public-key encryption. Cryptology ePrint Archive, Report 2009/090, 2009. http://eprint.iacr.org/.
56. Diehard Battery of Tests of Randomness CD, 1995. http://i.cs.hku.hk/~diehard/.
57. W. Diffie. The first ten years of public-key cryptography. *Innovations in Internetworking*, pages 510–527, 1988.
58. W. Diffie and M. E. Hellman. New directions in cryptography. *IEEE Transactions on Information Theory*, IT-22:644–654, 1976.
59. W. Diffie and M. E. Hellman. Exhaustive cryptanalysis of the NBS Data Encryption Standard. *COMPUTER*, 10(6):74–84, June 1977.
60. Whitfield Diffie, Paul C. Van Oorschot, and Michael J. Wiener. Authentication and authenticated key exchanges. *Des. Codes Cryptography*, 2(2):107–125, 1992.
61. Hans Dobbertin. Alf swindles Ann. *CRYPTOBYTES*, 3(1), 1995.
62. Hans Dobbertin. The status of MD5 after a recent attack. *CRYPTOBYTES*, 2(2), 1996.
63. Saar Drimer, Tim Güneysu, and Christof Paar. DSPs, BRAMs and a Pinch of Logic: New Recipes for AES on FPGAs. *IEEE Symposium on Field-Programmable Custom Computing Machines (FCCM)*, 0:99–108, 2008.
64. Morris Dworkin. Recommendation for Block Cipher Modes of Operation: The CCM Mode for Authentication and Confidentiality, May 2004. http://csrc.nist.gov/publications/nistpubs/800-38C/SP800-38C_updated-July20_2007.pdf.

65. Morris Dworkin. Recommendation for Block Cipher Modes of Operation: The CMAC Mode for Authentication, NIST Special Publication 800-38D, May 2005. `http://csrc.nist.gov/publications/nistpubs/800-38D/SP-800-38D.pdf`.

66. Morris Dworkin. Recommendation for Block Cipher Modes of Operation: Galois Counter Mode (GCM) and GMAC, NIST Special Publication 800-38D, November 2007. `http://csrc.nist.gov/publications/nistpubs/800-38D/SP-800-38D.pdf`.

67. H. Eberle and C.P. Thacker. A 1 GBIT/second GaAs DES chip. In *Custom Integrated Circuits Conference*, pages 19.7/1–4. IEEE, 1992.

68. AES Lounge, 2007. `http://www.iaik.tu-graz.ac.at/research/krypto/AES/`.

69. eSTREAM—The ECRYPT Stream Cipher Project, 2007. `http://www.ecrypt.eu.org/stream/`.

70. The Side Channel Cryptanalysis Lounge, 2007. `http://www.crypto.ruhr-uni-bochum.de/en_sclounge.html`.

71. Thomas Eisenbarth, Sandeep Kumar, Christof Paar, Axel Poschmann, and Leif Uhsadel. A Survey of Lightweight Cryptography Implementations. *IEEE Design & Test of Computers – Special Issue on Secure ICs for Secure Embedded Computing*, 24(6):522 – 533, November/December 2007.

72. S. E. Eldridge and C. D. Walter. Hardware implementation of Montgomery's modular multiplication algorithm. *IEEE Transactions on Computers*, 42(6):693–699, July 1993.

73. T. ElGamal. A public-key cryptosystem and a signature scheme based on discrete logarithms. *IEEE Transactions on Information Theory*, IT-31(4):469–472, 1985.

74. C. Ellison and B. Schneier. Ten risks of PKI: What you're not being told about public key infrastructure. *Computer Security Journal*, 16(1):1–7, 2000. See also `http://www.counterpane.com/pki-risks.html`.

75. M. Feldhofer, J. Wolkerstorfer, and V. Rijmen. AES implementation on a grain of sand. *Information Security, IEE Proceedings*, 152(1):13–20, 2005.

76. Amos Fiat and Adi Shamir. How to prove yourself: practical solutions to identification and signature problems. In *CRYPTO '86: Proceedings of the 6th Annual International Cryptology Conference, Advances in Cryptology*, pages 186–194. Springer, 1987.

77. Federal Information Processing Standards Publications — FIPS PUBS. `http://www.itl.nist.gov/fipspubs/index.htm`.

78. Electronic Frontier Foundation. Frequently Asked Questions (FAQ) About the Electronic Frontier Foundation's DES Cracker Machine, 1998. `http://w2.eff.org/Privacy/Crypto/Crypto_misc/DESCracker/HTML/19980716_eff_des_faq.html`.

79. J. Franke, T. Kleinjung, C. Paar, J. Pelzl, C. Priplata, and C. Stahlke. SHARK — A Realizable Special Hardware Sieving Device for Factoring 1024-bit Integers. In Josyula R. Rao and Berk Sunar, editors, *CHES '05: Proceedings of the 7th International Workshop on Cryptographic Hardware and Embedded Systems*, volume 3659 of *LNCS*, pages 119–130. Springer, August 2005.

80. Bundesamt für Sicherheit in der Informationstechnik. Anwendungshinweise und Interpretationen zum Schema (AIS). Funktionalitätsklassen und Evaluationsmethodologie für physikalische Zufallszahlengeneratoren. AIS 31, Version 1, 2001. `http://www.bsi.bund.de/zertifiz/zert/interpr/ais31.pdf`.

81. Oded Goldreich. *Foundations of Cryptography: Basic Tools*. Cambridge University Press, New York, NY, USA, 2000.

82. Oded Goldreich. Zero-Knowledge: A tutorial by Oded Goldreich, 2001. `http://www.wisdom.weizmann.ac.il/~oded/zk-tut02.html`.

83. Oded Goldreich. *Foundations of Cryptography: Volume 2, Basic Applications*. Cambridge University Press, New York, NY, USA, 2004.

84. Oded Goldreich. On post-modern cryptography. Cryptology ePrint Archive, Report 2006/461, 2006. `http://eprint.iacr.org/`.

85. Jovan Dj. Golic. On the security of shift register based keystream generators. In *Fast Software Encryption, Cambridge Security Workshop*, pages 90–100. Springer, 1994.

86. Tim Good and Mohammed Benaissa. AES on FPGA from the fastest to the smallest. *CHES '05: Proceedings of the 7th International Workshop on Cryptographic Hardware and Embedded Systems*, pages 427–440, 2005.

87. L. Grover. A fast quantum-mechanical algorithm for database search. In *Proceedings of the Twenty-eighth Annual ACM Symposium on Theory of Computing*, pages 212–219. ACM, 1996.

88. Tim Güneysu, Timo Kasper, Martin Novotny, Christof Paar, and Andy Rupp. Cryptanalysis with COPACOBANA. *IEEE Transactions on Computers*, 57(11):1498–1513, 2008.

89. S. Halevi and H. Krawczyk. MMH: message authentication in software in the Gbit/second rates. In *Proceedings of the 4th Workshop on Fast Software Encryption*, volume 1267, pages 172–189. Springer, 1997.

90. D. R. Hankerson, A. J. Menezes, and S. A. Vanstone. *Guide to Elliptic Curve Cryptography*. Springer, 2004.

91. M. Hellman. A cryptanalytic time-memory tradeoff. *IEEE Transactions on Information Theory*, 26(4):401–406, 1980.

92. Shoichi Hirose. Some plausible constructions of double-block-length hash functions. In *FSE: Fast Software Encryption*, volume 4047 of *LNCS*, pages 210–225. Springer, 2006.

93. Deukjo Hong, Jaechul Sung, and Seokhie Hong et al. Hight: A new block cipher suitable for low-resource device. In *CHES '06: Proceedings of the 8th International Workshop on Cryptographic Hardware and Embedded Systems*, pages 46–59. Springer, 2006.

94. International Organization for Standardization (ISO). ISO/IEC 15408, 15443-1, 15446, 19790, 19791, 19792, 21827.

95. International Organization for Standardization (ISO). ISO/IEC 9796-1:1991, 9796-2:2000, 9796-3:2002, 1991–2002.

96. International Organization for Standardization (ISO). ISO/IEC 10118-4, Information technology—Security techniques—Hash-functions—Part 4: Hash-functions using modular arithmetic, 1998. http://www.iso.org/iso/.

97. D. Kahn. *The Codebreakers. The Story of Secret Writing*. Macmillan, 1967.

98. Jens-Peter Kaps, Gunnar Gaubatz, and Berk Sunar. Cryptography on a speck of dust. *Computer*, 40(2):38–44, 2007.

99. A. Karatsuba and Y. Ofman. Multiplication of multidigit numbers on automata. *Soviet Physics Doklady (English translation)*, 7(7):595–596, 1963.

100. Ann Hibner Koblitz, Neal Koblitz, and Alfred Menezes. Elliptic curve cryptography: The serpentine course of a paradigm shift. Cryptology ePrint Archive, Report 2008/390, 2008. http://eprint.iacr.org/cgi-bin/cite.pl?entry=2008/390.

101. Neal Koblitz. *Introduction to Elliptic Curves and Modular Forms*. Springer, 1993.

102. Neal Koblitz. The uneasy relationship between mathematics and cryptography. *Notices of the AMS*, pages 973–979, September 2007.

103. Neal Koblitz, Alfred Menezes, and Scott Vanstone. The state of elliptic curve cryptography. *Des. Codes Cryptography*, 19(2-3):173–193, 2000.

104. Çetin Kaya Koç. *Cryptographic Engineering*. Springer, 2008.

105. S. Kumar, C. Paar, J. Pelzl, G. Pfeiffer, and M. Schimmler. Breaking ciphers with COPACOBANA—A cost-optimized parallel code breaker. In *CHES '06: Proceedings of the 8th International Workshop on Cryptographic Hardware and Embedded Systems*, LNCS. Springer, October 2006.

106. Matthew Kwan. Reducing the Gate Count of Bitslice DES, 1999. http://www.darkside.com.au/bitslice/bitslice.ps.

107. Ben Laurie. Seven and a Half Non-risks of PKI: What You Shouldn't Be Told about Public Key Infrastructure. http://www.apache-ssl.org/7.5things.txt.

108. Laurie Law, Alfred Menezes, Minghua Qu, Jerry Solinas, and Scott Vanstone. An efficient protocol for authenticated key agreement. *Des. Codes Cryptography*, 28(2):119–134, 2003.

109. Arjen K. Lenstra and Eric R. Verheul. The XTR public key system. In *CRYPTO '00: Proceedings of the 20th Annual International Cryptology Conference, Advances in Cryptology*, pages 1–19. Springer, 2000.

110. Rudolf Lidl and Harald Niederreiter. *Introduction to Finite Fields and Their Applications.* Cambridge University Press, 2nd edition, 1994.

111. Chae Hoon Lim and Tymur Korkishko. mCrypton–A lightweight block cipher for security of low-cost RFID tags and Sensors. In *Information Security Applications*, volume 3786, pages 243–258. Springer, 2006.

112. Yehuda Lindell. *Composition of Secure Multi-Party Protocols: A Comprehensive Study.* Springer, 2003.

113. Stefan Mangard, Elisabeth Oswald, and Thomas Popp. *Power Analysis Attacks: Revealing the Secrets of Smart Cards (Advances in Information Security).* Springer, 2007.

114. Mitsuru Matsui. Linear cryptanalysis method for DES cipher. In *Advances in Cryptology - EUROCRYPT '93*, 1993.

115. Mitsuru Matsui. How far can we go on the x64 processors? In *FSE: Fast Software Encryption*, volume 4047 of *LNCS*, pages 341–358. Springer, 2006.

116. Mitsuru Matsui and S. Fukuda. How to maximize software performance of symmetric primitives on Pentium III and 4 processors. In *FSE: Fast Software Encryption*, volume 3557 of *LNCS*, pages 398–412. Springer, 2005.

117. Mitsuru Matsui and Junko Nakajima. On the power of bitslice implementation on Intel Core2 processor. In *CHES '07: Proceedings of the 9th International Workshop on Cryptographic Hardware and Embedded Systems*, pages 121–134. Springer, 2007.

118. Ueli M. Maurer and Stefan Wolf. The relationship between breaking the Diffie–Hellman protocol and computing discrete logarithms. *SIAM Journal on Computing*, 28(5):1689–1721, 1999.

119. D. McGrew and J. Viega. RFC 4543: The Use of Galois Message Authentication Code (GMAC) in IPsec ESP and AH. Technical report, Corporation for National Research Initiatives, Internet Engineering Task Force, Network Working Group, May 2006. Available at http://rfc.net/rfc4543.html.

120. A. J. Menezes, P. C. van Oorschot, and S. A. Vanstone. *Handbook of Applied Cryptography.* CRC Press, Boca Raton, Florida, USA, 1997.

121. Ralph C. Merkle. Secure communications over insecure channels. *Commun. ACM*, 21(4):294–299, 1978.

122. Sean Murphy and Matthew J. B. Robshaw. Essential algebraic structure within the AES. In *CRYPTO '02: Proceedings of the 22nd Annual International Cryptology Conference, Advances in Cryptology*, pages 1–16. Springer, 2002.

123. David Naccache and David M'Rahi. Cryptographic smart cards. *IEEE Micro*, 16(3):14–24, 1996.

124. Block Cipher Modes Workshops. http://csrc.nist.gov/groups/ST/toolkit/BCM/workshops.html.

125. NIST test suite for random numbers. http://csrc.nist.gov/rng/.

126. National Institute of Standards and Technology (NIST). Digital Signature Standards (DSS), FIPS186-3. Technical report, Federal Information Processing Standards Publication (FIPS), June 2009. Available at http://csrc.nist.gov/publications/fips/fips186-3/fips_186-3.pdf.

127. J. Nechvatal. Public key cryptography. In Gustavus J. Simmons, editor, *Contemporary Cryptology: The Science of Information Integrity*, pages 177–288. IEEE Press, Piscataway, NJ, USA, 1994.

128. Security Architecture for the Internet Protocol. http://www.rfc-editor.org/rfc/rfc4301.txt.

129. I. Niven, H.S. Zuckerman, and H.L. Montgomery. *An Introduction to the Theory of Numbers (5th Edition).* Wiley, 1991.

130. NSA Suite B Cryptography. http://www.nsa.gov/ia/programs/suiteb_cryptography/index.shtml.

131. Philippe Oechslin. Making a Faster Cryptanalytic Time-Memory Trade-Off. In *CRYPTO '03: Proceedings of the 23rd Annual International Cryptology Conference, Advances in Cryptology*, volume 2729 of *LNCS*, pages 617–630, 2003.

132. The OpenSSL Project, 2009. http://www.openssl.org/.
133. European Parliament. Directive 1999/93/EC of the European Parliament and of the Council of 13 December 1999 on a Community framework for electronic signatures, 1999. http://europa.eu/eur-lex/pri/en/oj/dat/2000/l_013/l_01320000119en00120020.pdf.
134. D. Pointcheval and J. Stern. Security proofs for signature schemes. In U. Maurer, editor, *Advances in Cryptology — EUROCRYPT'96*, volume 1070 of *LNCS*, pages 387–398. Springer, 1996.
135. Axel Poschmann. *Lightweight Cryptography — Cryptographic Engineering for a Pervasive World*. PhD thesis, Department of Electrical Engineering and Computer Sciences, Ruhr-University Bochum, Germany, April 2009. http://www.crypto.ruhr-uni-bochum.de/en_theses.html.
136. B. Preneel, R. Govaerts, and J. Vandewalle. Hash functions based on block ciphers: A synthetic approach. *LNCS*, 773:368–378, 1994.
137. Bart Preneel. MDC-2 and MDC-4. In Henk C. A. van Tilborg, editor, *Encyclopedia of Cryptography and Security*. Springer, 2005.
138. Electronic Signatures in Global and National Commerce Act, United States of America, 2000.
139. Jean-Jacques Quisquater, Louis Guillou, Marie Annick, and Tom Berson. How to explain zero-knowledge protocols to your children. In *CRYPTO '89: Proceedings of the 9th Annual International Cryptology Conference, Advances in Cryptology*, pages 628–631. Springer, 1989.
140. M. O. Rabin. Digitalized Signatures and Public-Key Functions as Intractable as Factorization. Technical report, Massachusetts Institute of Technology, 1979.
141. W. Rankl and W. Effing. *Smart Card Handbook*. John Wiley & Sons, Inc., 2003.
142. RC4 Page. http://www.wisdom.weizmann.ac.il/~itsik/RC4/rc4.html.
143. R. L. Rivest, A. Shamir, and L. Adleman. A method for obtaining digital signatures and public-key cryptosystems. *Communications of the ACM*, 21(2):120–126, February 1978.
144. Ron Rivest. The RC4 Encryption Algorithm, March 1992. http://www.rsasecurity.com.
145. Dorothy Elizabeth Robling Denning. *Cryptography and Data Security*. Addison-Wesley Longman Publishing Co., Inc., 1982.
146. Matthew Robshaw and Olivier Billet, editors. *New Stream Cipher Designs: The eSTREAM Finalists*, volume 4986 of *LNCS*. Springer, 2008.
147. Carsten Rolfes, Axel Poschmann, Gregor Leander, and Christof Paar. Ultra-lightweight implementations for smart devices–security for 1000 gate equivalents. In *Proceedings of the 8th Smart Card Research and Advanced Application IFIP Conference – CARDIS 2008*, volume 5189 of *LNCS*, pages 89–103. Springer, 2008.
148. K. H. Rosen. *Elementary Number Theory, 5th Edition*. Addison-Wesley, 2005.
149. Public Key Cryptography Standard (PKCS), 1991. http://www.rsasecurity.com/rsalabs/node.asp?id=2124.
150. Claus-Peter Schnorr. Efficient signature generation by smartcards. *Journal of Cryptology*, 4:161–174, 1991.
151. A. Shamir. Factoring large numbers with the TWINKLE device. In *CHES '99: Proceedings of the 1st International Workshop on Cryptographic Hardware and Embedded Systems*, volume 1717 of *LNCS*, pages 2–12. Springer, August 1999.
152. A. Shamir and E. Tromer. Factoring Large Numbers with the TWIRL Device. In *CRYPTO '03: Proceedings of the 23rd Annual International Cryptology Conference, Advances in Cryptology*, volume 2729 of *LNCS*, pages 1–26. Springer, 2003.
153. P. Shor. Polynomial-time algorithms for prime factorization and discrete logarithms. *SIAM Journal on Computing, Communication Theory of Secrecy Systems*, 26:1484–1509, 1997.
154. J. H. Silverman. *The Arithmetic of Elliptic Curves*. Springer, 1986.
155. J. H. Silverman. *Advanced Topics in the Arithmetic of Elliptic Curves*. Springer, 1994.
156. J. H. Silverman. *A Friendly Introduction to Number Theory*. Prentice Hall, 3rd edition, 2006.

157. Simon Singh. *The Code Book: The Science of Secrecy from Ancient Egypt to Quantum Cryptography.* Anchor, August 2000.

158. Jerome A. Solinas. Efficient arithmetic on Koblitz curves. *Designs, Codes and Cryptography,* 19(2-3):195–249, 2000.

159. J.H. Song, R. Poovendran, J. Lee, and T. Iwata. RFC 4493: The AES-CMAC Algorithm. Technical report, Corporation for National Research Initiatives, Internet Engineering Task Force, Network Working Group, June 2006. Available at `http://rfc.net/rfc4493.html`.

160. NIST Special Publication SP800-38D: Recommendation for Block Cipher Modes of Operation: Galois Counter Mode (GCM) and GMAC, November 2007. Available at `http://csrc.nist.gov/publications/nistpubs/800-38D/SP-800-38D.pdf`.

161. W. Stallings. *Cryptography and Network Security: Principles and Practice.* Prentice Hall, 4th edition, 2005.

162. Tsuyoshi Takagi. Fast RSA-type cryptosystem modulo $p^k q$. In *CRYPTO '98: Proceedings of the 18th Annual International Cryptology Conference, Advances in Cryptology,* pages 318–326. Springer, 1998.

163. S. Trimberger, R. Pang, and A. Singh. A 12 Gbps DES Encryptor/Decryptor Core in an FPGA. In Ç. K. Koç and C. Paar, editors, *CHES '00: Proceedings of the 2nd International Workshop on Cryptographic Hardware and Embedded Systems,* volume 1965 of *LNCS,* pages 157–163. Springer, August 17-18, 2000.

164. Trivium Specifications. `http://www.ecrypt.eu.org/stream/p3ciphers/trivium/trivium_p3.pdf`.

165. Walter Tuchman. A brief history of the data encryption standard. In *Internet Besieged: Countering Cyberspace Scofflaws,* pages 275–280. ACM Press/Addison-Wesley, 1998.

166. Annual Workshop on Elliptic Curve Cryptography, ECC. `http://cacr.math.uwaterloo.ca/conferences/`.

167. Digital Signature Law Survey. `https://dsls.rechten.uvt.nl/`.

168. Henk C. A. van Tilborg, editor. *Encyclopedia of Cryptography and Security.* Springer, 2005.

169. Ingrid Verbauwhede, Frank Hoornaert, Joos Vandewalle, and Hugo De Man. ASIC cryptographical processor based on DES, 1991. `http://www.ivgroup.ee.ucla.edu/pdf/1991euroasic.pdf`.

170. SHARCS — Special-purpose Hardware for Attacking Cryptographic Systems. `http://www.sharcs.org/`.

171. WAIFI — International Workshop on the Arithmetic of Finite Fields. `http://www.waifi.org/`.

172. Andre Weimerskirch and Christof Paar. Generalizations of the Karatsuba algorithm for efficient implementations. Cryptology ePrint Archive, Report 2006/224. `http://eprint.iacr.org/2006/224`.

173. D. Whiting, R. Housley, and N. Ferguson. RFC 3610: Counter with CBC-MAC (CCM). Technical report, Corporation for National Research Initiatives, Internet Engineering Task Force, Network Working Group, September 2003.

174. M.J. Wiener. Efficient DES Key Search: An Update. *CRYPTOBYTES,* 3(2):6–8, Autumn 1997.

175. Thomas Wollinger, Jan Pelzl, and Christof Paar. Cantor versus Harley: Optimization and analysis of explicit formulae for hyperelliptic curve cryptosystems. *IEEE Transactions on Computers,* 54(7):861–872, 2005.

Index

3DES, *see* triple DES

A5/1 cipher, 31
access control, 264
active attack, 225
Adleman, Leonard, 173
Advanced Encryption Standard, 57, 87, 88
  affine mapping, 103
  byte substitution layer, 90, 101
  diffusion layer, 90, 103
  hardware implementation, 115
  key addition layer, 90, 106
  key schedule, 106
  key whitening, 106
  layers of, 90
  MixColumn, 90, 103, 104
  overview, 89
  S-Box, 90, 101
  selection process, 88
  ShiftRows, 90, 103
  software implementation, 115
  state of, 90
  T-Box, 115
AES, *see* Advanced Encryption Standard
affine cipher, 19
affine mapping, 103
Alice and Bob, 4
anonymity, 264
asymmetric cryptography, *see* public-key
    cryptography
attack
  brute-force, *see* brute-force attack
  buffer overflow, 11
auditing, 264
authenticated channel, 342, *see* channel
authenticated encryption, 143
authentication tag, 320

availability, 264
avalanche effect, 66

baby-step giant-step method, 221
Biham, Eli, 75, 76
binary extended Euclidean algorithm, 168
birthday attack, 299
birthday paradox, 299
bit-slicing, 82
block cipher, 30
  confusion, 57
  diffusion, 57
block ciphers
  and hash functions, 305
Blowfish, 81, 307
brute-force attack, 7, 136
  for discrete logarithms, 220
BSI, 22

CA, *see* certification authority
Caesar cipher, *see* shift cipher
cardinality, *see* group
Carmichael number, 189
CAST, 81
CBC, *see* cipher block chaining mode
CBC-MAC, 143, 325
CC, *see* Common Criteria
CCM, 327
certificate, 155
  chain of, 350
certificate revocation list, 350
  delta CRL, 350
certificates, 345
certification authority, 345
CFB, *see* cipher feedback mode
chain of trust, 347, 350
challenge-response protocol, 340

Printed in the United States
By Bookmasters